TRACE METALS
IN HEALTH AND DISEASE

Trace Metals in Health and Disease

New Roles of Metals in Biochemistry,
the Environment, and Clinical/Nutritional Studies

An Intra-Science Research Foundation Symposium

Editor

Norman Kharasch, Ph.D.
*Professor of Biomedicinal Chemistry
School of Pharmacy
University of Southern California
Los Angeles, California*

Raven Press ■ New York

Raven Press, 1140 Avenue of the Americas, New York, New York 10036

© 1979 by Raven Press Books, Ltd. All rights reserved. This book is protected by copyright. No part of it may be reproduced, stored in a retrieval system, or transmitted, in any form or by any means, electronic, mechanical, photocopying, recording, or otherwise, without the prior written permission of the publisher.

Made in the United States of America

Library of Congress Cataloging in Publication Data

Main entry under title:

Trace metals in health and disease

     Proceedings of the 12th annual Intra-Science Research Foundation sumposium, held in Santa Monica, Calif., Nov. 30 – Dec. 1, 1978.
     Includes bibliographies and index.
     1. Metals – Physiological effect – Congresses.
2. Trace elements – Physiological effect – Congresses.
I. Kharasch, Norman. II. Intra-Science Research Foundation. [DNLM: 1. Trace elements – Physiology – Congresses. 2. Environmental health – Congresses. QU130.3 I161t 1978]
QP 532.T72   615.9'2   76-64431
ISBN 0-89004-389-2

This volume is dedicated to Dr. Klaus Schwarz 1914–1978, whose pioneering experimental works and perceptive deductions about trace metal research are the very basis of the now well-established science of Trace Metal Research in Health and Disease

Preface

The role of trace metals in health and disease is currently an extremely important area of research. Our understanding of the impact that the presence or absence of trace quantities of metals has on metabolic processes is increasing at a rapid rate. This volume is based on the presentations delivered at the Twelfth Annual Symposia of the Intra-Science Research Foundation at which biochemists, biophysicists, nutritionists, and pharmacologists came together to discuss and evaluate the present state of knowledge of the roles of trace metals in biochemistry, in the environment, and in clinical/nutritional applications.

Topics ranging from new aspects of the impact of metals in the environment on human health to the roles of metals in genetic information transfer are discussed. Particular attention is focused on the role of metals in carcinogenesis, and the possible role of metal-induced mutagenesis in the initiation of cancer is explored. The interdisciplinary range of this volume makes it of interest to scientists and clinicians in the areas of environmental biochemistry, nutrition, pharmacology, cancer research toxicology, and medicine.

Norman Kharasch

Acknowledgments

This volume contains the Proceedings of the Twelfth Annual Intra-Science Research Foundation Symposium on "Trace Elements in Health and Disease: New Roles of Metals in Biochemistry, the Environment, and Clinical/Nutritional Studies," held on November 29, 30, and December 1, 1978 in Santa Monica, California.

The foundation wishes to acknowledge the special help of Prof. Roslyn Alfin-Slater, Professor and Division Head, Environmental and Nutritional Sciences, UCLA, who served as general chairman of the symposium. The cooperation and assistance of Dr. Seymour Siegal, Chairman of the Foundation's Science Council, and of Mr. Milton Golden, Chairman of the Executive Committee, are deeply appreciated.

Intra-Science Research Foundation

700 State Drive, Exposition Park, P.O. Box 18589, Los Angeles, California 90018

Board of Directors and Officers

Theodore A. Venia
Chairman
(Vice President, E. T. Horn Co.)

Maurice Samson
President
(President, Samson Raw Materials, Co.)

Milton M. Golden
Chairman, Executive Committee
(Vice President, Standard Brands Paint Co., Inc.)

Joseph Schultz, Ph.D.
Vice-Chairman

Frank D. Davis
Treasurer

Richard Brenneman, Ph.D.
Secretary
(Industrial Sociologist)

Norman Kharasch, Ph.D.
Scientific Director
(Professor, Biomedical Chemistry, USC)

Seymour Siegel, Ph.D.
Chairman, Science Council
(Director, Chemistry & Physics Laboratory, The Aerospace Corporation)

Sam Chess
(Technical Director, CPR Division, The Upjohn Company)

Ray R. Coulter, Pharm. D., LL.B.
(General Counsel, Lear Siegler, Inc.)

Robert Graham
(Founder, Chairman of the Board Armorlite, Inc.)

Science Council

Seymour Siegel, Ph.D.
Chairman

Roslyn Alfin-Slater, Ph.D.
(Professor and Division Head Environmental and Nutritional Sciences, UCLA)

Stuart Eriksen, Ph.D.
(Director of Research, Allergan Pharmaceutical Co.)

Milton M. Golden

Norman Kharasch, Ph.D.

Thomas Lincoln, M.D.
(Associate Professor, Pathology USC, Cancer Center)

Maurice Samson

Joseph Schultz, Ph.D.

Theodore A. Venia

Walter Wolf, Ph.D.
(Professor and Director, Radiopharmacy Program, USC)

Morris Wolfred, Ph.D.
(Prescription Centers)

Contents

SIGNIFICANCE OF ENVIRONMENTAL PATHWAYS IN POLLUTANT TOXICOLOGY

1   Metabolic Response of Microbiota to Chromium and Other Metals
H. Drucker, T. R. Garland, and R. E. Wildung

27   Biomedical and Environmental Significance of Siderophores
J. B. Neilands

43   Some Bio-Inorganic Chemical Reactions of Environmental Significance
J. M. Wood, Y.-T. Fanchiang, and P. J. Craig

55   Quantitative Mammalian Cell Mutagenesis and a Preliminary Study of the Mutagenic Potential of Metallic Compounds
Abraham W. Hsie, Neil P. Johnson, D. Bruce Couch, Juan R. San Sebastian, J. Patrick O'Neill, James D. Hoeschele, Ronald O. Rahn, and Nancy L. Forbes

71   In Vitro Clonal Growth Assay for Evaluating Toxicity of Metal Salts
M. E. Frazier and T. K. Andrews

ROLE OF METALS IN CARCINOGENESIS AND IN SELECTED CANCER DRUGS

83   Problems in Metal Carcinogenesis
Arthur Furst

93   Trace Element Interactions in Carcinogenesis
Gerald L. Fisher

109   Metals as Mutagenic Initiators of Cancer
C. Peter Flessel

ROLE OF METALS IN GENETIC INFORMATION TRANSFER

123   Essential and Deleterious Effects in the Interaction of Metal Ions with Nucleic Acids
G. L. Eichhorn, Y. A. Shin, P. Clark, J. Rifkind, J. Pitha, E. Tarien, G. Rao, S. J. Karlik, and D. R. Crapper

135   Metals, DNA Polymerization, and Genetic Miscoding
Richard A. Zakour, Lawrence A. Loeb, Thomas A. Kunkel, and R. Marlene Koplitz

155 Heavy Atom Labeling in Atomic Microscopy
Michael Beer, Christian Stoeckert, Rex Hjelm, Jr., James Resch, David Tunkel, Robert Hyland, and J. W. Wiggins

NEW ASPECTS OF THE ROLE OF ZINC AS A TRACE METAL

167 The Role of Zinc in Prenatal and Neonatal Development
Lucille S. Hurley

177 The Role of Zinc in the Biochemistry of the *Euglena gracilis* Cell Cycle
Kenneth H. Falchuk

ROLES OF CALCIUM IN THE ACTION AND DESIGN OF DRUGS

189 Vitamin D Metabolism and Function
Hector F. DeLuca

217 A Review of the Basic and Applied Pharmacology of a New Group of Calcium Antagonists: The 2-Substituted 3-Dimethylamino-5,6-Methylenedioxyindenes
Ralf G. Rahwan and Donald T. Witiak

227 Calcium Antagonists: Mechanisms of Action with Special Reference to Nifedipine
Philip D. Henry

235 Verapamil: Mechanisms of Pharmacologic Actions and Therapeutic Applications
Bramah N. Singh

ROLES OF SELENIUM AND RELATED ELEMENTS IN BIOCHEMICAL, ENVIRONMENTAL, AND CLINICAL STUDIES. A SYMPOSIUM IN MEMORIAM OF PROFESSOR KLAUS SCHWARZ

251 Klaus Schwarz, 1914–1978. Commemoration of a Leader in Trace Element Research
G. N. Schrauzer

263 The Glutathione Peroxidase Reaction: A Key to Understand the Selenium Requirement of Mammals
L. Flohé, W. A. Günzler, and G. Loschen

287 Antioxidants, Cancer, and the Immune Response
Werner A. Baumgartner

307 Subject Index

Contributors

T. K. Andrews
Molecular Biology and Biophysics Section
Pacific Northwest Laboratory
Battelle Memorial Institute
Richland, Washington 99352

Werner A. Baumgartner
Radioimmunoassay and In Vitro Labs
Nuclear Medicine Service
Veterans Administration
 Wadsworth Hospital Center
Los Angeles, California 90073

Michael Beer
Professor and Chairman
Department of Biophysics
Faculty of Arts and Sciences
Johns Hopkins University
Baltimore, Maryland 21218

P. Clark
Laboratory of Cellular and Molecular
 Biology
Gerontology Research Center
National Institute on Aging
National Institutes of Health
Baltimore City Hospitals
Baltimore, Maryland 21224

D. Bruce Couch
Chemical Industry Institute of Toxicology
Research Triangle Park,
Durham, North Carolina 27709

P. J. Craig
Gray Freshwater Biological Institute
Department of Biochemistry
University of Minnesota
Navarre, Minnesota 55392

D. R. Crapper
Department of Physiology
Medical Sciences Building
University of Toronto
Toronto, Ontario M5S 1A8
Canada

H. Drucker
Biology Department
Pacific Northwest Laboratory
Battelle Memorial Institute
Richland, Washington 99352

G. L. Eichhorn
Laboratory of Cellular and Molecular
 Biology
Gerontology Research Center
National Institute on Aging
National Institutes of Health
Baltimore City Hospitals
Baltimore, Maryland 21224

Kenneth H. Falchuk
The Howard Hughes Medical
 Institute and
Biophysics Research Laboratory and
Department of Medicine
Harvard Medical School and
Division of Medical Biology
Affiliated Hospitals Center, Inc.
Boston, Massachusetts 02115

Y.-T. Fanchiang
Gray Freshwater Biological Institute
Department of Biochemistry
University of Minnesota
Navarre, Minnesota 55392

Gerald L. Fisher
Radiobiology Laboratory
University of California
Davis, California 95616

C. Peter Flessel
Air and Industrial Hygiene Laboratory
 Section
Calfornia Department of Health Services
Berkeley, California 94704

L. Flohé
Grünenthal GmbH
Center of Research
D 5100 Aachen, West Germany

Nancy L. Forbes
Biology Division
Oak Ridge National Laboratory and
University of Tennessee—Oak Ridge
  Graduate School of Biomedical
  Sciences
Oak Ridge, Tennessee 37830

M. E. Frazier
Molecular Biology and Biophysics Section
Pacific Northwest Laboratory
Battelle Memorial Institute
Richland, Washington 99352

Arthur Furst
Institute of Chemical Biology
University of San Francisco
San Francisco, California 94117

T. R. Garland
Ecosystems Department
Pacific Northwest Laboratory
Battelle Memorial Institute
Richland, Washington 99352

W. A. Günzler
Grünenthal GmbH
Center of Research
D 5100 Aachen, West Germany

Philip D. Henry
Cardiovascular Division
Barnes Hospital
Washington University School of Medicine
St. Louis, Missouri 63110

Rex Hjelm, Jr.
Department of Biophysics
Johns Hopkins University
Baltimore, Maryland 21218

James D. Hoeschele
Biology Division
Oak Ridge National Laboratory and
University of Tennessee—Oak Ridge
  Graduate School of Biomedical
  Sciences
Oak Ridge, Tennessee 37830

Abraham W. Hsie
Biology Division
Oak Ridge National Laboratory and
University of Tennessee—Oak Ridge
  Graduate School of Biomedical
  Sciences
Oak Ridge, Tennessee 37830

Lucille S. Hurley
Department of Nutrition
University of California
Davis, California 95616

Robert Hyland
Department of Biophysics
Johns Hopkins University
Baltimore, Maryland 21218

Neil P. Johnson
Postdoctoral Fellow
Biology Division
Oak Ridge National Laboratory and
University of Tennessee—Oak Ridge
  Graduate School of Biomedical
  Sciences
Oak Ridge, Tennessee 37830

S. J. Karlik
Department of Physiology
Medical Sciences Building
University of Toronto
Toronto, Ontario M5S 1A8
Canada

R. Marlene Koplitz
Department of Pathology and
Gottstein Memorial Cancer Research
  Laboratory
University of Washington
Seattle, Washington 98195

Thomas A. Kunkel
Department of Pathology and
Gottstein Memorial Cancer Research
  Laboratory
University of Washington
Seattle, Washington 98195

Lawrence A. Loeb
Department of Pathology and
Director, Gottstein Memorial Cancer
  Research Laboratory
University of Washington
Seattle, Washington 98195

G. Loschen
Grünenthal GmbH
Center of Research
D 5100, West Germany

J. B. Neilands
Biochemistry Department
University of California
Berkeley, California 94720

CONTRIBUTORS

J. Patrick O'Neill
Biology Division
Oak Ridge National Laboratory and
University of Tennessee—Oak Ridge
  Graduate School of Biomedical
  Sciences
Oak Ridge, Tennessee 37830

J. Pitha
Laboratory of Cellular and Molecular
  Biology
Gerontology Research Center
National Institute on Aging
National Institutes of Health
Baltimore City Hospitals
Baltimore, Maryland 21224

Ronald O. Rahn
Health and Safety Research Division
Oak Ridge National Laboratory and
University of Tennessee—Oak Ridge
  Graduate School of Biomedical
  Sciences
Oak Ridge, Tennessee 37830

Ralf G. Rahwan
Divisions of Pharmacology and Medicinal
  Chemistry
College of Pharmacy
The Ohio State University
Columbus, Ohio 43210

G. Rao
Laboratory of Cellular and Molecular
  Biology
Gerontology Research Center
National Institute on Aging
National Institutes of Health
Baltimore City Hospitals
Baltimore, Maryland 21224

James Resch
Department of Biophysics
Johns Hopkins University
Baltimore, Maryland 21218

J. Rifkind
Laboratory of Cellular and Molecular
  Biology
Gerontology Research Center
National Institute on Aging
National Institutes of Health
Baltimore City Hospitals
Baltimore, Maryland 21224

Juan R. San Sebastian
Postdoctoral Fellow
Biology Division
Oak Ridge National Laboratory and
University of Tennessee—Oak Ridge
  Graduate School of Biomedical
  Sciences
Oak Ridge, Tennessee 37830

G. N. Schrauzer
Department of Chemistry
University of California at San Diego
Revelle College
La Jolla, California 92093

Y. A. Shin
Laboratory of Cellular and Molecular
  Biology
Gerontology Research Center
National Institute on Aging
National Institutes of Health
Baltimore City Hospitals
Baltimore, Maryland 21224

Bramah N. Singh
Department of Cardiology
Wadsworth Veteran's Administration
  Hospital and
Department of Medicine
UCLA School of Medicine
Los Angeles, California 90073

Christian Stoeckert
Department of Biophysics
Johns Hopkins University
Baltimore, Maryland 21218

E. Tarien
Laboratory of Cellular and Molecular
  Biology
Gerontology Research Center
National Institute on Aging
National Institutes of Health
Baltimore City Hospitals
Baltimore, Maryland 21224

David Tunkel
Department of Biophysics
Johns Hopkins University
Baltimore, Maryland 21218

J. W. Wiggins
Department of Biophysics
Johns Hopkins University
Baltimore, Maryland 21218

R. E. Wildung
Ecosystems Department
Pacific Northwest Laboratory
Battelle Memorial Institute
Richland, Washington 99352

Donald T. Witiak
Divisions of Pharmacology and Medicinal
  Chemistry
College of Pharmacy
Ohio State University
Columbus, Ohio 43210

J. M. Wood
Gray Freshwater Biological Institute
Department of Biochemistry
University of Minnesota
Navarre, Minnesota 55392

Richard A. Zakour
Department of Pathology and
Gottstein Memorial Cancer Research
  Laboratory
University of Washington
Seattle, Washington 98195

Opening Remarks

Theodore Venia
Chairman, Board of Directors
Intra-Science Research Foundation

It is a special privilege for me to open this Twelfth Annual Intra-Science Symposium on Trace Metals in Health and Disease—New Roles of Metals in Biochemistry, the Environment, and Clinical/Nutritional Studies. Our aim in sponsoring these Symposia has always been to seek out a topic that is ready to "come of age" and to provide an in-depth analysis of it that will help make this happen.

We hope that your meeting here will add to new understanding of scientific problems, to the making of new contacts, and exposure to new ideas. All of us also hope that you will enjoy yourselves and that your interaction with Intra-Science will be a stimulating and enduring one.

I wish to acknowledge the assistance of the many people who have helped to make this event possible. Although the list is incomplete, I wish to especially acknowledge our Chairman, Dr. Roslyn B. Alfin-Slater, Dr. Seymour Siegel, Chairman of our Science Council, Professor Norman Kharasch, Founder and Scientific Director of Intra-Science, and Mr. Milton Golden, Chairman of the Foundation's Executive Committee.

I also wish to express appreciation to the following organizations which have supported our 1978 symposium:

- American Cyanamid Co.
- American Hoechst Corporation
- Hoffman LaRoche Inc.
- ICN Pharmaceuticals
- McGaw Laboratories
- Pennwalt Corporation
- Soltec Corporation
- Syntex Research

Metabolic Response of Microbiota to Chromium and Other Metals

*H. Drucker, **T. R. Garland, and **R. E. Wildung

*Biology Department and **Ecosystems Department, Battelle Pacific Northwest Laboratories, Richland, Washington 99352

ABSTRACT

Soil microflora may play a key role in modification of heavy metals added to soil as a result of man's industrial and agricultural activity. For such modification to occur, soils must contain populations of microbiota resistant to the toxic effects of a given metal and capable of altering the chemical form of the added metal. We use chromium here as an example of our approach to the problem of soil metabolism of heavy metals. We determined the effects of chromium as a function of its concentration on soil populations or aerobic and anaerobic spore-forming and nonspore-forming bacteria, fungi, and actinomycetes. In the same experimental framework, we examined chromium effects on soil respiration, measured as carbon dioxide evolution from a starch carbon source. Ten ppm chromium affected soil microbial populations; 1 ppm chromium affected soil respiration. The effect of chromium on respiration appeared to be a result of altered sugar metabolism rather than toxicity to microbial populations per se. Similar studies with other toxic metals suggested that metals could be ranked in terms of their toxic effects to soil microflora. We determined that the order of effect was (from greatest effect to least): Ag, Hg, Cr, Cd, Cu, Ni, Zn, Tl, Fe, Sn, W, Mo, Mn, As, Co, Sb, Pb.

Mixed microbial cultures derived from soil, and their ability to alter the chemical form of chromium were examined. Based upon results from gel permeation chromatography, and thin layer electrophoresis and chromatography, these cultures appear to be capable of modifying chromium form. The most likely mechanism of chromium modification appears to be interaction of chromium with microbial metabolites produced during the growth of chromium-resistant microflora.

Through the years, toxicologists have developed a large body of data that considers the biological effect of heavy metals in the chemical forms in which they enter commerce; that is, as chlorides, sulfates, nitrates, etc. Although these data are of great use in setting occupational health standards, and in determining effects that can be expected as a result of accidental exposures of the general populace to these compounds, they may not reflect the toxic potential of these metals for environmentally exposed populations over long periods of time. The obvious reason for this last observation is that the toxicity of any element is, to a great extent, a function of its chemical form: Mercuric chloride in the workplace differs in toxicity and biological properties from methyl mercury which might be present in food (1); sodium arsenate is not equivalent to trimethyl

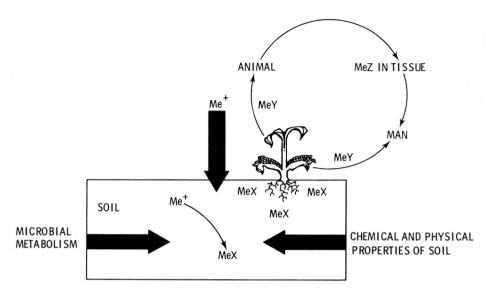

FIG. 1. Postulated path of metals in the environment.

arsine, in terms of its biological properties (2). Determining potential effects on man of heavy metals in the environment requires knowledge of the chemical form to which man might be exposed.

With this in mind, we will consider one form of exposure of man to metals—through food chains. What are the factors determining chemical form (Fig. 1)? If we consider this problem in a systematic fashion, there are six basic areas requiring analysis:

1. The chemical form of the metal as it enters into soils.
2. The physical and chemical properties of the soil as they relate to the chemistry and solubility of the added metal.
3. The biochemical potential of soil as it relates to the "metabolism" of the added metal (primarily a function, we believe, of soil microbiota).
4. The transport and translocation of the most soluble or biologically available form of the metal into plant roots.
5. The plant metabolism as it relates to the translocated metal (if the plant component is consumed by man, the chemical form in the ingested tissue is of importance here).
6. Transfer through gut and animal metabolism of the plant-incorporated metal. (Here, form in the animal tissue determines the hazard of the metal to man.)

For the past several years, we have been considering metal transfer in food chains through research in areas 1 to 5 listed above. In this article, we will discuss our efforts in the third area of concern: microbial modification of metal form. We will describe, as representative of our approach to the problem, work we have done on the metal chromium.

Initially, we wished to establish conditions of soil culture that would yield microbial-stressed populations which might be resistant to metals because of an ability to "detoxify" them—that is, to change their chemical form. To determine this, we examined the effect of chromium on microbial numbers and types, and on respiration (CO_2) evolution from a typical agricultural soil amended with a carbon and nitrogen source. We then examined the effect of chromium on metabolism of the soil cultures by measuring stoichiometry of glucose utilization versus CO_2 evolution. Our studies with chromium were part of a study determining the effects of 17 metals on soil microflora and metabolism. With the quantitative data base developed by these comparative studies, we utilized two numerical ranking systems in order to develop a comparative index of metal effect on soil microbial populations.

We then examined the ability of cultures derived from soil to modify the chemical form of added chromium. We hoped to determine whether soil microorganisms modified chromium either as a consequence of metabolism or as the result of specific detoxification mechanisms. Chromium modification, therefore, was analyzed as a function of growth of soil enrichment cultures on a range of carbon sources and in the presence and absence of the metal during growth phase of the cultures.

MATERIALS AND METHODS

Preparation of Soil and Metal Solution

A Ritzville surface soil, 15 cm, taken on the Hanford Reservation, was utilized for plant and microbial studies. After sampling in sufficient quantity, the soil was air-dried (approximately 8% moisture), sieved (0.3 cm), thoroughly mixed and stored (air-dry) for further use. Subsamples were removed for subsequent microbial experiments. The results of all analyses are reported on the basis of oven-dry (110°C) soil.

Standard solutions of unlabeled metals were made gravimetrically after drying in a desiccator under vacuum, or were diluted from commercial standards. Chromium was added to soil as K_2CrO_4; other metals were added as chlorides, nitrates, or sulfates. We attempted to use metal compounds which were soluble and stable at neutral or soil pH. Concentrations of metals in solution were checked by atomic absorption spectrophotometry.

Microbial Studies

Two methods were employed to determine the effect of metals on the number and types of microorganisms in the soil, including measurements of (a) viable organisms using plate counts and (b) metabolic activity as indicated by CO_2

TABLE 1. *Media and conditions used for isolation of representative soil microbiota*

Organism type	Media	Condition for selectivity
Bacteria Aerobic and microaerophilic non-sporeformers	Soil extract agar[a]	Aerobic incubation heat inactivation of non-sporeformers
Sporeformers Anaerobic and facultative anaerobic nonsporeformers Sporeformers	Soil extract agar[a]	Anaerobic incubation heat inactivation of nonsporeformers
Fungi	Oxgall agar[b]	Streptomycin (30 ppm)
Actinomycetes	Glycerol-asparagenate[c]	Substrate, identification

[a]From ref. 3. [b]From ref. 4. [c]From ref. 5.

evolution rate. Measurements were conducted concurrently, using the same incubation system. Soils (100 g dry weight) containing 0.5 g NH_4NO_3 and 1.0 g of starch to insure sufficient microbial activity for an accurate assessment of metal effects on microbial growth rate, H_2O (22%), and sufficient heavy metal standard solution to attain 1.0, 10.0, and 100.0 $\mu g/g$ metal on a dry weight basis, were incubated in duplicate in glass incubation cells (250 cc) in the dark at 30°C for up to 14 days. The cells were stoppered and continuously flushed with moistened (H_2O bubbler), CO_2-free (Ascarite) air. The soils were subsampled periodically (0, 2 and/or 6, 10, and 13 days of incubation) by insertion and withdrawal of a glass tube through a sampling port in the top of the incubation cell. The subsample of soil (1 g dry weight) was placed in 1 liter of sterile distilled H_2O. Appropriate soil dilutions were made from this stock inoculum. Aliquots of the diluted stock solutions (1 ml) were subsequently cultured in triplicate on selective media (Table 1; references 3–5), and the number of organisms was determined by plate count after incubation for 6 days at 28°C.

The CO_2 in the soil effluent gases, an index of respiration rate, was absorbed in NaOH contained in Pettenkofer tubes and analyzed (6) by titration of the unneutralized NaOH after precipitation of $CO_3^=$ as the Ba salt. A continuously recording automatic titrimeter was used. To improve the sensitivity of the method, the concentrations of NaOH used for collection of CO_2 were adjusted according to expected CO_2 evolution rate. Concentrations of HCl used for titration of unneutralized base were adjusted accordingly.

Measurement of Glucose in Soil

The determination of glucose in soil was accomplished by adaptation of the glucose oxidase method for the determination of glucose in blood and urine (7–9). In the modified procedure, the glucose was extracted from soil (1–5 g) with two 5-ml portions of heated (80–90°C) H_2O. The combined H_2O extracts were eluted through an anion exchange resin (20–50 mesh Agl-X4, Cl form) to remove chromium that interfered with colorimetric analysis. The chromium-free solution was deproteinized by addition of 0.3 M NaOH (1 ml) and 0.15 M $ZnSO_4$ (1 ml) and centrifugation for 20 min at 8,000 rpm. The clarified solution was taken to pH 7 by titration with NaOH. Reduced chromogen and glucose oxidase were added, and the reaction was taken to completion by heating the solution at 37°C for 45 min. The absorbance of the solution at 400 nm was compared to solutions of known glucose concentration, and the glucose content of the soil was calculated.

Enrichment Culturing of Metal-Resistant Mixed Fungal Cultures, Using Selective Mixed-Carbon Sources—Sugars and Organic Acids

A schematic diagram of the procedure used is shown in Fig. 2. One-hundred ml of standard mineral base medium (10) at pH 6.0, containing streptomycin sulfate (2 µg/ml) and 0.5% (wt/vol) each of sucrose, glucose, and fructose (mixed sugar media) or 0.5% (wt/vol) each of malate, succinate, lactate, and acetate (K-salts, mixed organic acid media), were inoculated in a 500-ml Erlenmeyer flask with 1.0 g of soil (mixed sugar media), or with 2.5 g of soil (mixed organic acid media). The flask was incubated aerobically at 28°C in a shaking incubator (200 rpm) for 48 hr. The resulting fungal mat was then removed and ground in a Dounce homogenizer to produce a homogenous slurry. An aliquot (5 ml) of this homogenous slurry was then reinoculated into fresh standard mineral-base medium and reincubated aerobically at 28°C in a shaking incubator (200 rpm for 48 hr). This cell suspension was used as the inoculum for enrichment culturing of metal-resistant mixed fungal cultures.

An aliquot (5 ml) of the homogenous cell suspension was inoculated into each of three Erlenmeyer flasks (500 ml) containing standard mineral-base medium (pH 6.0) with streptomycin sulfate (2 µg/ml), mixed sugar or mixed organic acid carbon sources and radiolabeled metal, ^{51}Cr (0.2 µCi/ml, as K_2CrO_4), at 10, 50, or 100 µg/ml. The flasks were then incubated aerobically at 28°C in a shaking incubator (200 rpm) for 120 hr (stationary phase).

An aliquot (5 ml) of the same homogenous cell suspension was inoculated (48 hr later) into an Erlenmeyer flask (500 ml) containing standard mineral-base medium at pH 6.0 with streptomycin sulfate (2 µg/ml) and either mixed sugar or mixed organic acids as carbon source. The flask was then incubated aerobically at 28°C in a shaking incubator (200 rpm) for 48 hr. At this time, sufficient radiolabeled metal was added to give a final metal concentration equal to the highest metal concentration permitting normal growth in the three previously described cultures, and a final radioactivity concentration of 0.20 µCi/ml. The culture was then reincubated and allowed to grow to stationary phase (28°C, 200 rpm, 24 hr).

The culture initially grown in the presence of metal at the highest concentration permitting normal growth, and the culture initially grown in the absence of metal, were harvested at stationary phase. An aliquot (1 ml) of the stationary

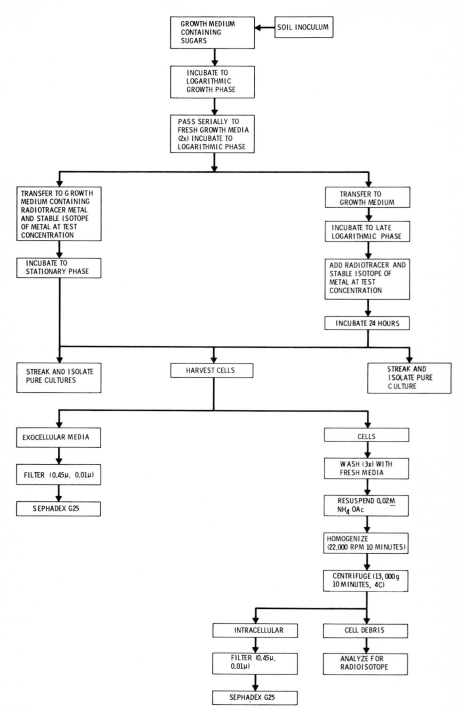

FIG. 2. Procedure for the isolation and enrichment of metal-resistant bacteria on either mixed organic acids or mixed sugar carbon sources.

cultures was saved for isolation of metal-resistant and nonresistant soil microorganisms.

The stationary-phase cultures were harvested by vacuum filtration onto a 0.45-μm Millipore filter. The resulting supernatant solution was decanted. This fraction was filtered through a 0.01-μm Sartorius filter, and stored at 4°C for biochemical analyses. Both exocellular media fractions were analyzed for radioactivity, using total gamma counting.

Cells on the 0.45-μm Millipore filter were washed three times with standard mineral-base medium (50 ml), pH 6.0, containing mixed sugars or mixed acids and unlabeled metal at the test concentration. A fraction of each wash was saved and assayed for radioactivity. The washed cells were resuspended in 0.02 M ammonium acetate (10 ml), pH 7.0. An aliquot (0.5 ml) of the cells was assayed for radioactivity.

Enrichment Culturing of Metal-Resistant Mixed Bacterial Cultures Utilizing Selective Mixed Carbon Sources—Sugars and Organic Acids

The procedure used for metabolic selection of metal-resistant bacteria on sugars and organic acids was the same as that used for fungal cultures, except that: *(a)* streptomycin sulfate (used in the fungal cultures to prevent bacterial growth) was omitted, and *(b)* cultures were incubated from 48 to 96 hr (to late logarithmic phase), instead of from 96 to 120 hr, as in the case of fungi.

Biochemical Analyses

Methods for gel permeation chromatography using Sephadex G-25, thin-layer chromatography (TLC) and electrophoresis (TLE) were previously described (11); however, visualization of chromium on TLE and TLC plates was by staining rather than by autoradiography.

Since the decay scheme of ^{51}Cr is primarily by electron capture and the emission of a 0.32-MeV γ, the sensitivity for film autoradiography on the TLC plates is low. This factor, coupled with the 27.7 day half-life of this isotope prompted the search for a technique whereby chromium could be visualized on the plate through the use of a spray reagent. McGonigle (12) stated that a 0.25% solution of diphenyl-carbazide in methylated spirits: glacial acetic acid (9:1) gave a magenta color with chromium using TLC with silica gel plates. The apparent low sensitivity (3.5–12 μg) of this assay led to investigation of the effect of chromium concentration on the intensity of spots visualized after TLC on cellulose. Two stock solutions, $CrCl_3$ in 2 M HCl and K_2CrO_4 in water (1,000 μg Cr/ml), and solutions containing 0.1, 0.5, 1.0, 5.0, and 25 μg/ml Cr^{+3} and CrO_4^{-2}, were prepared. From each solution, 0.005 ml was added to a cellulose plate (0.1-mm layer) and sprayed with the test reagent. Solutions containing Cr^{+3} appeared much less sensitive to color development than solutions of CrO_4^{-2}, especially after solvent development. The plates were therefore sprayed with 25% NH_4OH to raise the pH into the alkaline range, then sprayed with 15% H_2O_2, which oxidized Cr^{+3} to CrO_4^{-2}. This procedure was adopted for use on samples from microbial cultures. The sensitivity of this assay was 0.05 µg metal/spot for CrO_4^{-2}, and 0.5 µg metal/spot for Cr^{+3}.

RESULTS

Effects of Chromium on Soil Cultures

To examine microbial metabolic potentials in soil relative to metal modification, one should first define conditions under which metal-modifying bacteria might be isolated from soil. We postulated that bacteria capable of metal modification might be more prevalent (present in higher numbers) in soils stressed with a given metal than in soils to which no metal was added. We therefore had to determine concentrations of metal which showed some "stress" effect on the soil microbial population, but which did not eliminate the soil microflora. Two methods were employed for this aspect of our research: (a) We examined the effect of metals and metal concentration on numbers and classes of soil microflora to obtain a relatively specific idea of changes in microbial populations as a function of metal; (b) we examined respiration (CO_2 evolution) as a measure of the effects of metals on microbial metabolic activity.

Effects of Chromium on Soil Microbial Populations

We performed microbial counts on the selected classes of organisms at 0, 2, and 6 days of incubation with 0, 1, 10, and 100 ppm Cr present (Table 2). Total aerobic bacteria were found to decrease in number (by almost an order of magnitude) at 100 ppm chromium, versus no-metal-added control, both at 2 and 6 days of incubation. At this chromium concentration, numbers of organisms decreased markedly from day 0 to 2: the metal was exerting a pronounced killing effect. Concentrations of 1 and 10 ppm chromium may have been slowing growth rate over the interval 0 to 2 days, but at 2 to 6 days, numbers of total aerobic bacteria were almost equivalent to a no-metal-added control.

Growth rates of anaerobic bacteria were affected at 1 ppm chromium, but not markedly. At 10 and 100 ppm chromium, there were obvious bactericidal effects from 0 to 2 days of incubation, with some growth of the organisms occurring in the interval 2 to 6 days.

Aerobic spore-forming bacteria were not affected by chromium at 1 ppm; minimally, there was a pronounced bacteriostatic effect on this class of organisms at 10 ppm chromium; and an apparent bactericidal effect at 100 ppm chromium, with little or no recovery of the organisms over the interval of incubation.

Spore-forming anaerobic bacteria were not grossly affected in growth rate between 0 and 6 days of incubation in the presence of 1 ppm chromium. This class of organisms decreased in numbers by an order of magnitude relative to

TABLE 2. Microbial populations in Cr-stressed soils[a]

Metal concentration	Total aerobes —0 time	Organisms/g soil × 10^6 2 days	Organisms/g soil × 10^6 6 days	Total anaerobe	Organisms/g soil × 10^6 2 days	Organisms/g soil × 10^6 6 days
0	160	840	1,600	0.84	7.2	7.1
1	150	400	1,300	0.78	3.9	5.5
10	140	450	1,100	1.1	0.78	1.2
100	130	35	190	0.77	0.13	0.28

Metal concentration	Aerobe spore-forming	Organisms/g soil × 10^6 2 days	Organisms/g soil × 10^6 6 days	Anaerobe spore-forming	Organisms/g soil × 10^6 2 days	Organisms/g soil × 10^6 6 days
1	0.44	4.1	5.2	0.53	5.7	5.5
1	0.56	3.5	4.9	0.50	4.8	3.4
10	0.66	0.46	1.4	0.51	0.053	0.37
100	0.79	0.42	0.43	0.55	0.058	0.082

Metal concentration	Actinomycetes	Organisms/g soil × 10^6 2 days	Organisms/g soil × 10^6 6 days	Fungi	Organisms/g soil × 10^6 2 days	Organisms/g soil × 10^6 6 days
0	25	200	290	0	0.23	0.29
1	21	160	200	1	0.18	0.22
10	25	150	180	10	0.32	0.35
100	14	7.7	59	100	0.20	0.12

[a] Populations as a function of Cr concentrations were determined as described in Materials and Methods.

numbers at 0 day of incubation, and by two orders of magnitude relative to no-metal-added control at 10 and 100 ppm metal. There was little apparent recovery from the bactericidal effect of the metal at 100 ppm chromium.

Actinomycetes decreased in numbers at 1 and 10 ppm chromium relative to control throughout the 6-day incubation period. The data suggest that this may be an effect on growth rate. At a chromium concentration of 100 ppm, bactericidal effects were observed at 2 days of incubation, with growth of this class apparently resuming in the interval from 2 to 6 days of incubation.

Of all the classes of organisms measured, fungi appeared to be least sensitive to chromate toxicity. While they do not appear to grow very well in soil incubation, absolute numbers of these organisms in soil decreased only at 100 ppm chromium.

These data suggest, overall, that chromium has both bactericidal and bacteriostatic effects on soil microflora. These effects are a function of chromium concentration, duration of exposure of the biota to the metal, and class of microorganism.

Effects of Chromium on Soil Respiration

Carbon dioxide evolution was measured from soil cultures amended with 0, 1, 10, or 100 ppm chromium over a period of 13 days (Fig. 3). The lower chromium levels (1 and 10 ppm chromium) did not appear to affect CO_2 pro-

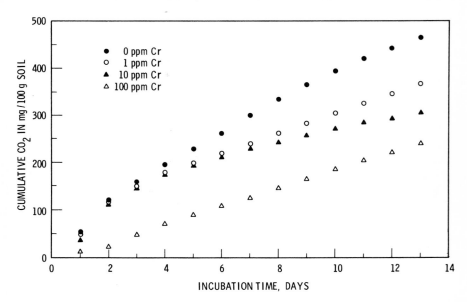

FIG. 3. Cumulative respiration (CO_2) evolution as a function of incubation time on soil cultures amended with 0, 1, 10, or 100 ppm chromium. Soil culture preparation and method of CO_2 measurement are described in Materials and Methods.

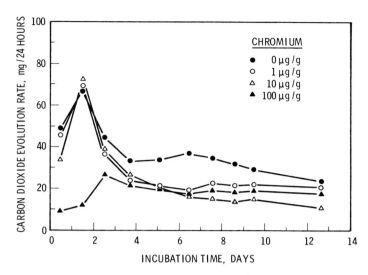

FIG. 4. Rate of soil respiration (CO_2 evolution) as a function of incubation time in soil cultures amended with 0, 1, 10, and 100 ppm chromium. Data were derived from results of experiment described in Fig. 3.

duction until approximately day 3 of incubation. The 1-ppm chromium culture appeared to show a greater degree of CO_2 evolution than the 10-ppm-amended culture; both, however, are significantly lower than the control culture in CO_2 production. The culture amended with 100 ppm chromium appears to be almost qualitatively different from the other three soil cultures. Whereas the 0-, 1-, and 10-ppm-amended cultures appear to produce CO_2 at an initial rapid rate, followed by a slower rate (with this slower rate apparently being a function of chromium concentration), respiration in the 100-ppm culture appears to progress at the same relatively slow rate throughout the incubation period.

This can be more clearly demonstrated by examining rates of carbon dioxide evolution as a function of incubation time (Fig. 4, a derivative plot of the data shown in Fig. 3). There are clearly defined optima in respiration rates in the 0-, 1-, and 10-ppm chromium-amended cultures at 1.5 days of incubation. The values of the optimal rate are about the same in all three cultures; rates then fall, and become essentially constant for the rest of the incubation period. There is no clearly defined maximum for respiration from the 100-ppm chromium culture; rate of respiration remains approximately constant throughout the incubation period.

A Potential Mechanism for Cr Effect on Soil Cultures

While effects on respiration at the lower chromium concentrations could be explained by slower rates of carbon metabolism (the nature of this metabolism

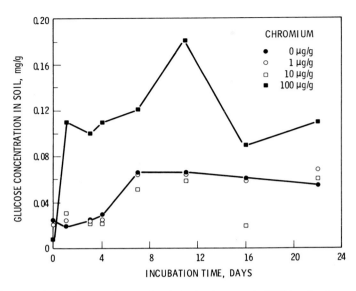

FIG. 5. Influence of chromium on glucose levels in starch-amended (1%) soil cultures containing chromium at 0, 1, 10, or 100 ppm. Cultures were incubated as described for respiration experiments (see Fig. 3).

not changing from soil containing no added chromium), the effect of 100-ppm chromium suggested that a radical change in carbon metabolism might be occurring. In this situation, "carbon metabolism" refers to the metabolic processing of glucose released from the carbon source, starch, by the action of exocellular amylotic enzymes. A simple explanation of this apparent effect could be that chromium, added as CrO_4, inhibits these enzymes, resulting in less available glucose. Carbon dioxide evolution is thus limited in rate by the availability of the sugar. To examine this possibility, we determined glucose levels by glucostat analyses in soils amended with the starch substrate and chromium at 0, 1, 10 and 100 ppm, and then incubated as in the previous experiments (Fig. 5). Results were the converse of what would be expected if chromium inhibited glucose formation from starch: Throughout the period of incubation, there were higher levels of glucose in the soil from the 100-μg/g culture than in the 0-, 1-, or 10-ppm-amended soils. There were essentially no differences in free glucose levels in the 0, 1, and 10 ppm chromium soil cultures.

Given these data, we decided to measure glucose utilization in cultures where glucose, not starch, was used as carbon source. We could then measure moles of sugar utilized versus moles of carbon dioxide evolved, a crude (but reasonable) measure of changes in carbon metabolic potential as a function of chromium concentration. We found, as was the case in starch- and chromium-amended soils, that small differences in soil respiration occurred at 1 and 10 ppm chromium; large changes occurred at 100 ppm chromium (Fig. 6). Data on glucose

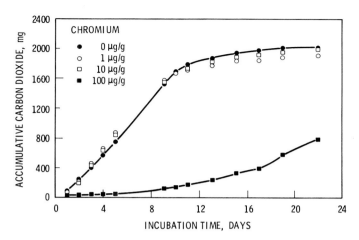

FIG. 6. Influence of chromium on cumulative carbon dioxide evolution from glucose-amended (1%) soil cultures with chromium at 0, 1, 10, or 100 µg/g day soil. Conditions of culture were the same as for starch-amended cultures (see Materials and Methods; Fig. 3).

utilization (Fig. 7) also mirrored results from starch-amended soil cultures, in that 0, 1, and 10 µg/g chromium-added cultures showed little difference in soil glucose levels as a function of incubation time, whereas the 100-ppm-amended culture showed significantly less glucose use/unit time relative to the 0 metal control.

A good index of what kind of change might be occurring in these cultures

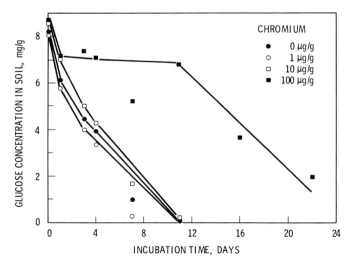

FIG. 7. Influence of chromium glucose levels in glucose-amended (1%) soil cultures containing 0, 1, 10, or 100 µg/g soil cultures. Data were from cultures described in Fig. 6.

TABLE 3. *Influence of chromium on microbial utilization of glucose relative to soil respiration (CO_2 evolution)*

Level of chromium addition ($\mu g/g$)	Total^{a,b} glucose utilized (mM)	Total CO_2^{a,b} evolved from soil (mM)	mM CO_2/mM glucose
0	9.2	41.5	4.5
1	9.1	39.5	4.3
10	9.0	40.3	4.5
100	5.5	13.1	2.4

Data were derived from experiments described in Figs. 6 and 7.
^{a} Soil weight, 200 g.
^{b} Calculated for the incubation period 3–11 days except at the 100 $\mu g/g$ addition level which was calculated for the period 11–22 days.

is provided by the ratios of carbon dioxide evolved per mole glucose utilized (Table 3). Total glucose utilized and total CO_2 produced do not change as a function of chromium in the 0-, 1-, and 10-ppm-amended cultures. The 100-ppm culture uses less glucose and evolves less CO_2. The ratio mM CO_2/mM glucose utilized is approximately 4.5:1 for 1 and 10 ppm chromium. It drops to 2.4:1 at 100 ppm chromium. These results suggest that there is a change in carbon metabolism for soil cultures grown at the high chromium dose. While the cultures at lower concentrations of chromium appear to metabolize glucose in an oxidative fashion (e.g., ratios of CO_2/glucose utilized are what might be expected of sugars metabolized through the tricarboxylic acid cycle), the soil cultures at 100 ppm chromium appear to be metabolizing the sugar in almost fermentative fashion (the ratio CO_2/glucose utilized is what might be expected of metabolism through, for example, an acetate or ethanol formation).

Earlier in this chapter, we presented data suggesting large quantitative changes in soil microbial population. The metabolic data just described suggest that this change may be qualitative as well as quantitative; that is, unique soil microflora, of altered metabolic potential relative to microbiota present in soils containing no soluble chromium, have been selected in relatively large numbers by the presence of chromium.

Comparative Evaluation of Chromium Versus Other Toxic Metals

We have stated that effects on soil microbiota and respiration rate were determined for 17 metals. The next question is, how does chromium compare, in effects on soil metabolism and microbial populations, with the other metals examined? We can compare the effects of metals on the measured parameters in a qualitative fashion. Table 4 shows where statistically significant effects occurred on a given parameter, and at what concentration of metal. A metal-

TABLE 4. *Summary of the effects of metals on respiration rate and on the relative distribution of microorganisms in soil*

Metal	Respiration rate	Relative effect on[a,b] numbers of total aerobic bacteria	Distribution of other microorganisms relative to total aerobic bacteria
Ag	+ (1, 10, 100)	+ (1, 10, 100)	+ (1, 10, 100)
As	+ (1, 10, 100)	0	+ (10, 100)
Cd	+ (10, 100)	+ (100)	+ (1, 10, 100)
Co	0	+ (100)	+ (100)
Cr	+ (1, 10, 100)	+ (10, 100)	+ (10, 100)
Cu	+ (10, 100)	+ (100)	+ (100)
Fe	0	+ (1, 10, 100)	+ (1, 10, 100)
Hg	+ (1, 10, 100)	+ (1, 10, 100)	+ (1, 10, 100)
Mn	0	+ (1, 10, 100)	+ (1, 10, 100)
Mo	+ (100)	+ (1, 10, 100)	+ (1, 10, 100)
Ni	+ (100)	+ (100)	+ (100)
Pb	0	0	0
Sb	+ (100)	+ (100)	+ (100)
Sn	0	0	+ (10, 100)
Tl	+ (100)	+ (1, 100)	+ (1, 10, 100)
W	0	0	0
Zn	0	+ (100)	+ (10, 100)

Data were from soil cultures with test metals added at 0, 1, 10, and 100 μg/g soil. Experimental procedures are described in Materials and Methods.

[a] The + denotes a pronounced metal-induced change relative to controls; the 0 indicates that pronounced metal-induced change did not occur.

[b] Parenthetical values indicate metal amendment levels in μg/g at which effects were observed.

induced effect, whether positive or negative, is scored +; zero means that no effect was observed over the metal concentration range employed: 0, 1, 10, and 100 ppm. Effects on other microorganisms were examined, relative to population of total aerobes, by determining changes in ratio to total aerobes.

One can see from the table that chromium does not affect the parameters measured as much as the metals Ag and Hg. On the other hand, chromium would appear to change soil microflora and metabolism, as a function of its concentration in soil, more than the metals W and Pb; and, perhaps, Sb and Ni. But, how does one rank chromium versus Mn, where changes on microbial populations occur at lower concentrations, but no effects on respiration are observed at any concentration; or, how does one rank Cr vs. Mo, etc.? To resolve this problem, we attempted to develop consistent numerical ranking systems. Achieving experimental consistency in work with soils appears to be considerably more difficult than achieving consistent results in other biological systems, such as the liver of the Wistar rat. In our experiments, we used one set of soil samples from one site, all amended with carbon and nitrogen in similar fashion, held at the same temperature and moisture content. It should follow that all the soil samples contain a reasonably constant amount of naturally occurring metal; this was indeed the case experimentally.

We found, however, that there was a wide degree of variability for controls run with different sets of soils, in spite of our attempts to maintain constant experimental conditions. We define a set as a subsample of soil from a large store, from which we determined effects of one or two metals on experimental parameters. We found pronounced variation in both CO_2 evolution and microbial numbers for the eight control soils run in parallel with the 17 metal-added soils. There was no real trend in this variability which might suggest cause; that is, time of storage, sets of soils drawn from different stores, etc., do not appear to correlate with the observed changes. There appeared to be no pronounced storage-dependent decreases in soil respiration rate or microbial numbers: Fresh soils, dug recently from the field, might give low respiration rates and microbial numbers, whereas soil stored for 6 months might yield high values for respiration, and/or microbial populations, versus the values observed when it was first used.

However, although variable results occurred, two general observations gave us some degree of confidence in our ability to define the effects of metals on soil microbial communities:

1. Replicate controls, run at the same time from the same subsamples, showed only small degrees of variation in measured parameters.

2. Effects for a given metal versus its control were consistent. That is, within different soil sets, increases or decreases in a parameter of similar magnitude (with respect to control) were observed.

These considerations led to the use of ratios in analyses of effects of a given metal on soil microbial populations and metabolism, and in determining a comparative ranking of metal effects.

Metal Ranking System I

In the first of two ranking systems (Table 5), we determined the value of the ratio of numbers of a class of organisms in soil plus metal after 6 days of incubation, versus numbers from a control incubated 6 days minus metal:

$$R_i = Nm_i/Nm_o \qquad \text{(i)}$$

where N is the mean number of a class of organisms in soil with m, amended metal concentration, at 1, 10, or 100 ppm; m_o = no metal added. If N_{mi} is less than N_{mo}, however, the reciprocal of ratio (i) was employed:

$$R_i = Nm_o/Nm_i \qquad \text{(ii)}$$

In this system, an attempt was made to compare all data on a common basis. Thus, an overall standard error (in percent) was determined for each class of organism. The standard error varied between 16.2 and 23.2%, with an average value of 19.6%.

The same procedure was used for respiration data: we calculated ratios of

TABLE 5. *Method of rank determination: System I*

(i) $R_i = Nm_i/Nm_o$
    N = mean number of organism in soil, i = added metal concentration, m_o is no metal added control

(ii) If $Nm_i < Nm_o$ then $R_i = Nm_o/Nm_i$
    Ratios for respiration data determined as in (i) and (ii) above

(iii) $MR_i = k_1 R_o + 4k_2 (R_A + R_{MR})$
    MR = Numerical value rank
    R_o = Average ratio all organism classes
    $k_1 = 1; k_2 = 0.5$
    R_A = Cumulative CO_2 at 12 days + metal/cumulative CO_2 at 12 days—Metal or reciprocal
    R_{MR} = Maximum respiration rate + metal/maximum respiration rate − metal or reciprocal

(iv) $\overline{OMR} = \dfrac{MR_{1\ ppm} + MR_{10\ ppm} + MR_{100\ ppm}}{3}$

maximal respiration rate plus metal, versus control, as well as ratios of cumulative CO_2 produced in soils plus metal at 12 days of incubation, versus control. The average standard error associated with respiration rate was ± 5% over all data.

This first ranking system attempted to statistically weigh the body of data and to "equalize" effects on respiration with effects on microbial numbers. We used the empirical equation shown here ($MR = k_1R_o + 4k_2(R_A + R_{MR})$, where MR is the numerical value of the metal rank at one of the test metal concentrations, R_o is the average ratio for all organism classes, R_A is the ratio of cumulative CO_2, and R_{MR} the ratio of maximum respiration rate. The constants, k_1 and k_2, are normalization factors with values of 1 and 0.5, respectively. They are used to give equivalent weights to the combined respiratory parameters and the parameter derived from microbial numbers. The multiplicand, 4 (for respiratory parameters), is used to compensate for the increased statistical significance of ratios derived from respiratory data, versus those derived from microbial numbers (that is, 20% average standard error for microbial numbers/5% standard error for respiratory data = 4). For overall metal rank, we used the average of metal rank values at the three test concentrations.

Metal Ranking System II

The second ranking system also uses the ratio of parameters measured in the presence of metal, versus the parameter measured in the absence of metal. In the first system, a large change in any one parameter will weight a metal's rank possibly even more than several smaller changes in a number of parameters. In order to correct for this, the second system has an arbitrary index of change assigned to certain ranges for ratio values. After indices of changes are assigned and appropriate normalization is performed, the indices values are added, resulting in a numerical rank.

TABLE 6. *Numerical definition of values for indices of change used in metal ranking System II*

A. For microbial numbers	
If R_i	Then indices =
1 to 1.5	0
1.5 to 3	1
3 to 5	2
5 to 100	3
above 100	4
B. For respiration data	
If R_i	Then indices =
1 to 1.13	0
1.13 to 1.5	1
1.5 to 3	2
above 3	3

TABLE 7. *Method of rank determination: System II*

(i) At metal level n_1
 $I_{tot_n} = \Sigma O_n + 3 \, (RR_n + Acc_n)$
 O_I = Index number individual microbial classes at [metal] = n
 RR_n = Index number for maximum respiratory rate at [metal] = n
 Acc_n = Index number for accumulated CO_2 at metal = [n]

(ii) $OMR = \Sigma I_{tot_n}$
 n = 1, 10, 100 ppm

The arbitrary index of change values is given in Table 6. The ratios were determined as previously described. The 0 levels of index were chosen as those levels representing no statistically significant change. The index of change numbers for respiratory data was weighted to approximate equivalence with the index of change numbers for microbial population by multiplying by a factor of three (Table 7). At metal level n, $I_{tot_n} = \Sigma O_n + 3 \, [RR_n + Acc_n]$, where O_n is the index number obtained from ratio values for the six organism classes, RR_n is the index number obtained from ratios of maximal respiration rates, and Acc_n is the index number obtained for cumulative CO_2 ratios. The I_{tot_n} for each level of amendment was used to determine overall metal rank (OMR) by simple addition:

$$OMR = I_{tot(1 \, ppm)} + I_{tot(10 \, ppm)} + I_{tot(100 \, ppm)}.$$

Comparison of Metal Toxicity Ranking

Both ranking systems are empirical and arbitrary. In the second system, the values assigned to the index of change, and the cut-off points used for increases or decreases in change values, can all, obviously, be manipulatable, resulting in significant changes in the numerical value of OMR. However, both System

TABLE 8. *Overall metal rank determined by ranking Systems I and II*

System I		System II	
Metal	OMR	Metal	OMR
Ag	53.30	Hg	71
Hg	37.30	Ag	66
Cr	15.10	Cr	49
Cd	9.96	Cd	33
Cu	8.19	Zn	29
Ni	8.03	Fe	25
Zn	7.58	Ni	22
Tl	6.96	Cu	20
Fe	6.90	Sn	19
Sn	6.19	Mn	14
W	5.93	Tl	14
Mo	5.87	As	11
Mn	5.86	Mo	10
As	5.84	W	10
Co	5.69	Co	6
Sb	5.62	Pb	5
Pb	5.55	Sb	4

I and System II, using absolute values for ratios, give similar rankings for the 17 metals in our study (Table 8).

The OMR obtained using ranking System I is an averaged number; that is, MR for the three test metal concentrations is divided by three. In ranking System II, the additive results of ranking values at the three metal concentrations can be obtained. For metals having no effect (either positive or negative) on the measured parameters, we would obtain a numerical rank of 5 by using ranking System I. Either positive or negative effects, in this ranking system, could give an infinitely high value to the rank. In System II, the rank value for "no effect" would be 0; a metal showing maximal effect on all parameters would, using our indices, score 126 for overall metal rank value.

Comparing the overall metal ranks, we see that rankings for the first four metals (Ag, Hg, Cr, Cd) are common to both ranking systems. Positional translocations are never greater than three ranks, and the numerical value of rank, after about the first nine metals, becomes rather uniformly low.

While examining this table of overall ranks, it may be worthwhile to discuss the trends in data observed for metal ranks at the three test concentrations used in our studies. The 17 metals could be grouped into three classes: nutrilites, toxicants, and metals having little or no effect. Metals behaving as nutrilites (Fe, Sn, Zn) caused relatively great changes in measured parameters, thus scoring higher in rank at low metal concentrations (1 and 10 ppm). The magnitude of rank value for these metals did not change greatly at the 100-ppm level of amendment. Toxicants which affected all measured parameters (such as Ag, Hg, Cd, Cr, Ni) increased in magnitude of effect as concentration of metal increased. Rank change as a function of metal concentration was, naturally, least consistent for those metals having little effect.

Both ranking systems gave similar ranking of metals at all three metal concentrations. Total agreement between the two systems was best at the 100-ppm level of amendment, where positional correspondence occurred for 11 of the 17 test metals. This is most likely due to the greater effect of most metals at this high concentration.

Metals causing great effect (such as Ag, Hg, and Cr) appear to correspond in OMR. The same is true for metals that have little effect (Pb, Sb, W). The majority of the tested metals showed rather selective effects on specific classes of microorganisms, or on respiration. With the exception of the nutrilite metals, the effect increased with increasing metal concentrations (rate of respiration or numbers of organisms decreased).

Even though the ranking systems are not absolute, they reflect the effect of chromium relative to other metals in soil: while it does not affect measured parameters as much as the most toxic elements (Ag and Hg), chromium has a much greater effect than the other test metals (Cd to Pb). It appears to be one of the more toxic metals to soil microflora.

Biochemical Analyses: Chromium in Mixed Soil Microbial Cultures

We desired to perform a limited biochemical analysis of mixed liquid cultures derived from chromium-stressed soils. We isolated these mixed cultures from soil, based on chromium resistance and ability to grow and metabolize different carbon sources.

Location of Chromium in Mixed Microbial Cultures

To determine whether the cultures transported, occluded, or associated with the metal, we determined the distribution of chromium in the cellular and exocellular compartments of the liquid cultures. We examined potential to gel permeation chromatography (GPC), TLC, and TLE. These techniques generally separate materials on the basis of molecular size, solubility properties, and charge.

As shown in Table 9, we first examined the quantity of soluble chromium (present initially as chromate) in the exocellular media of mixed fungal and bacterial cultures, isolated from soil, and grown in the presence of mixed sugars or organic acids, with metal added both at the stationary-growth phase (to identify metabolites from normal metabolism), and at the beginning of growth (to identify metabolites formed on continuous metal exposure and stress). These experiments utilized ^{51}Cr as a tracer. In sterile media controls (not shown), the chromium remained 100% soluble as the chromate. However, comparing the values for fungi and bacteria, it is apparent that the quantity of chromium soluble in the exocellular media was lower, regardless of time of metal exposure, when sugars were used as a carbon source, as compared to organic acids.

TABLE 9. *Chromium in the exocellular media of microbial mixed cultures*

	Concentration of chromium (%) soluble in exocellular media[a]			
	Fungi		Bacteria	
Time of exposure	Mixed sugars	Organic acids	Mixed sugars	Organic acids
Stationary phase	46	77	72	83
Continuously	72	97	73	99

Bacterial and fungal mixed cultures, derived as described in Materials and Methods (see Fig. 2), were tagged with ^{51}Cr as CrO_4^{2-} and distribution of metal in cultures was determined.
[a] Cultures not replicated. Analytical precision was $< \pm 10\%$.

Comparing the values for fungi, it is apparent that solubility was lower when the cultures were equilibrated with the metal at the stationary phase, compared to continuous exposure of the organisms to the metal. This was true only for the bacteria grown on organic acids, not for the bacteria grown on sugars.

It would appear that solubility of chromium in the culture, which was perhaps determined by the chemical form of chromium, was a function of organism type, carbon source, and period in the life of the culture when exposed to chromium.

Modification of Chromium: Gel Permeation Chromatography

As an illustration of microbially mediated modification of chromium, we examined the results of biochemical analyses for the exocellular component of liquid mixed-fungal cultures grown on the mixed sugar carbon source. Gel permation chromatography (Sephadex G-25) of ^{51}Cr-tagged materials verified that changes in the solubility of chromium in the exocellular media were accompanied by changes in the form of chromium, as compared to the medium alone (Fig. 8).

The upper exclusion limit for the column employed was 4,000, so components with equivalent molecular weights greater than 4,000 should have exited the column at the eluant volume. The dotted line on the right in Fig. 8 represents the position of chromate in the medium alone. It is apparent, given the mobility of chromium as chromate in sterile media (Fig. 8A), that growth of microorganisms altered the form of chromium, both through simple interaction with metabolites at the stationary phase (Fig. 8B), and, perhaps, by modes of modification more specific for the presence of the metal insult (Fig. 8C). There appeared to be some residual chromate in the exocellular fraction from the single exposure at the stationary phase (Fig. 8B), but none remained under continuous exposure conditions (Fig. 8C), the treatment which resulted in the greatest chromium solubility.

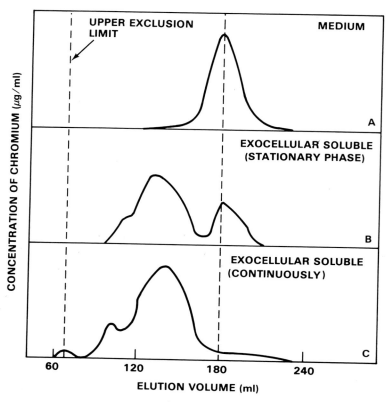

FIG. 8. Separation of soluble chromium complexes in mixed fungal liquid cultures by gel permeation chromatography on Sephadex G-25. **A.** Medium control. Chromium added as CrO_4^{2-} plus $^{51}CrO_4^{2-}$ tracer to sterile mixed sugar media and equilibrated with media for 96 hr. **B.** Stationary-phase addition of chromium. Mixed fungal cultures grown on mixed sugar carbon source received CrO_4 plus $^{51}CrO_4$ as tracer at stationary growth phase. The exocellular soluble fraction of this culture (passes 0.01-μm filter) was placed on the column. **C.** Chromium present throughout culture growth. As above, except CrO_4^{2-} plus $^{51}CrO_4^{2-}$ tracer was added at time culture was inoculated. See Materials and Methods for further description.

Modification of Chromium: Thin Layer Chromatography and Electrophoresis

Thin-layer chromatography was applied to the material subjected to gel permeation chromatography (Fig. 9). Chromium-containing materials were visualized by staining with diphenylcarbazide. Chromium in water and in sterile media was present as a single component, verified chemically, prior to chromatography, to be chromate. The slight differences in migration for chromate in water versus chromate in sterile media are probably due to the location of a nonchromium-containing component of the medium, which is located (as visualized by UV) immediately in front of the chromium component on the plate, thus restricting its migration. Thin-layer chromatography tended to verify the solubility, and

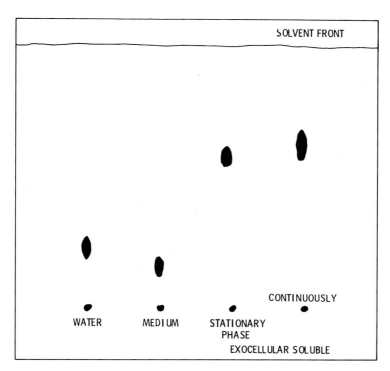

FIG. 9. Thin-layer chromatograph of soluble (passes 0.01-μm filter) exocellular fraction of a mixed fungal culture grown on mixed sugar carbon source. Chromatography matrix was cellulose and solvent was NH_4OH:ethanol (1:4). CrO_4^{2-} added at either stationary phase or present throughout growth. Chromium-containing spots visualized with diphenylcarbazide (see Materials and Methods). Controls are CrO_4^{2-} in water or sterile medium equilibrated 46 hr.

GPC analyses, in that the exocellular media from both growth conditions contained components with chromatographic mobility different from that of the controls and, perhaps, different from each other.

Thin-layer electrophoresis (Fig. 10) also verified that alteration of chromium form occurred on microbial growth. The chromate in both the water and the medium was present in a single negatively charged spot, migrating toward the positive pole. The exocellular soluble fraction of the fungal culture exposed to the chromium at the stationary phase contained three chromium complexes, while only two were present in the continually exposed culture. All complexes formed in the exocellular media were positively charged, migrating to the negative pole.

These results suggest that soil microbiota, at least in mixed culture, are capable of changing the chemical form of chromium (added as chromate) in the media. They suggest that intermediates in the metabolism of these mixed cultures (chromium form changed upon addition to stationary-phase cultures) are able to

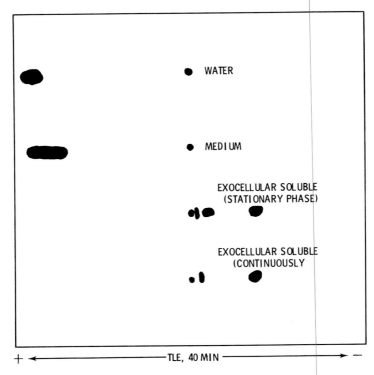

FIG. 10. Thin-layer electropherogram of soluble (passes 0.01-μm filter) exocellular fraction of a mixed fungal culture grown on mixed sugar carbon source. Plate matrix was cellulose, and buffer was 0.01M HEPES, pH 7.4. Chromium-containing spots visualized with diphenylcarbazide (see Materials and Methods). CrO_4^{2-} added at either stationary phase or present throughout growth. Controls are CrO_4^{2-} in water or sterile medium equilibrated 96 hr.

modify the form of the metal, and that cultures stressed throughout growth by the presence of chromium may be capable of different (or further) modification of the added metal.

DISCUSSION

We are in the process of analyzing some 110 cultures of single isolates of chromium-resistant soil microbiota for their ability to modify chromium, as judged by GPC, TLC, and TLE. Early results suggest that these purified strains, like the mixed cultures, are capable of modifying chromium (added to their media as CrO_4), both as a consequence of metabolism and, perhaps, as a result of more specific mechanisms of modification.

One of our more interesting general observations on chromium, and one that appears to apply to other metals we are presently analyzing, is that metabolites produced as a result of growth of chromium-resistant cultures appear to be capable of modifying the form of chromium added to the milieu. These results suggest that one may not have to postulate and then search for exotic classes of compounds to determine the potential for microbially mediated chromium modification and mobilization from soils. It may be that some very common materials, such as amino acids, peptides, or saccharides, produced as a normal consequence of growth of chromium-resistant microbiota, may serve as vectors for chromium mobilization through soils to the roots of plants, to plant tissue, and then to man.

ACKNOWLEDGMENT

This work was supported by the National Institute for Environmental Health Sciences under contract no. 2311100844.

REFERENCES

1. Bremner, I. (1974): Heavy metal toxicities. *Q. Rev. Biophys.*, 7:75–124.
2. Frost, D. V. (1967): Arsenicals in biology—retrospect and prospect. *Fed. Proc.*, 26:194–208.
3. Lochhead, A. G. (1940): Qualitative studies of soil microorganisms. III. Influence of plant growth on the character of the bacterial flora. *Can. J. Res., Sect. C.*, 18:42–53.
4. Littman, M. L. (1947): A culture medium for the primary isolation of fungi. *Science*, 106:109–111.
5. Conn, H. J. (1921): The use of various culture media in characterizing actinomycetes. *N.Y. Agr. Exp. Sta. Tech. Bull.*, 83, 26 pp.
6. Stotzky, G. (1965): Microbial respiration. In: *Methods of Soil Analysis, Part 2,* edited by C. A. Black, pp. 1550–1558. American Society of Agronomy, Inc., Madison, Wisconsin.
7. Salomon, L. L., and Johnson, J. E. (1959): Enzymatic microdetermination of glucose in blood and urine. *Anal. Chem.*, 31:453–456.
8. Keston, A. S. (1956): Specific colorimetric enzymatic reagents for glucose. In: *Abstracts of Papers, 129th Meeting of the American Chemical Society,* April 1956, p. 31.
9. Washko, M. E., and Rice, E. W. (1961): Determination of glucose by an improved enzymatic procedure. *Clin. Chem.*, 7:542–545.
10. Stanier, R. Y., Palleroni, N. J., and Doudoroff, M. (1966): The aerobic *Pseudomonads:* A taxonomic study. *J. Gen. Microbiol.*, 43:159–271.
11. Robinson, A. V., Garland, T. R., Schneiderman, G. S., Wildung, R. E., and Drucker, H. (1977): Microbial transformation of a soluble organoplutonium complex. In: *Biological Implications of Metals in the Environment,* edited by H. Drucker and R. E. Wildung, pp. 52–62. NTIS, Springfield, Virginia.
12. McGonigle, E. J. (1975): Thin layer chromatographic analysis in waste chemistry. In: *Chromatographic Analysis of the Environment,* edited by R. L. Grob, Chapter 19. Marcel Dekker, New York.

Biomedical and Environmental Significance of Siderophores

J. B. Neilands

Biochemistry Department, University of California, Berkeley, California 94720

INTRODUCTION

Microbial iron assimilation has lately become a popular topic for research and much is written on the subject in the current literature (5,18,26). The present review is largely speculative and will explore the subject from a new angle, namely, the possibility of transforming basic research in this field into a useful applied technology.

Without appearing to be overly enthusiastic, it is my belief that basic research on microbial iron metabolism is pregnant with possibilities for practical application. Before launching into this matter, however, a brief review of the current level of understanding of the fundamental process of microbial iron uptake is in order.

Iron occurs in both heme and non-heme proteins and is *probably* a universal nutrient for all living cells (14). The element is found in redox catalysts with potentials ranging from ferredoxin to cytochrome oxidase. Even in cells acquiring energy via the non-iron utilizing processes of glycolysis and fermentation, the element will still be required, at least in species as disparate as *Escherichia coli* and man, for synthesis of ribotide reductase (30). The latter enzyme is responsible for generation of the deoxyribotide precursors of DNA. However, in order to make these critically important biocatalysts, the iron must first be assimilated. Herein lies a difficulty. Dating from the advent of O_2 evolving "plant type" photosynthesis, the surface iron of the planet has existed predominantly in the ferric form, which exhibits a profound insolubility at neutral pH ($K_{sol} = <10^{-38}$ M) (23).

During the past decade, an impressive body of knowledge has accumulated on the mechanism of iron assimilation by aerobic and facultative anaerobic microorganisms (24). It has been discovered that such species are equipped with two mechanisms for acquisition of iron, namely, systems which have been dubbed *low affinity* and *high affinity* (26).

Low Affinity Iron Transport

In the presence of sufficiently high levels of inorganic iron, the element becomes available to the cells by a nonspecific process. The latter does not readily lend itself to genetic manipulation and hence little is yet known about how the system operates. The low affinity process will hold in all circumstances where cells are grown in complex laboratory media or in synthetic media in which the available iron concentration is ca. micromolar or higher.

High Affinity Iron Transport

When the iron concentration is less than about micromolar or when iron is bound in a form which is unavailable to the cell, the high affinity system is invoked. The latter consists of two parts: namely soluble, relatively low molecular weight, ferric-specific ligands collectively designated *siderophores,* and the matching membrane receptor for the iron-laden form of the siderophore. Both components of the high affinity system are repressed by iron. The high affinity system has evidently been retained through evolutionary time as a means of assuring survival of microorganisms under environmental conditions where iron is limiting, such as within host tissue, in water, and in certain soils.

TABLE 1. *Distribution of siderophores among bacteria, yeast, and fungi*[a]

Microorganism	Siderophore
Bacteria	
Escherichia coli	enterobactin (enterochelin)
Salmonella typhimurium	
and other enteric bacteria	
Azotobacter vinelandii	2-*N*,6-*N*-di-(2,3-dihydroxybenzoyl)-L-lysine
Bacillus subtilis	2,3-dihydroxybenzoylglycine
Klebsiella oxytoca	2,3-dihydroxybenzoylthreonine
Paracoccus denitrificans	2,3-dihydroxybenzoyl spermidine compounds
Bacillus megaterium	schizokinen
Anabaena sp.	schizokinen
Arthrobacter pascens	arthrobactin
Pseudomonas fluorescens	ferribactin
Streptomyces sp.	ferrioxamines
Yeast	
Rhodotorula and related yeasts	rhodotorulic acid
Fungi	
Ascomycetes	ferrichromes
Basidiomycetes	
Fungi imperfecti	
Aspergilli	
Neurospora	
Penicillia	
Fusaria	fusarinines
Neurospora	coprogen
Penicillia	coprogen

[a] For a more complete list, see the article published by G. C. Rodgers and J. B. Neilands in *Handbook of Microbiology,* A. I. Laskin and H. A. Lechevalier, eds., CRC Press, 1973, p. 823.

The high affinity iron transport system has been most thoroughly studied in enteric bacteria such as *E. coli* and *Salmonella typhimurium* (26). The wealth of genetic information which has been accumulated with these species is reflected in our substantial knowledge of the genetics and biochemistry of their iron metabolism. The enteric bacteria synthesize their own siderophore, called enterobactin (or enterochelin), plus cognate membrane receptor; these species also made efficient use of siderophores such as ferrichrome (20). The latter is not made by enteric bacteria but is a typical fungal siderophore; e.g., all *Penicillia* examined thus far produce ferrichrome. To date several dozen siderophores have been chemically characterized from microorganisms and most of them fall into one of two classes, the catechols and the hydroxamic acids (Table 1) (25). Enterobactin (Fig. 1) and ferrichrome (Fig. 2) are prototypes of the catechol and hydroxamate type siderophores, respectively.

The data recorded in Table 1, which comprise only a partial list, predict that siderophores will probably be found in all aerobic and facultative anaerobic organisms which are critically investigated for their presence. The ubiquitous presence of siderophores in typical soil organisms is worthy of special note. Strict anaerobes have never been reported to contain siderophores, possibly because the ferrous ion is sufficiently soluble to satisfy their growth requirements. However, certain anaerobic organisms have a high requirement for iron and

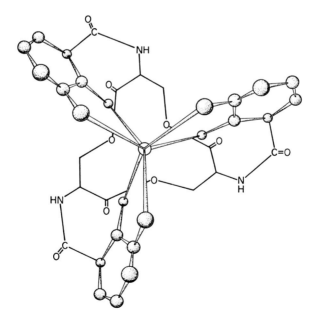

FIG. 1. Ferric enterobactin, the ferric complex of cyclo-tri-2,3-dihydroxybenzoyl serine. This is the common siderophore of enteric bacteria and is the prototype of the catechol members of the series of microbial iron carriers.

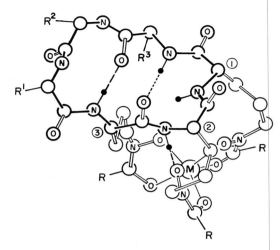

FIG. 2. Ferrichrome, the prototype of the hydroxamate type of siderophores. Ferrichrome is a ferric trihydroxamate complex of a cyclohexapeptide. It is produced commonly by fungi, such as the *Penicillia*, but is capable to feeding iron to *Escherichia coli* mutants deprived of their native high-affinity carrier (enterobactin).

at the same time generate H_2S. Thus, such species must be equipped with a device for obtaining soluble iron ions from ferrous sulfide.

As chemical entities, siderophores are generally all-oxygen, hexadentate ligands (25). They are virtually ferric-specific and exhibit formation constants ranging from 10^{30} to 10^{50} M. They penetrate to the cytoplasm, the ferric ion is released or transferred by reduction, and the ligand is then returned to the environment to participate in a new round of iron transport (19). In vesicles prepared from the cytoplasmic membrane of *E. coli*, ferrichrome is transported by a process resembling nonproton symport (22).

Little is yet known about the molecular mechanics of siderophore transport except that in the case of both ferric enterobactin and ferrichrome, outer membrane receptors are needed for the most efficient utilization of the molecule. The size of the typical siderophore is 500 to 1,000 daltons, a number apparently too large to allow free diffusion through the small water-filled pores of the sealed outer membrane of Gram-negative bacteria (26).

It has been demonstrated in several laboratories that the receptors for ferrichrome and ferric enterobactin also serve as the binding sites for certain specific phages and bacteriocins (11,35). Apparently these noxious agents have "learned" how to exploit siderophore receptors as a means of gaining entry to the cell. This observation affords a satisfactory explanation for the biochemical function of receptors for *E. coli* phages such as T1, T5, and Φ80 (26).

In the author's laboratory, a major research effort is now concentrated upon the cloning of the iron operon in *E. coli*. Hopefully, this will lead ultimately to identification of the structures of all of the gene products involved in the high affinity iron assimilation pathway, and its regulation. Armed with this detailed knowledge of iron assimilation in one living cell, we may now ask how such basic information could be adapted to the biomedical and other fields of applied science.

Simple iron deficiency anemia is the single most important nutritional disease in our society. Since iron is never completely lacking in the diet, this may be ascribed, in large measure, to inefficient utilization of the element. There is no biological mechanism for excretion of iron, and the repression-type control, which appears to hold for microbes, plants, and animals, operates to admit just enough iron to balance losses of "wear and tear." If we could understand how this regulatory system works in *E. coli,* we would have a model for iron assimilation in higher organisms, including man. In eukaryotes, the mitochondria, undoubtedly the evolutionary descendants of bacteria, contain much of the functional iron of the cell.

Improved iron nutrition of the animal organisms might be the first anticipated practical application of knowledge of microbial iron transport gleaned from basic research in the field. In fact, the work has diverged into a number of additional areas such as chelation therapy. The theoretical basis for these applications will now be enumerated. To repeat: Iron is probably a universal requirement of all living cells, the lactic acid bacteria constituting perhaps the only exception to this rule. Siderophores are produced by all aerobic and facultative anaerobic microbial species but, apparently, these substances do not occur in plant or animal tissues. There is one recent, as yet unconfirmed, report that a siderophore-like compound occurs in cells in tissue culture (7). Should the growth of pathogens depend upon the high affinity iron transport system, then this provides a point of attack with a selective drug which, since it occurs so generally, is apt to lack a profound and generalized toxicity to all cells.

INFECTION AND IMMUNITY

Animals

There is now abundant evidence that iron is one of the prime determinants of virulence in organisms afflicting man. The subject has been investigated *in extenso* by Lankford, Kochan, Bullen and others and this work has recently been reviewed by Weinberg (36). The list of organisms in this category includes, but is not limited to, *E. coli, S. typhimurium, Yersinia pestis, Pseudomonas aeruginosa, Neisseria gonorrhoeae* and *N. meningitidis.* Using a strain of *S. typhimurium* believed to be deficient only in its ability to synthesize its siderophore, enterobactin, Yancy et al. (37) showed that the altered organism had greatly diminished virulence for the mouse.

Plants

In extending the work with animals to plants, Ong et al. (28) asked if certain common plant pathogens could be induced to form siderophores on low iron media. *Agrobacterium tumefaciens,* the crown gall organism, was shown to produce a new siderophore of the catechol type, named agrobactin. It is closely related to the catechol type siderophore isolated from *Paracoccus denitrificans* by Tait (33) except that it contains an additional residue of 2,3-dihydroxybenzoic acid in place of salicylic acid and the threonyl moiety is present in an oxazoline ring.

In living tissue, the iron will be locked up in the storage proteins ferritin and hemosiderin, in transferrin, lactoferrin, heme, and in iron sulfur clusters. None of this iron can be called readily available and so it makes good sense that the invading pathogenic species should be capable of switching on their high affinity iron-gathering apparatus. However, in no case has it yet been shown that an infected tissue actually contains a siderophore. Work with *Neurospora crassa* has demonstrated that the organism contains two types of siderophores, an internal form and a related exported form (13). Thus, one must be cognizant of the possibility that the siderophore detected extracellularly may not be the whole story and that the type of carrier produced may be influenced by the habitat.

ANTIMETABOLITES

Mycobactin (Fig. 3) was the first member of the class of compounds we now call siderophores to be chemically characterized and knowledge of its structure greatly facilitated identification of subsequently discovered members of the hydroxamate series, although aspergillic acid retains the distinction of being the first microbial product shown to contain the -CON(OH)-grouping. Interestingly enough, the study of mycobactin, a growth factor for *Mycobacterium johnei* produced by *Mycobacterium phlei,* was undertaken by chemists at the Imperial Chemical Industries facility in hopes of developing an anti-TB drug.

Figure 4 shows some of the possible lines of attack which could be used for the rational development of antimetabolites based on known iron assimilation pathways in microorganisms. Use of a modified siderophore depends, of course,

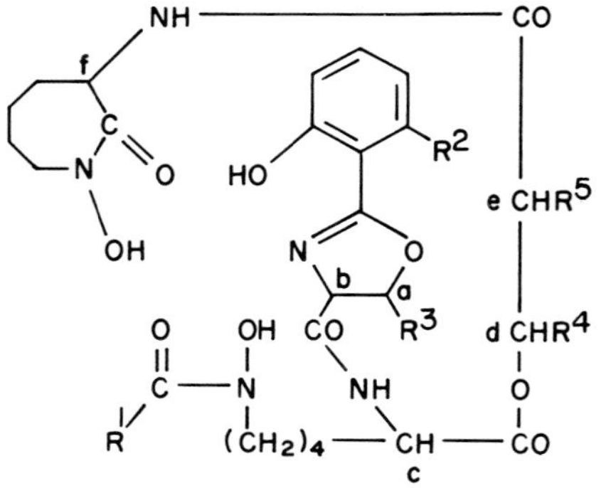

FIG. 3. Mycobactin, the siderophore of mycobacterial species. Depending on the source organisms, various different substituents occur at R^1–R^5. The sites a–f are centers of potential optical activity.

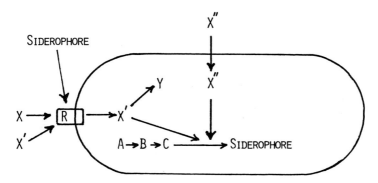

FIG. 4. Possible scenarios for the rational design of drugs and anti-metabolites based on the interdiction of the high affinity iron transport system. X, a siderophore analog, competes for a specific cell surface receptor (R), but is not transported. X', a siderophore analog, is "illicitly" transported and blocks either a vital cellular process (Y) or the biosynthesis of the siderophore. X" is a drug specifically blocking the biosyntheic path to the siderophore near its ultimate terminus. Elimination of the high affinity iron uptake system may require supplementation with a second agent, such as deferriferrichrome A, in order to deprive the cell of the low affinity assimilation pathway. Albomycin is a natural anti-metabolite utilizing the ferrichrome receptor.

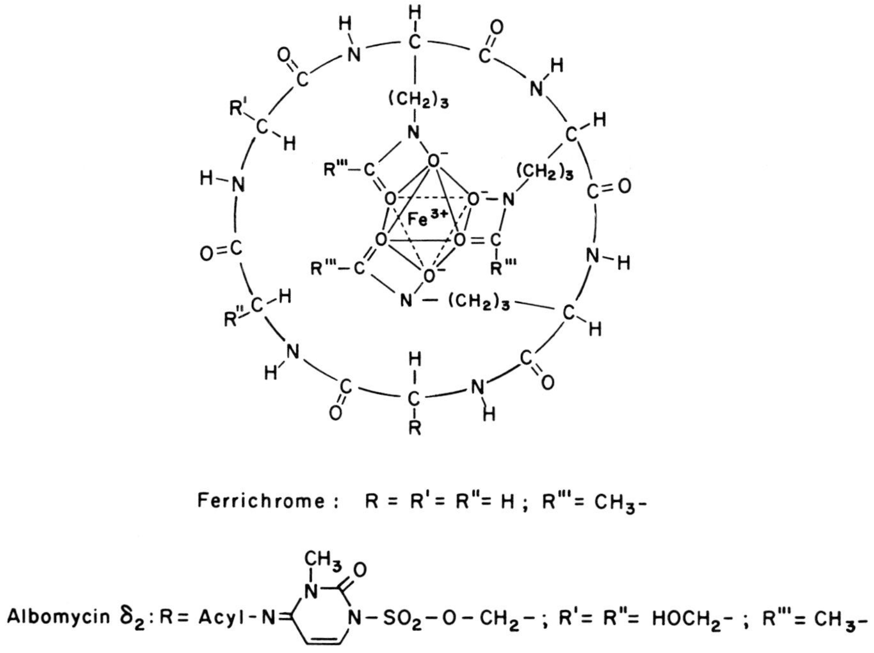

Ferrichrome: $R = R' = R'' = H$; $R''' = CH_3-$

Albomycin δ_2: $R =$ Acyl-N=⟨...⟩N-SO_2-O-CH_2-; $R' = R'' = HOCH_2-$; $R''' = CH_3-$

FIG. 5. Structures of ferrichrome and albomycin. The formula shown for the latter is probably not quite correct and may require some revision; however, it is certainly closely related to ferrichrome in its general constitution.

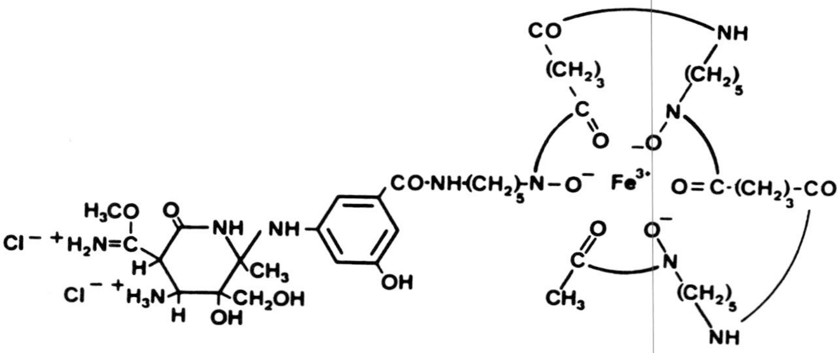

Ferrimycin A_1

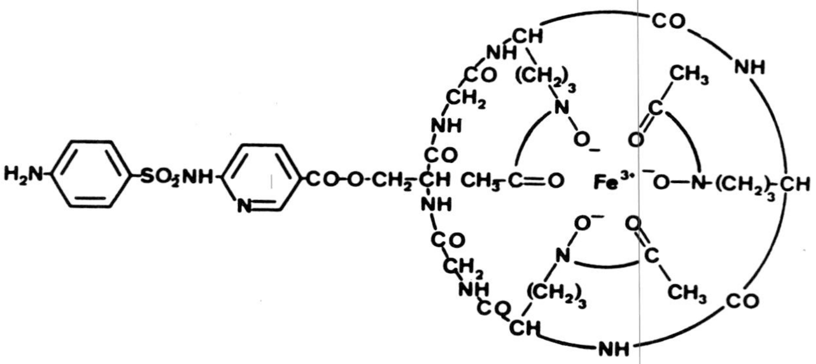

6-Sulfanilamido-nicotinic acid-ferricrocinyl ester

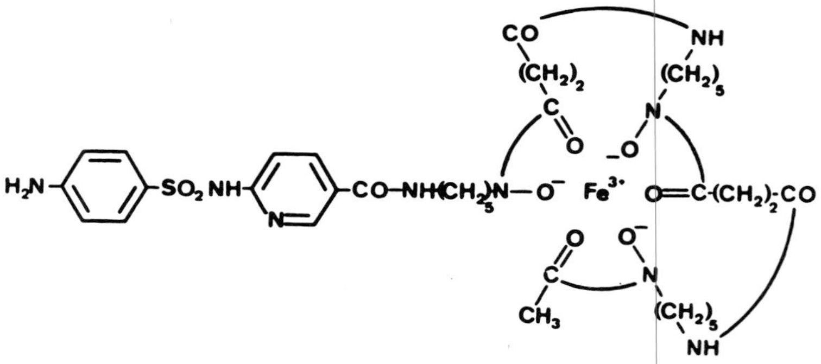

N-(6-Sulfanilamido-nicotinoyl)-ferrioxamin B

FIG. 6. Sulfonamide derivatives of ferricrocin and ferrioxamine B, the toxicities of which for *Staphylococcus aureus* can be antagonized by the corresponding unsubstituted siderophores (Naegeli, 21). Also shown is ferrimycin A_1, an antibiotic related to the ferrioxamine family of siderophores.

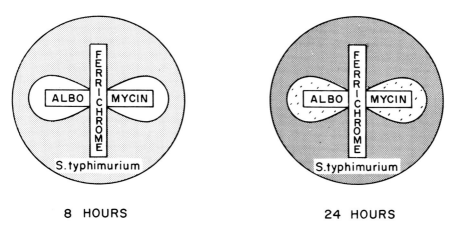

FIG. 7. The "cross test" for competition for membrane receptors. Filter paper strips are soaked in 0.1 mM test solution and placed on the surface of seeded plates.

on some flexibility in the specificity requirements of the surface receptor. There is promise that this hope will be realized since, for example, the ferrichrome receptor recognizes and transports close structural analogs of the siderophore, including albomycin (Fig. 5). Recently, the Swiss workers have reported (21) the preparation of new antimetabolites based on ferricrocin, an analog of ferrichrome in which the triglycine residue is replaced by a residue of glycylserylglycine. The latter affords an hydroxyl group available for derivation which, in the case of the Swiss investigators, was employed for the attachment of antifolate drugs (Fig. 6). Since the semi-synthetic ferricrocins antagonized ferrichrome in the "cross test" (Fig. 7) with *Staphylococcus aureus,* it is obvious that the former are able to utilize the ferrichrome receptor. The latter also accepts the coordination isomer of ferrichrome, namely, enantioferrichrome (21).

There are several advantages inherent in this approach to drug development. The surface receptors for siderophores are known to have a very great affinity for the ligand and hence the drugs should be active at low dose. Furthermore, the specificity and selectivity of the antimetabolites should be high since they are designed for the microbial target. Since the pathogen *in situ* is probably iron-starved, thanks to host defense mechanisms, its high affinity biosynthetic machinery should be derepressed and the siderophore receptor should be present in abundant quantities. Since the organism must have iron for growth and proliferation, its possibilities for development of resistance are limited. Finally, exploitation of the siderophore receptors provides an opportunity to overcome the permeability limit of the outer membrane of gram-negative bacteria, a barrier which precludes the application of some of the most useful antibiotics to these species.

Admittedly, there are certain problems with the above approach. The surface receptor must not be so specific as to exclude interaction with the siderophore analogs. This may not be the major problem, since in addition to the ferrichrome-albomycin-ferricrocin analogs series just described must be added the observation that the ferric enterobactin receptor accepts the ferric complexes of the purely synthetic compounds 1,3,5-tris(N,N',N"-2,3-dihydroxybenzoyl)aminomethylbenzene (34) and cis-1,5,9-tris(2,3-dihydroxybenzamido)cyclododecane (12). Once inside the cell, the drug will operate by a mechanism unrelated to iron metabolism, which requires that the siderophore analog must suffer modification in order that the toxic moiety can be released. To be really effective, such drugs would probably have to be given in combination with a second nontransportable iron-scavenging agent which would block out the low affinity iron assimilation pathway. Deferriferrichrome A seems to be the best candidate for this purpose since its ferric complex is not taken up even by the source organism, the iron being shunted to ferrichrome which then readily enters the cell.

The generalized ability of microorganisms to synthesize siderophores and the high affinity of the latter for ferric ion will, in most instances, defeat attempts to starve the pathogen through external application of a competing ligand for iron. However, King et al. (17) have been able to control certain dermatophytes by topical application of apotransferrin.

Relatively little is known about the biosynthesis of most siderophores, but this should be a fruitful area of inquiry for the development of new antimicrobial agents. Such agents should preferably act near the terminus of the biosynthetic chain since early intermediates will undoubtedly be required for construction of other molecules essential to the cell.

DEMETALLATION DRUGS

Although people generally suffer from an insufficiency of iron, a not insignificant number are plagued with the opposite problem of iron overload. Here we have the most obvious application of the siderophore in the clinic and the whole area of siderophore-promoted deferration is now being investigated actively, goaded on by the Iron Chelation Program of the National Institutes of Health.

Using the "cross test" as an assay, the Ciba Pharmaceutical Company and their collaborators at the Eidgenössischen Technischen Hochschule in Zurich, isolated a number of siderophores from fungal sources (16). Since the substances were as active, or even more active, than penicillin, their original hope was to find new antibiotics. This research culminated in the characterization of the ferrioxamines, comprised of a dozen or so molecular species based on repeating units of 1-amino-ω-hydroxyaminoalkane, succinic acid, and acetic acid. One member of the series, ferrimycin, is an antibiotic and bears the relationship to the ferrioxamines that albomycin bears to ferrichrome. The best known ferrioxamine is component B (Fig. 8). It has been synthesized chemically but the procedure based on low-iron growth of *Streptomyces pilosus* is the method employed for production of the clinical product, which is a salt of the ligand, viz., deferriox-

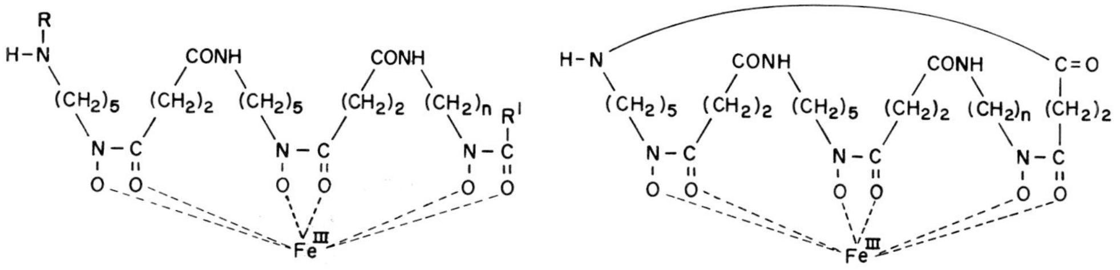

FIG. 8. The ferrioxamine family of siderophores. These substances occur commonly among the streptomyces and actinomyces sp. Deferrioxamine B, R = H; Mesylate salt = Desferal.

amine B mesylate (Desferal). It is stocked in the emergency room for use in cases of acute accidental iron poisoning in children. However, with the advent of childproof containers for iron tablets, the more significant use for Desferal and certain other siderophores lies in the treatment of the chronic iron poisoning resulting from the long-term supportive therapy, repeated transfusions, for thalassemia (Cooley's anemia) and aplastic anemia (1). Patients afflicted with these diseases can be managed over many years but eventually succumb to accumulated deposits of iron granules in the vital organs of the body.

After some initial promise for the effective treatment of transfusion-induced siderosis, Desferal fell into a decline owing to a combination of factors among which were its inability to remove really gross quantities of iron, certain toxic side reactions, and the requirement that the drug must be injected. Recently, with the development of an infusion pump, there has been renewed interest in Desferal (29). Among the hydroxamate type siderophores, rhodotorulic acid (Fig. 9) seems to be somewhat more effective than Desferal and has the advantage of economy of production (8). The source organism, *Rhodotorula pilimanae,* can be induced to excrete several grams per liter on low iron media, and since the product can be recrystallized from water, it is somewhat cleaner than the commercial preparations of Desferal which have been available to us. Rhodotorulic acid forms a neutral binuclear complex with Fe-III (6).

Within the hydroxamate class of siderophores, it has been generally assumed

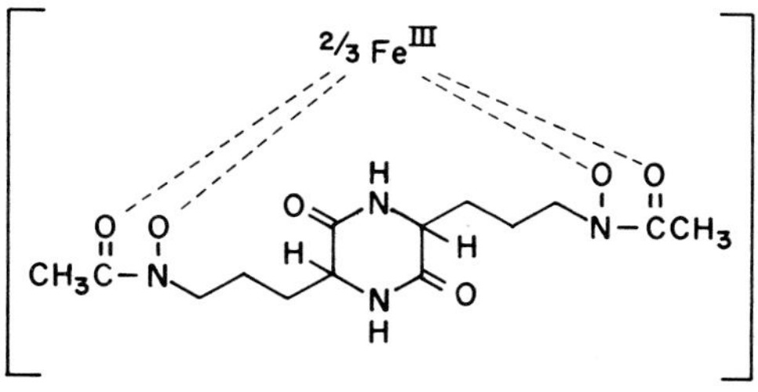

FIG. 9. Rhodotorulic acid, a siderophore produced by *Rhodotorula pilimanae* and related yeasts.

that neither ligand nor iron complex can be transported across the intestinal mucosa, although ferric triacetohydroxamate has recently been advocated as a source of iron for the anemic animal (3). This has fueled the quest for deferration drugs which would be effective via the oral route, and some success has attended the use of the catechol series of siderophores in this way. The instability of the ester bonds of enterobactin would require that it also be injected, and when administered via this route it proved capable of releasing relatively large fractions of the iron stores in the mouse (10). Enterobactin has the advantage of extreme stability of binding to Fe-III (log $K_s \cong 51$, ref. 2) and there is little doubt that it can remove the ferric ion from iron transferrin. The carbocyclic analog and the triscatechol derivative of 1,3,5-aminomethylbenzene have yet to be evaluated in the hypertransfused laboratory animal. The former easily removes iron from ferrichrome while the latter exhibits marginal ability to do so and must have a stability constant close to that of ferrichrome itself, namely, $10^{29.3}$. Cholylhydroxamic acid has proven to be the most orally efficacious drug examined thus far (9).

The search for a deferration drug is fraught with a variety of problems, not the least of which is the difficulty in extrapolation from laboratory animal to the human subject. Hyperferremia is known to lead to increased incidence of infections and the administration of a siderophore might compound this undesirable response. Thus, in contrast to the wild type, a double mutant of *S. typhimurium* blocked in enterobactin synthesis and unable to either transport or utilize Desferal showed no increased virulence in the iron-loaded mouse treated with the drug (15).

Conceivably, siderophore research could lead to the development of an anticancer nostrum. Except in neoplastic tissue, there is a very slow turnover of DNA and hence a drug aimed at polydeoxyribonucleotide synthesis might control—if not cure—cancer. An attractive possibility is the biosynthetic pathway leading to deoxyribotides, a reaction sequence which has been well documented in *E. coli* and which occurs also in human tissue. Briefly, the catalytic component of ribotide reductase is an iron-containing enzyme, and as such, affords a possible target for an iron-binding drug (32). Hydroxyurea, a hydroxamic acid, inhibits this enzyme but appears to do so through a free radical scavenging mechanism.

It is well known that certain ions in the transuranic series have coordination characteristics closely parallel to those of Fe-III. This is a consequence of the comparable ratio of charge/ionic radius. Thus Pu-IV displays stability constant with simple ferric hydroxamic acids which are of the same order of magnitude as those given by Fe-III. Bulman (4) has shown that rhodotorulic acid and 2,3-dihydroxybenzoylglycine can promote the extrusion of Pu-IV from a gel permeation column. The possible role of siderophores, which occur in soil and are produced generally by soil microorganisms in moving transuranic elements through the food chain, is being investigated by scientists at Battelle Northwest in Richland, Washington. Mixed ligand chelate therapy has been proposed for treatment of plutonium and cadmium poisoning (31).

AGRICULTURAL

Although synthetic iron chelates are routinely used for improved iron nutrition of plants, there have been no reports of the application of siderophores for this purpose. It is doubtful that the latter could compete on an economic basis with simple synthetics, such as ethylenediaminetetraacetate (EDTA), in spite of the virtue of selectivity and biodegradability. In tests with tomato plants growing in hydroponic solution, deferriferrichrome and deferriferrichrome A proved to be superior to EDTA in providing iron in slightly alkaline media (J. Orlando and J. B. Neilands, *unpublished*). Hydroxamic acids are inhibitors of urease and the application of acetohydroxamic acid to soil has been proposed as a means of preventing nitrogen loss through the action of urease.

Nitrogen fixation in the biosphere, a reaction upon which all life depends, is a purely bacterial process (prokaryotic). Ferredoxin, a low molecular weight protein rich in iron, is the usual source of the energetic electrons required for the reaction

$$N_2 + 6H^+ + 6\ e \rightleftarrows NH_3$$

Nitrogenase proper consists of an iron protein and an iron-molybdenum protein, and is uniformly the same enzyme regardless of the source. Hydrogenase, another iron-containing enzyme, is always associated with nitrogenase. Leghemoglobin is present in nodules, where it may play a role in modulating the oxygen supply to the bacteroids. After nitrogen has been fixed as ammonia and organically bound as glutamine, the iron-sulfur protein glutamate synthase is required for movement of the N into amino acids and other cell metabolites. Thus, in both symbiotic and free living species, iron is involved at multiple stages in the fixation of N_2. It would hence be pertinent to investigate the role of the high affinity iron assimilation system in such likely subjects as *Klebsiella pneumoniae* and *Azotobacter vinelandii*. Enterobactin has been isolated from the former and bis-2,3-dihydroxybenzoyl lysine from the latter, but the presence of these catechols has not been correlated with the capacity of the organism to fix N_2.

MICROBIAL BIOGEOCHEMISTRY

Although, as already mentioned, the presence of siderophore activity in soil extracts can be readily demonstrated, there are as yet no systematic studies reported on the ability of these natural chelating agents to promote the dissolution of minerals.

FOOD PRODUCTS

Mold-ripened blue cheese has been reported to contain a ferrichrome equivalent of 5 mg per kilo (27). *Penicillium roquefortii* and an isolate from blue cheese have been shown to form the siderophore coprogen when the organisms are cultured on low iron laboratory media. Sake contains small quantities of ferrichrome-type siderophores. In no instance has the nutritional or public health significance of the presence of siderophores in foods been established.

ACKNOWLEDGMENT

Basic research in the author's laboratory on microbial iron transport has been supported by NIH Grants AI 04156 and AM 17146.

REFERENCES

1. Anderson, W. F., and Hiller, M. C., editors (1976): *Development of Iron Chelators for Clinical Use.* U.S. Department of Health, Education and Welfare Publication No. (NIH) 76–994.
2. Avdeef, A., Sofen, S. R., Bregante, T. L., and Raymond, K. N. (1978): Stability constants for catechol models of enterobactin. *J. Am. Chem. Soc.,* 100:5362–5370.
3. Brown, D. A., Chidambaram, M. V., Clarke, J. J., and McAleese, D. M. (1978): Design of iron (III) chelates in oral treatment of anemia: Solution properties and absorption of iron (III) acetohydroxamate in anemic rats. *Bioinorganic Chem.,* 9:255–275.
4. Bulman, R. A. (1978): Chemistry of plutonium and the transuranics in the biosphere. *Structure & Bonding,* 34:39–77.
5. Byers, B. R., and Arceneaux, J. E. L. (1977): Microbial transport and utilization of iron. In: *Microorganisms and Minerals,* edited by E. D. Weinberg, pp. 215–249. Marcel Dekker, New York.
6. Carrano, C. J., and Raymond, K. N. (1978): Characterization of the complexes of rhodotorulic acid, a dihydroxamate siderophore. *J. Am. Chem. Soc.,* 100:5371–5374.
7. Fernandez-Pol, J. A. (1978): Siderophore-like growth factor synthesized by SV-40 transformed cells adapted to picolinic acid stimulates DNA synthesis in cultured cells. *FEBS Letters,* 88:345–348.
8. Grady, R. W., Graziano, J. H., Akers, H. A., and Cerami, A. (1976): The development of new iron chelating drugs. *J. Pharmacol. Exp. Ther.,* 196:478–485.
9. Grady, R. W., Graziano, J. H., White, G. P., Jacobs, A., and Cerami, A. (1978): The development of new iron chelating drugs. II. *J. Pharmacol. Exp. Ther.,* 205:757–765.
10. Guterman, S. K., Morris, P. M., and Tannenberg, W. J. K. (1978): Feasibility of enterochelin as an iron chelating drug. *Gen. Pharmac.,* 9:123–127.
11. Hantke, K., and Braun, V. (1975): Membrane receptor dependent iron transport in *Escherichia coli. FEBS Letters,* 49:301–305.
12. Hollifield, W. C., and Neilands, J. B. (1978): Ferric enterobactin transport system in *Escherichia coli* K-12. *Biochemistry,* 17:1922–1928.
13. Horowitz, N. H., Charlang, G., Horn, G., and Williams, N. P. (1976): Isolation and identification of the conidial germination factor of *Neurospora crassa. J. Bacteriol.,* 127:135–140.
14. Jacobs, A., and Worwood, A. P., editors (1974): *Iron in Biochemistry and Medicine.* Academic Press, London.
15. Jones, R. L., Peterson, C. M., Grady, R. W., Kumbaraci, T., and Cerami, A. (1977): Effects of iron chelators and iron overload on *Salmonella* infection. *Nature,* 267:63–65.
16. Keller-Schierlein, W. (1964): Natural iron (III) trihydroxamate complexes. *Prog. Chem. Org. Nat. Prod.,* 22:279–322.
17. King, R., Khan, H., Foye, J., Greenberg, J., and Jones, H. (1975): Transferrin, iron and dermatophytes. *J. Lab. Clin. Med.,* 86:204–212.
18. Kochan, I. (1977): Role of siderophores in nutritional immunity and bacterial parasitism. In: *Microorganisms and Minerals,* edited by E. D. Weinberg, pp. 252–288. Marcel Dekker, New York.
19. Leong, J., and Neilands, J. B. (1976): Mechanisms of siderophore iron transport in enteric bacteria. *J. Bacteriol.,* 126:823–830.
20. Luckey, M., Pollack, J. R., Wayne, R., Ames, B. N., and Neilands, J. B. (1972): Iron transport in *Salmonella typhimurium. J. Bacteriol.,* 111:731–738.

21. Naegeli, H.-U. (1978): Synthetic models for the antagonism between sideromycins and sideramines. Doctoral Dissertation, ETH, Juris Druck & Verlag, Zurich, Switzerland.
22. Negrin, R. S., and Neilands, J. B. (1978): Ferrichrome transport in inner membrane vesicles of *Escherichia coli* K-12. *J. Biol. Chem.,* 253:2339–2342.
23. Neilands, J. B. (1972): Evolution of biological iron binding centers. *Structure & Bonding,* 11:145–170.
24. Neilands, J. B., editor (1974): *Microbial Iron Metabolism.* Academic Press, New York.
25. Neilands, J. B. (1976): Microbial iron transport compounds (siderophores). In: *Development of Iron Chelators for Clinical Use,* edited by W. F. Anderson and M. C. Hiller. U.S. Department of Health Education and Welfare Publication No. (NIH) 76–994.
26. Neilands, J. B. (1977): Siderophores: Biochemical ecology and mechanism of iron transport in enterobacteria. In: *Bioinorganic Chemistry-II,* edited by K. N. Raymond, pp. 3–32. American Chemical Society, Washington, D.C.
27. Ong, S. A., and Neilands, J. B. (1979): *J. Agr. Food Chem. (in press).*
28. Ong, S. A., Peterson, T., and Neilands, J. B. (1979): Agrobactin, a siderophore from *Agrobacterium tumefaciens. J. Biol. Chem.,* 254:1860–1865.
29. Propper, R. D., Shurin, S. B., and Nathan, D. G. (1976): Reassessment of desferrioxamine B in iron overload. *N. Engl. J. Med.,* 294:1421–1423.
30. Reichard, P. (1968): The biosynthesis of deoxyribotides. *Eur. J. Biochem.,* 3:259–266.
31. Schubert, J., and Derr, S. K. (1978): Mixed ligand chelate therapy for plutonium and cadmium poisoning. *Nature,* 275:311–313.
32. Sjöberg, B.-M., Reichard, P., Gräslund, A., and Ehremberg, A. (1978): The tyrosine free radical in ribonucleotide reductase from *Escherichia coli. J. Biol. Chem.,* 253:6863–6866.
33. Tait, G. H. (1975): The identification and biosynthesis of siderochromes formed by *Micrococcus denitrificans. Biochem. J.,* 146:191–204.
34. Venuti, M. C., Rastetter, W. H., and Neilands, J. B. (1979): 1,3,5-Tris (N, N',N''-2,3-dihydroxybenzoyl)aminomethylbenzene, a synthetic iron chelator related to enterobactin. *J. Med. Chem.,* 22:123–124.
35. Wayne, R., and Neilands, J. B. (1975): Evidence for common binding sites for ferrichrome compounds and bacteriophage Φ80 in the cell envelope of *Escherichia coli. J. Bacteriol.,* 121:497–503.
36. Weinberg, E. D. (1978): Iron and infection. *Microbiol. Rev.,* 42:45–66.
37. Yancy, R. J., Breeding, S. A. L., and Lankford, C. E. (1976): Enterochelin: a virulence factor for *Salmonella typhimurium.* 76th Annual Meeting, American Society for Microbiology.

Some Bio-Inorganic Chemical Reactions of Environmental Significance

J. M. Wood, Y.-T. Fanchiang, and P. J. Craig

Gray Freshwater Biological Institute, Department of Biochemistry, University of Minnesota, Navarre, Minnesota 55392

In the 1930's, Challenger discovered the biomethylation of arsenic, and provided us with the first example of how biological systems possess the capability for synthesizing very toxic organoarsenic compounds from less toxic inorganic substrates (2,17). Even in those early days, Challenger recognized that biomethylation could only present a local environmental hazard, if the methylated product is produced in significant concentration so as to exert its toxic effect and if the methylated product is stable to hydrolysis.

In 1968, we discovered that mercury could be biomethylated to give methylmercury as the major product (12,27). This discovery has proved to be important in two respects: (a) it provided us with the recognition that biological systems are capable of synthesizing very toxic organometallic compounds from relatively innocuous inorganic precursors, and (b) it opened the door to the study of organometallic compounds produced by biological mechanisms in the aqueous environment. Since this initial discovery, a great deal of research has been done on the biomethylation of elements other than mercury (16,27–30). Detailed biochemical studies have provided us with five alternative mechanisms for the biomethylation of metals and metalloids (3,4). Recently, mechanistic studies have been used to predict the physical conditions which must be met for biomethylation to occur in the environment (16). In fact, on the basis of oxidation-reduction chemistry, we are now able to formulate the conditions for biomethylation and show that oxidation state properties are important to the biomethylation of metals.

Once synthesized, these methylated metals are invariably more toxic than their inorganic substrates. This toxicity is probably due to the nonpolar nature of many organometallic compounds which allows them to diffuse rapidly into and through cell membranes. During the last 6 months, we have learned a great deal about the kinetics for transport of methylmercury into the cell, but at the present time we have little information on macromolecular interactions of methylmercury after this molecule penetrates cell membranes (28). Interactions of methylated-metals with sulfhydryl groups are undoubtedly important in cases of acute toxicity. Also, the catalytic interaction of methylmercury with phospholipids in the central nervous system, such as the interaction with plasmalogens, may cause cell lysis (20). However, at this point a reasonable biochemical mechanism has to be found in order to explain the acute neurotoxicity of methylmercury.

Dynamic aspects of these methylation reactions are of critical importance, since even though most methylated metals are thermodynamically unstable in water, many of them are kinetically stable. In fact, it is well known that metals which are lower in their periodic groups form metalalkyls which are kinetically more stable. For example, mercury, platinum, and possibly lead offer potentially stable systems whereas palladium, chromium, and cadmium do not.

In this review, we examine the different mechanisms for B_{12}-dependent methyl-transfer to a selected group of toxic elements placing special emphasis on "redox" conditions. Five mechanisms have been elucidated for B_{12}-dependent methyl-transfer to date: (a) heterolytic cleavage of the Co-C bond with the transfer of a carbanion to the attacking metal ion; (b) heterolytic cleavage of the Co-C bond with the transfer of a carbonium ion to an attacking nucleophile; (c) homolytic cleavage of the Co-C bond with transfer of a methyl-radical to an attacking free radical; (d) "Redox-Switch", a mechanism where metal ions complex with the corrin macrocycle to labilize the Co-C bond to attack by weak electrophiles; and (e) cleavage caused by single electron "outer-sphere" oxidation of methylcobalamin.

The transfer of CH_3^- or $CH_3\cdot$ have been found to be the most predominant reaction mechanisms for a number of metals and metalloids.

ELECTROPHILIC ATTACK ON THE Co-C BOND OF METHYL-B_{12}

Methyl-B_{12} reacts rapidly with a number of inorganic metal ions, in aqueous media under aerobic conditions, to give metal-alkyls and aquo-B_{12} as the reaction products. For example, the reaction between methyl-B_{12} and mercuric acetate involves carbanion transfer; this mechanism is presented in Scheme 1. We have elucidated detailed mechanisms for the biosynthesis of methylmercury and dimethylmercury (3). The second-order rate constant for the biosynthesis of methylmercury from mercuric acetate and methyl-B_{12} has been determined as 3.7×10^2 sec^{-1}M^{-1}, and therefore under optimum conditions the kinetics for methylmercury synthesis have been shown to be extremely rapid (3). The biosynthesis of methylmercury in sediments and by bacteria isolated from sediments is well established (32 and references therein).

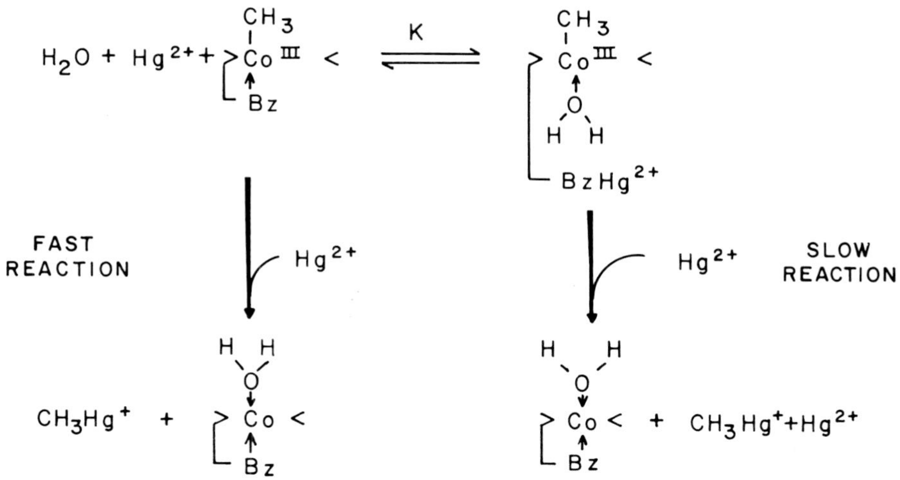

$$H_2O + Hg^{2+} + \begin{bmatrix} & CH_3 \\ & | \\ > & Co^{III} < \\ & \uparrow \\ & Bz \end{bmatrix} \underset{K}{\rightleftharpoons} \begin{bmatrix} & CH_3 \\ & | \\ > & Co^{III} < \\ & \downarrow \\ & O \\ & HH \\ & BzHg^{2+} \end{bmatrix}$$

FAST REACTION $\quad \downarrow Hg^{2+} \qquad\qquad \downarrow Hg^{2+} \quad$ SLOW REACTION

$$CH_3Hg^+ + \begin{bmatrix} & HH \\ & \searrow O \swarrow \\ > & Co < \\ & \uparrow \\ & Bz \end{bmatrix} \qquad \begin{bmatrix} & HH \\ & \searrow O \swarrow \\ > & Co < \\ & \uparrow \\ & Bz \end{bmatrix} + CH_3Hg^+ + Hg^{2+}$$

Scheme 1

The rate of methylmercury biosynthesis can be enhanced by adding B_{12} to certain bacterial cultures (25).

The reaction between mercuric ion and methyl corrinoids is an example of carbanion methyl-transfer. Because mercuric ion is a good electrophile, it also coordinates to the nitrogen of the 5,6-dimethylbenzimidazole base to give a mixture of "base off" and "base on" methyl-B_{12}. The "base on" species reacts 1,000 times faster than the "base off" species to give methylmercury as the product (3). Other metals which are known to react with methyl-B_{12} by a similar mechanism to mercuric ion are lead (Pb^{IV}), thallium (Tl^{III}), and palladium (Pd^{II}) (1,23,32).

In 1975, three publications appeared in the literature addressing the subject of biomethylation of both organo-lead and inorganic lead compounds in the aqueous environment (11,21,26). Since lead is certainly the most widely distributed toxic heavy metal in advanced industrial society, then it is critically important to establish whether or not lead is biomethylated in the aqueous environment. Gasoline additives, tetra-ethyl and tetra-methyl lead, have been shown to be unstable in the aqueous environment, with dealkylation leading to the formation of the corresponding trialkyl and dialkyl products (13). These trialkyl and dialkyl lead compounds appear to be quite stable in water, presumably due to the ionic nature of these species, but do not bioaccumulate to any appreciable extent in the food chain. (13). Even so, tetra-ethyl lead and to a lesser extent tetra-methyl lead are very nonpolar compounds which may be taken up rapidly by aquatic organisms, probably by diffusion-controlled processes. The uptake of tetra-alkyl lead compounds can be compared with the uptake of dimethylmercury or other nonpolar organomercurials which are similarly nonpolar in the aquatic environment. Methylmercuric chloride has been shown to diffuse through a 40 Å membrane in 20×10^{-9} seconds (30). This is close to the rate of diffusion of methylmercuric chloride through an equivalent 40 Å of an organic solvent such as toluene. Therefore, it seems likely that if organo-lead compounds are to be taken up by aquatic organisms, then the concentration of tetra-alkyl lead in water will determine the extent of contamination. Once the tetra-alkyl lead has been taken up by a living organism, then it may be dealkylated to a charged chemical species before it can react at the molecular level inside the cell. It is to be expected that diffusion and partition of tetra-alkyl lead compounds, through and into cell membranes, will represent important processes prior to conversion to less toxic reactive molecules such as triethyl-lead and trimethyl-lead salts (13).

Methylation of stable organo-lead compounds has been demonstrated in sediments (26) by mixed microbial populations (21) and by a B_{12}-dependent reaction (32). This reaction can be shown to occur under anaerobic conditions, though for the B_{12}-mechanism, aerobic conditions are favored. By similarity with mercuric ion, methyl-transfer to plumbic ion occurs by electrophilic attack on the Co-C bond of methyl-B_{12} (Scheme 2).

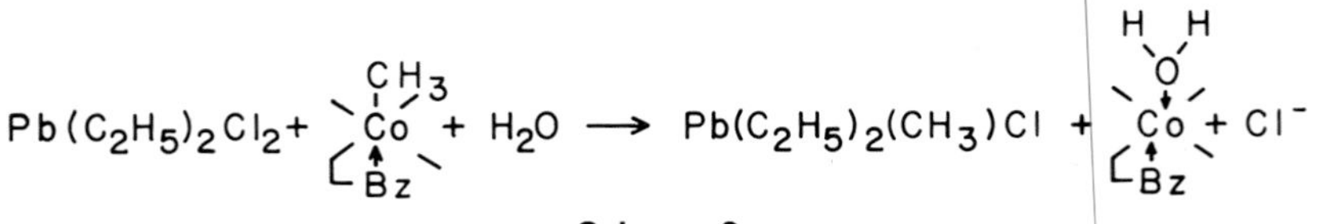

$$Pb(C_2H_5)_2Cl_2 + \underset{Bz}{\overset{CH_3}{Co}} + H_2O \longrightarrow Pb(C_2H_5)_2(CH_3)Cl + \underset{Bz}{\overset{H\diagup O\diagdown H}{Co}} + Cl^-$$

Scheme 2

In the case of inorganic lead salts, we have shown that electrophilic attack on the Co-C bond of methyl-B_{12} occurs to displace the methyl group as CH_3^-, but the resulting monomethyl-lead species is very unstable and is hydrolyzed in milliseconds to regenerate the inorganic lead salt. Taylor and Hanna have shown that prolonged incubation of ^{14}C-methyl-B_{12} with fine suspensions of Pb^{IV} oxides results in demethylation of B_{12}, but the resulting ^{14}C lead complex is unstable (24). In these experiments, the ^{14}C was volatilized from reaction mixtures. In Taylor's laboratory, and in our own, we have shown that there is no reaction with Pb^{II} salts (24,26,28).

The reactions described above all involve the displacement of a carbanion from the cobalt atom of methyl-B_{12}. Many of these reactions occur under aerobic conditions with half-lives in the order of milliseconds. It seems reasonable that metals which react by electrophilic attack on the Co-C bond (SE_2 mechanism) occur with the more oxidized state of the metal (e.g., Pb^{IV}, Tl^{III}, Hg^{II}, Pd^{II}, etc.).

FREE RADICAL ATTACK ON THE Co-C BOND OF METHYL-B_{12}

Free radical reactions involve homolytic cleavage of the Co-C bond of methylcobalamin with the transfer of $CH_3{}^\bullet$ and the production of Co(II)cobalamin.

The attacking metal or metalloid species contains an unpaired free electron which when coupled with $CH_3\cdot$ results in a formal oxidation of the element. One example of a free radical reaction is the methylation of thiols by methylcobalamin. In the case of ethane-thiol sulfonic acid or mercaptoethanol (10), the reaction has been shown to involve thiol free radicals. Methylation requires the presence of oxygen for the generation of the radical species and is characterized by a lag period during which a steady-state concentration of attacking radicals is generated.

Reactions of this kind require the generation of a radical intermediate. This radical intermediate can be produced under anaerobic conditions in the laboratory either by adding a single electron oxidant to a metal ion in the reduced state of a redox couple (e.g., $Sn^{II} \rightarrow Sn^{III} + e$) or by adding a single electron reductant to a metal ion in the oxidized state of a redox couple (e.g., $Au^{III} + e \rightarrow Au^{II}$). In the case of tin, we believe that reductive Co-C bond cleavage of methyl-B_{12} occurs by a transient Sn^{III} radical which is generated by one equivalent oxidation of Sn^{II} (Scheme 3) (4). In the case of gold, we have shown that cleavage of the Co-C bond requires the generation of Au^{II} by single electron reduction of Au^{III} (Scheme 4) (6). For this reaction, preincubation of the Au^{III} salt with B_{12r} (CoII) or Fe is required before a reaction with methyl-B_{12} proceeds.

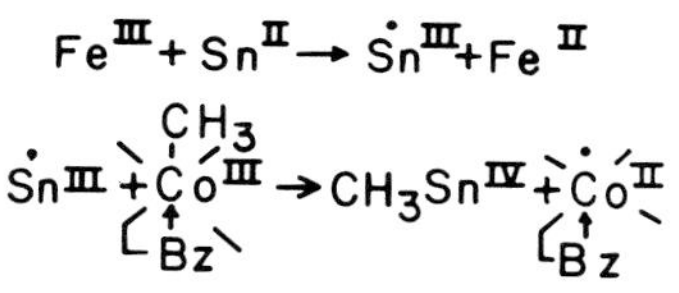

$$Fe^{III} + Sn^{II} \rightarrow \dot{S}n^{III} + Fe^{II}$$

$$\dot{S}n^{III} + \underset{Bz}{\overset{CH_3}{Co^{III}}} \rightarrow CH_3Sn^{IV} + \underset{Bz}{Co^{II}}$$

Scheme 3

$$Fe^{II} + Au^{III} \longrightarrow \dot{A}u^{II} + Fe^{III}$$

$$\dot{A}u^{II} + \underset{Bz}{\overset{CH_3}{Co^{III}}} \longrightarrow CH_3Au^{III} + \underset{Bz}{Co^{II}}$$

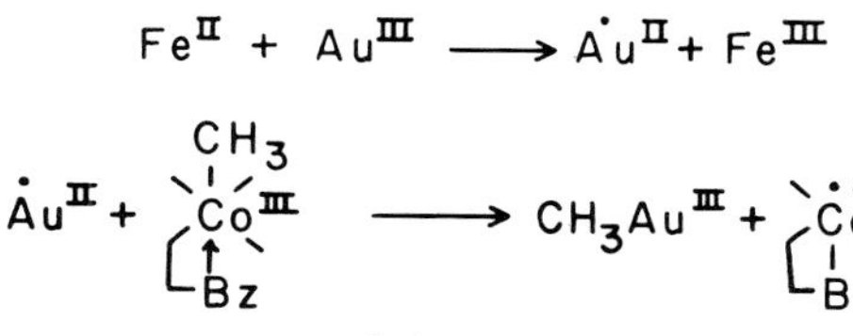

Scheme 4

The biomethylation of tin is quite a significant reaction in the environment, because methyl-tin compounds are produced biologically and have been detected in urine, tissues, as well as in the aquatic food-chain. A similar mechanism to that demonstrated for $RS\cdot$, Sn^{III}, and Au^{II} has been reported for Cr^{II} (5). It is likely that those metals with a high reduction potential (oxidizing agents) react by an electrophilic mechanism (Type I) and those with a low reduction potential (reducing agents) react by a reductive mechanism (Type II) (Table 1). This connection between standard reduction potential and mechanism for biomethylation seems highly rational, because $E°$ describes the relative thermodynamic tendency for the metals involved to accept or donate electrons.

TABLE 1. *Relationship between standard reduction potential ($E°$) and the mechanism of methylation for selected elements*

Redox couple	$E°$ (volts)	Mechanism of methylation
Pb(IV)/Pb(II)	+1.46	Type I
Tl(III)/Tl(I)	+1.26	Type I
Se(VI)/Se(IV) acid	+1.15	
Pd(II)/Pd(0)	+0.987	Type I
Hg(II)/Hg(0)	+0.854	Type I
Pt(IV)/Pt(II)	+0.760	Redox switch
As(V)/As(III) acid	+0.559	
Au(III)/Au(II)	+0.50[a]	Type II
Sn(IV)/Sn(II)	+0.154	Type II
Se(VI)/Se(IV) base	+0.05	
Cys-S-S Cys/2Cys-SH	−0.22	Type II
Cr(III)/Cr(II)	−0.41	Type II
As(V)/As(III) base	−0.67	

[a] The Au(III)/Au(II) couple is estimated (19).

The electrophilic mechanism occurs with inorganic compounds having a standard reduction potential ($E°$) of +0.85 volts or higher, while the free radical mechanism occurs with inorganic compounds having $E°$ of +0.50 volts and lower (7). Platinum is especially interesting because the $E°$ for Pt^{IV}/Pt^{II} couple is +0.76 volts and cannot be classified with either those elements that react electrophilically nor those which react by a free radical mechanism. Therefore, a study of the biomethylation of platinum is of considerable mechanistic interest since both Pt(IV) and Pt(II) oxidation states are required for methyl-transfer to occur (8).

REACTIONS OF PLATINUM WITH METHYL-B_{12}

Agnes et al. (1) were the first to report that the methylation of platinum by methylcobalamin required the addition of platinum in both oxidation states (Pt^{II} and Pt^{IV}). A pathway for this reaction was formulated and called the "redox switch" reaction (1). Later, Taylor and Hanna showed that under their conditions, methyl-Pt^{IV} was a product of this reaction, and they obtained some preliminary kinetic data to support a "redox switch" reaction mechanism (24). Spectrophotometric measurements showed the consumption of one mole of $Pt^{IV}Cl_6^{2-}$ per mole of MeB_{12}, with $Pt^{II}Cl_4^{2-}$ required in only catalytic quantities (8). Aquocobalamin (aquo-B_{12}) and methyl platinum were shown to be the initial products of the reaction. The overall stoichiometry for the demethylation of MeB_{12} by platinum complexes is expressed in Scheme V.

$$MeB_{12} + Pt^{IV}Cl_6^{2-} + H_2O \xrightarrow{(Pt^{II}Cl_4^{2-})} aquo\text{-}B_{12} + MePt^{IV}Cl_5^{2-} + Cl^-$$

Scheme 5

Recent experiments in our laboratory have shed some light on the "redox switch" mechanism first described by Agnes et al. (1). We have shown that $PtCl_4^{2-}$ forms a one-to-one complex with methyl-B_{12} by forming an adduct with the corrin macrocycle. There is no interaction with benzimidazole as determined by 270 MHz NMR. For this platinum-B_{12} complex, we observe a down field shift of the methyl resonance on the cobalt atom, indicating a change in electron density on the methyl group. The formation of this Pt^{II}-B_{12} complex appears to labilize the Co-C bond to attack by platinum, probably in the Pt^{IV} oxidation state, to give methyl-platinum species as the product (8), as demonstrated in Scheme 6.

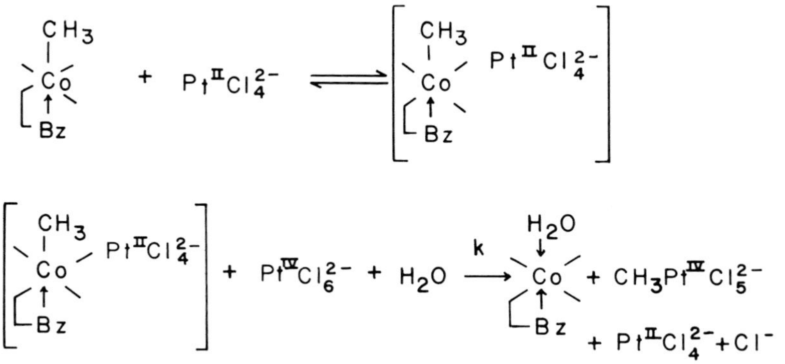

$$\begin{bmatrix} CH_3 \\ | \\ Co \\ \uparrow \\ Bz \end{bmatrix} + Pt^{II}Cl_4^{2-} \rightleftharpoons \begin{bmatrix} CH_3 \\ | \quad Pt^{II}Cl_4^{2-} \\ Co \\ \uparrow \\ Bz \end{bmatrix}$$

$$\begin{bmatrix} CH_3 \\ | \quad Pt^{II}Cl_4^{2-} \\ Co \\ \uparrow \\ Bz \end{bmatrix} + Pt^{IV}Cl_6^{2-} + H_2O \xrightarrow{k} \begin{matrix} H_2O \\ \downarrow \\ Co \\ \uparrow \\ Bz \end{matrix} + CH_3Pt^{IV}Cl_5^{2-} + Pt^{II}Cl_4^{2-} + Cl^-$$

Scheme 6

Although we do not know the detailed structure of the Pt^{II}-methyl B_{12} complex, it is clear that a third mechanism exists for methyl-transfer to metals which have standard reduction potentials close to that for molecular oxygen. This research provides us with the first example of "activation" of the Co-C σ bond of methyl-B_{12} through the interaction of a charged species with the corrin macrocycle, except for the interaction of the benzimidazole base with the Co-atom.

The coordination of benzimidazole to the cobalt atom in methyl-B_{12} is very important in determining the kinetics for [CH_3^-] methyl-transfer to metals. In the case of mercury, six coordinate "base on" methyl-B_{12} has a mercury methylation rate 1,000 × faster than the five coordinate "base off" species. A balance exists between the capacity of a metal ion for methylation and its ability to hinder methylation by coordination to benzimidazole. For mercuric ion, rapid methylation predominates. The coordinating nitrogen atom of benzimidazole, according to Pearson's definition, is a "hard" Lewis base. Therefore, maximum coordination to benzimidazole would be expected to occur with a "hard" Lewis acid (15) (i.e., small size, highly charged, oxidizing agent). Metals with "hard" acid properties convert methyl-B_{12} to the "base off" species which is a very weak methylating agent. In the light of this property, one would expect "soft" acids such as Hg^{II}, Pd^{II}, and Tl^{III} to be biomethylated. Experimental data confirm this expectation. A survey of other metals indicates that Bi^{III} and Sb^{III}, which are intermediary between "hard" and "soft" acids, may function as methyl-acceptors from methyl-B_{12}.

NUCLEOPHILIC ATTACK ON THE Co-C BOND OF METHYL-B_{12}

Schrauzer (22) was the first to postulate that B_{12}-dependent methionine synthesis occurs by nucleophilic attack with homocysteine thiolate anion on the cobalt-carbon bond (Scheme 7).

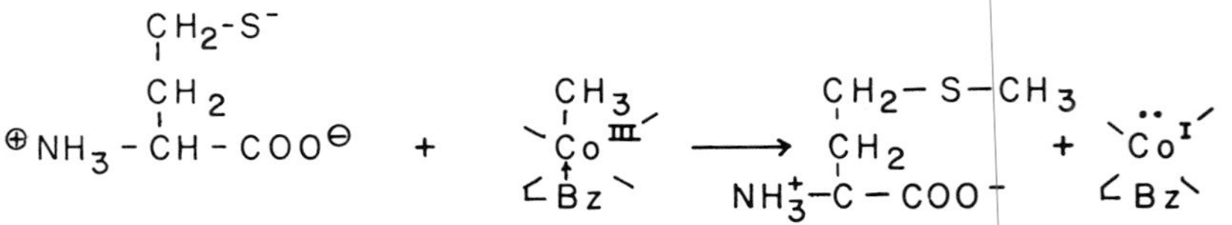

$$\overset{\oplus}{N}H_3-\underset{\underset{CH_2-S^-}{|}}{\overset{|}{CH}}-COO^{\ominus} + \underset{\underset{Bz}{\angle}}{\overset{CH_3}{\underset{|}{Co^{III}}}} \longrightarrow \underset{\overset{+}{N}H_3-\underset{|}{\overset{|}{C}}-COO^-}{\overset{CH_2-S-CH_3}{\underset{|}{CH_2}}} + \underset{\angle Bz}{\overset{..}{Co^I}}$$

Scheme 7

This mechanism was proposed because nonenzymatic methyl-transfer from methylcobalamin, or the B_{12}-model compound methylcobaloxime, was found to occur upon the addition of a solution of homocysteine at pH 14.0. It was suggested that a basic site in the protein would generate the thiolate ion in the vicinity of the active site, and that this nucleophile then could cleave the cobalt-carbon bond heterolytically. Two years later, Taylor and Hanna found evidence in support of Schrauzer's mechanism by identifying (B_{12}-s) as the B_{12}-product formed at the active site of methionine synthetase following alkyl-transfer to homocysteine. Ruediger (18), Ruediger and Jaenicke (19) have now performed extensive studies on the B_{12}-dependent methionine synthetase, and their results support Schrauzer's mechanism.

In model studies, similar to those conducted by Schrauzer, Agnes et al. (1) found evidence for thiol-promoted homolytic cleavage of methylcobalamin, but no evidence could be found for nucleophilic attack on the Co-C bond by thiolates or cyanide. Frick et al. (10) obtained conclusive evidence for a free radical mechanism in a study of methyl-transfer from methyl-B_{12} to biologically important thiols. Since methylcobalamin has been found to be extremely stable to stronger nucleophiles than thiolate, and since there is now overwhelming support for the homolytic cleavage mechanism with a number of free radicals, it appears that further studies on the mechanism of methionine synthetase are in order before the nucleophilic mechanism is universally acceptable. Although 5′-deoxyadenosylcobalamin is susceptible to nucleophilic attack by cyanide ion, the methyl-Co bond has been found to be very stable to nucleophiles, but very labile to electrophiles. This stability is not too difficult to rationalize when one considers the required stability of methyl-B_{12} coenzymes in the strong reducing conditions in which anaerobic bacteria function.

"OUTER SPHERE" OXIDATION OF METHYL-B_{12}

Halpern has reported that hexachloro-iridate (IV), a substitution inert "outer sphere" oxidant, is capable of oxidizing Co^{III} complexes to Co^{IV} complexes (14). The alkyl-Co^{IV} complexes are susceptible to nucleophilic attack leading to carbonium ion transfer and Co^{II} formation, as shown in Scheme 8.

$$R\ Co^{III} + Ir^{IV}\ Cl_6{}^{2-} \longrightarrow R\text{—}Co^{IV} + Ir^{III}\ Cl_6{}^{3-}$$
$$R\ Co^{IV} + X^- \longrightarrow RX + Co^{II}$$

Scheme 8

X^- = nucleophile.

Recently we have shown that methyl-B_{12} is oxidized by $Ir^{IV}Cl_6{}^{2-}$ to give an analogous complex which is rapidly demethylated. At the present time, we are working on the kinetics and mechanism of this interesting reaction (9).

From the time that biological methylation of mercury was discovered in 1968, we have come a long way in our understanding of the biochemical cycles for toxic metals. The discovery of five alternative mechanisms for methyltransfer has facilitated an understanding of the important role played by molecular oxygen and transition metals in these reactions. However, we lack information on the coordination chemistry of these metals in the mixed ligand environments we find in nature. If we are to continue to move and distribute metals such as lead, cadmium, nickel, copper, mercury, and tin in amounts vastly in excess of their natural geological flux rates, then we must have a better understanding of the fates of these inorganic complexes (33). At the present time in advanced industrial society, we are forcing microbial populations to adapt to elevated concentrations of heavy metals. What are the evolutionary consequences to life in a heavy metal-rich environment? What is the impact of the microbial world on the world of higher organisms? The recognition that microorganisms must synthesize our daily intake of Vitamin B_{12} shows us how delicate the relationship is between microorganisms and man.

ACKNOWLEDGMENTS

Our research work is supported by Public Health Service Grant (AM 18101), National Science Foundation Grant (PCM 17318), and a grant from the Northwest Area Foundation. We thank R. L. Thrift for his help with 270 MHz NMR experiments as well as the contributions of our colleagues L. J. Dizikes, H. J. Segall, R. E. DeSimone, F. S. Kennedy, A. Cheh, T. Frick, W. P. Ridley, and J. Pignatello.

REFERENCES

1. Agnes, G., Hill, H. A. O., Pratt, J. M., Ridsdale, S. C., Kennedy, F. S., and Williams, R. J. P. (1971): B_{12}-dependent methyl-transfer to metals. *Biochim. Biophys. Acta,* 252:207–210.
2. Challenger, F. L. (1945): The biological methylation of arsenic. *Chemistry Reviews,* 36:315–362.
3. DeSimone, R. E., Penley, M. W., Charbonneau, L., Smith, S. G., Wood, J. M., Hill, H. A. O., Pratt, J., Ridsdale, S., and Williams, R. J. P. (1973): The kinetics and mechanism of methyl and ethyl-transfer to mercuric ion. *Biochim. Biophys. Acta,* 304:851–863.
4. Dizikes, L. J., Ridley, W. P., and Wood, J. M. (1978): A mechanism for the biomethylation of tin by reductive Co-C bond cleavage in alkylcobalamins. *J. Am. Chem. Soc.,* 100:1010–1012.
5. Espenson, J. H., and Seelers, T. D. (1974): Free radical reactions with ethylcobalamin. *J. Am. Chem. Soc.,* 96:94–97.
6. Fanchiang, Y. -T., Ridley, W. P., and Wood, J. M. (1978): The biochemistry of toxic elements. *Quarterly Reviews of Biophysics,* edited by J. Eisinger and W. Blumberg.
7. Fanchiang, Y. -T., Ridley, W. P., and Wood, J. M. (1978): Mechanisms for biomethylation. Proceedings of an ACS Symposium on Organometallic Chemistry, edited by F. Brinckman. *ACS Monograph.*
8. Fanchiang, Y. -T., Ridley, W. P., and Wood, J. M. (1979): The methylation of platinum complexes by methylcobalamin. *J. Am. Chem. Soc.,* 101:1442–1447.
9. Fanchiang, Y. -T., and Wood, J. M. (1978): Unpublished results.
10. Frick, T., Francia, M. D., and Wood, J. M. (1976): Mechanism for the interaction of thiols with methylcobalamin. *Biochim. Biophys. Acta,* 428:808–818.
11. Jarvie, A. W. P., Markall, R. N., and Potter, H. R. (1975): Chemical alkylation of lead. *Nature,* 255:217–218.
12. Jensen, S., and Jernelöv, A. (1969): Biological methylation of mercury in aquatic organisms. *Nature,* 223:753–754.
13. Maddock, B. G., and Taylor, D. (1978): The acute toxicity and bioaccumulation of some lead alkyl compounds in marine animals. Proceedings of an International Conference on Lead in the Marine Environment, edited by M. Branica. Yugoslavia Chemical Society, Rovinj, Yugoslavia.
14. Magnusen, R. H., and Halpern, J. (1978): Stereochemistry of the nucleophilic cleavage of cobalt-carbon bonds in organocobalt(IV) compounds. *Chem. Communs.,* 44–46.
15. Pearson, R. G. (1963): Hard and soft acids. *J. Am. Chem. Soc.,* 85:3533–3539.
16. Ridley, W. P., Dizikes, L. J., and Wood, J. M. (1977): Biomethylation of toxic elements in the environment. *Science,* 197:329–332.
17. Ridley, W. P., Dizikes, L. J., Cheh, A., and Wood, J. M. (1977): Recent studies on the biomethylation and demethylation of toxic elements. *Environmental Health Perspectives,* 19:43–46.
18. Ruediger, H. (1971): Vitamin B_{12}-dependent methionine synthetase. Cycle of transmethylation. *Eur. J. Biochem.,* 21(2):264–268.
19. Ruediger, H., and Jaenicke, L. (1973): Biosynthesis of methionine. *Mol. Cell. Biochem.,* 1(2):157–168.
20. Segall, H. J., and Wood, J. M. (1974): Reaction of methylmercury with plasmalogens suggests a mechanism for the neurotoxicity of metal-alkyls. *Nature,* 248:456–458.
21. Schmidt, U., and Huber, F. (1975): Methylation of organo lead and lead(II) compounds to $(CH_3)_4$ Pb by microorganisms. *Nature,* 259:157–158.
22. Schrauzer, G. N. (1968): Organocobalt chemistry of vitamin B_{12} model compounds (cobaloximes). *Acc. Chem. Res.,* 1:97.
23. Scovell, W. H. (1974): The mechanism for B_{12}-dependent methyl-transfer to palladium. *J. Am. Chem. Soc.,* 96:3451–3456.
24. Taylor, R. T., and Hanna, M. L. (1976): The methylation of lead and platinum by MeB_{12}. *J. Environ. Sci. Health Part A Env. Sci., Eng.,* 3:201.
25. Vonk, I., and Sjipersteijn, A. (1973): Microbial methylation of mercury. *Antonie van Leeuwenhoek, J. Microbiol. Serol.,* 39:505.
26. Wong, P. T. S., Chan, Y. K., and Luxon, P. L. (1975): Methylation of lead in the environment. *Nature,* 253:263–264.

27. Wood, J. M., Kennedy, F. S., and Rosen, C. G. (1968): The synthesis of methylmercury compounds by methanogenic bacteria. *Nature,* 220:173–175.
28. Wood, J. M. (1974): Biological cycles for toxic elements in the environment. *Science,* 183:1049–1054.
29. Wood, J. M. (1976): Les metaux toxiques dans l'environment. *LaRecherche* 70:711–716.
30. Wood, J. M., Cheh, A., Dizikes, L. J., Ridley, W. P., Rakow, S., and Lakowicz, J. R. (1978): Biological cycles for toxic elements. *Fed. Proc. Fed. Am. Soc. Exp. Biol.,* 37:1, 16–21.
31. Wood, J. M. (1976): *Biochemical and Biophysical Perspectives in Marine Biology,* Volume 3, edited by D. C. Malins and J. R. Sargent, pp. 407–431. Academic Press, New York.
32. Wood, J. M. (1977): Lead in the marine environment: Some biochemical considerations. Proceedings of an International Conference on Lead in the Marine Environment, edited by M. Branica. Yugoslavia Chemical Society, Rovinj, Yugoslavia.
33. Wood, J. M., and Goldberg, E. D. (1977): *Impact of Metals on the Biosphere. Dahlem Konferenzen Global Chemical Cycles and Their Alterations by Man,* edited by W. Stumm. Chapter 10, pp. 137–154. Abakon Verlagsgesellschaft, Berlin.

Quantitative Mammalian Cell Mutagenesis and a Preliminary Study of the Mutagenic Potential of Metallic Compounds*

Abraham W. Hsie, **Neil P. Johnson, **,†D. Bruce Couch, **Juan R. San Sebastian, J. Patrick O'Neill, James D. Hoeschele, Ronald O. Rahn, and Nancy L. Forbes

Biology Division, and Health and Safety Research Division, Oak Ridge National Laboratory, and University of Tennessee—Oak Ridge Graduate School of Biomedical Sciences, Oak Ridge, Tennessee 37830

As science and technology have progressed, increased amounts of toxic metallic compounds have been released into our environment. While many of the short-term toxic effects of metals on plants and animals have been documented, their long-range consequences remain largely unknown. The carcinogenic effects of metallic compounds containing Be, Cd, Co, Cr, Fe, Ni, Pb, Ti, and Zn have been demonstrated in experimental animals. Based on epidemiological studies, As, Cd, Cr, and Ni compounds have been implicated as human carcinogens (37). These carcinogenic metals have been reported to enhance susceptibility to viral transformation in hamster embryo cells in culture (3). Teratogenic effects of metals have also been demonstrated experimentally (10). Evidence for a mutagenic effect of metals includes the induction of chromosome and mitotic aberrations (12), the direct metal—DNA interactions (25,33), the enhancement of error in *in vitro* DNA synthesis (34,35), and the induction by certain metals of mutations in microorganisms (11).

Recent studies of mutagenesis and DNA repair in microorganisms, especially *Salmonella typhimurium* and *Escherichia coli*, have established that approximately 90% of chemical carcinogens induce mutations or cause DNA damage in these bacteria (23,36). Such findings imply that the bacterial tests are useful for identification not only of potential mutagens but also of carcinogens in the environment. However, studies of mutagenesis in prokaryotes may not reveal

---

* By acceptance of this article, the publisher or recipient acknowledges the right of the U.S. Government to retain a nonexclusive royalty-free license in and to any copyright covering the article.
  ** Postdoctoral Fellow.
  † Presently affiliated with Chemical Industry Institute of Toxicology, Research Triangle Park, North Carolina.

some fundamental mechanisms of mutagenesis in mammals, because mammals differ from prokaryotes in their level of organization and repair of DNA, mechanisms of metabolism of chemicals, and other related functions. In addition, it is well known that chromosomal abnormality, one of the major causes of inheritable human diseases, is often associated with the process of malignancy. The great majority of chemical carcinogens are also known to induce chromosomal aberrations (1,21), or sister-chromatid exchanges (1). Clearly, mammalian cell systems offer advantages over bacterial assays for study of genetic toxicity at the chromosome and chromatid level.

Within the last decade, several mammalian cell mutation systems (4), especially those utilizing resistance to purine analogs such as 8-azaguanine and 6-thioguanine (TG)[1] as a genetic marker, have been developed for study of mechanisms of mammalian mutation and for assessment of the mutagenic hazard of environmental agents to humans. The selection for mutation induction to purine analog resistance is based on the fact that the wild-type cells containing hypoxanthine-guanine phosphoribosyl transferase (HGPRT) activity are capable of converting the analog to toxic metabolites, leading to cell death; the presumptive mutants, by virtue of the loss of HGPRT activity, are incapable of catalyzing this detrimental metabolism and, hence, escape the lethal effect of the purine analog.

The near-diploid Chinese hamster ovary (CHO) cell line has been chosen for our study because a mutation assay, referred to as CHO/HGPRT system (16,19), has been developed. We have used CHO cells because these are perhaps the best characterized mammalian cells genetically. They exhibit high cloning efficiency, achieving nearly 100% under normal growth conditions, and are capable of growing in a relatively well-defined medium on a glass or plastic substratum or in suspension with a population doubling time of 12 to 14 hr. In addition, the cells have a stable, easily recognizable karyotype of 20 or 21 chromosomes (depending on the subclone) and are thus suitable for studies of mutagen- or carcinogen-induced chromosome and chromatid aberrations and sister-chromatid exchange.

In this article, we first summarize our work on the development of a quantitative specific gene mutation system, CHO/HGPRT, and the application of this mutation assay to studies of chemical mutagenesis with reference to the interrelationships of cellular lethality and mutagenesis, the apparent exposure dose, the structure-activity (mutagenicity) relationship, and the mutagenicity of carcinogens. We then present our preliminary findings on the mutagenic potential of compounds containing metals such as Pt and Mn.

---

[1] *Abbreviations:* TG, 6-thioguanine; TG^r, TG-resistant or TG-resistance; HGPRT, hypoxanthine-guanine phosphoribosyl transferase; CHO cells, Chinese hamster ovary cells; EMS, ethyl methanesulfonate; MNNG, *N*-methyl-*N'*-nitro-*N*-nitrosoguanidine; iPMS, isopropyl methanesulfonate; DES, diethylsulfate; ENNG, *N*-ethyl-*N'*-nitro-*N*-nitrosoguanidine; B*(a)*P, benzo*(a)*pyrene; 4-NQO, 4-nitroquinoline-1-oxide; DMS, dimethylsulfate; MMS, methyl methanesulfonate; MNU, *N*-methyl-*N*-nitrosourea; ENU, *N*-ethyl-*N*-nitrosourea; BNU, *N*-butyl-*N*-nitrosourea; DMN, dimethylnitrosamine; DDP, dichlorodiammine Pt(II) [Pt(NH$_3$)$_2$Cl$_2$].

MATERIALS AND METHODS

Cell Culture

All studies to be described have employed a subclone of CHO-K$_1$ cells designated as CHO-K$_1$-BH$_4$ (14) which was isolated following selection in F12 medium containing aminopterin (10 μM). Cells are routinely cultured in Ham's F12 medium (K. C. Biological Co.) containing 5% heat-inactivated (56°C, 30 min), extensively dialyzed fetal calf serum (K. C. Biological Co.) (medium F12FCM5) in plastic tissue culture dishes (Falcon or Corning Glass Works) under standard conditions of 5% CO$_2$ in air at 37°C in a 100% humidified incubator. These cells grow in medium which contains aminopterin as well as in regular medium with 5 or 10% dialyzed fetal calf serum with a population doubling time of 12 to 14 hr. Cells are removed with 0.05% trypsin for subculture, and the number is determined with a Coulter counter (model B, Coulter Electronics).

Treatment with Chemicals

We have established treatment procedures which are found to be suitable for various chemicals (14,26). Briefly, CHO cells are plated at 5 × 10^5 cells/25 cm^2 bottle in medium F12FCM5. After a 16- to 24-hr growth period (cell number = ~1.0 to 1.5 × 10^6 cells/plate), the cells are washed twice with saline G, and sufficient serum-free F12 medium is added to bring the final volume to 5 ml after the addition of various amounts of microsome preparation (up to 1 ml) and 50 μl of chemical, usually dissolved in dimethyl sulfoxide. Chemicals and/or microsomes are omitted from some plates to provide controls. The microsomal preparation is prepared in this laboratory according to the method of Ames et al. (2) from livers of Aroclor 1254-induced male Sprague-Dawley rats; the microsome mix for biotransformation contains (per ml) 30 μmoles KCl, 10 μmoles MgCl$_2$, 10 μmoles CaCl$_2$, 4 μmoles NADP, 5 μmoles glucose-6-phosphate, 50 μmoles phosphate buffer (pH 8.0), and 0.1 ml microsome fraction (which contains 3 to 4 mg protein). Cells are then incubated for 5 hr and washed three times with saline G before 5 ml of F12FCM5 is added. After they are incubated overnight, cells are trypsinized and plated for cytotoxicity and specific gene mutagenesis to be described below.

Cytotoxicity

The effect of chemicals on the cellular cloning efficiency is determined by use of the treated cells described above. For an expected cloning efficiency higher than 50%, 200 well-dispersed single cells are plated, and for an expected survival lower than this, the number of cells plated is adjusted accordingly to yield 100 to 200 surviving colonies after standard incubation in medium F12FCM5 for 7 days. At the end of the incubation period, the plates are fixed with 3.7%

formalin and stained with a dilute crystal violet solution before the colonies are enumerated. A cluster of more than 50 cells growing within a confined area is considered to be a colony. Control cells, which do not receive treatment with mutagen, usually give 80% or higher plating efficiency under this condition. Neither the solvent-microsome mix nor these agents individually affect the cellular cloning efficiency. The effect of a chemical on the cloning efficiency is expressed as percent survival relative to the untreated controls.

Specific Gene Mutagenesis

The CHO/HGPRT system has been defined in terms of medium, TG concentration, optimal cell density for selection (and, hence, recovery of the presumptive mutants), and expression time for the mutant phenotype (14,26). For the determination of mutation induction, the treated cells are allowed to express the mutant phenotype in F12 medium for 7 to 9 days, at which time mutation induction reaches a maximum which is maintained thereafter (as long as 35 days examined) for several agents [ethyl methanesulfonate (EMS), N-methyl-N'-nitro-N-nitrosoguanidine (MNNG), ICR-191, X-ray, and UV light] irrespective of concentration or intensity of the mutagen (26–28,30–32). Routine subculture is performed at 2-day intervals during the expression period, and at the end of this time the cells are plated for selection in hypoxanthine-free F12FCM5 containing 1.7 μg/ml (10 μM) of TG at a density of 2.0×10^5 cells/100-mm plastic dish (Corning or Falcon), which permits 100% mutant recovery in reconstruction experiments (26). We find the use of dialyzed serum particularly important, presumably due to potential competition between hypoxanthine and TG for transport into the cells and for catalysis by HGPRT (26). After 7 to 8 days in the selective medium, the drug-resistant colonies develop; they are then fixed, stained, and counted. Such a protocol permits the maximum stable mutation induction by various physical and chemical agents of TG-resistant variants (TG^r), >98% of which have highly reduced HGPRT activity (26). Mutation frequency is calculated based on the number of drug-resistant colonies per survivor at the end of the expression period.

To analyze the nature of the dose-response curve, we fit data from several different dose ranges to the linear model by the method of weighted least-squares (14).

RESULTS AND DISCUSSIONS

Development of the CHO/HGPRT System and Its Application in Quantitative Mutagenesis

Development of a Protocol for Quantifying Specific Gene Mutagenesis

Many investigators have reported the induction and isolation of various mutants different from the phenotype of the parental mammalian cells (4). Generally, their procedures for mutation induction do not take into consideration the quantitative aspects of the mutagenesis, since the purpose of these studies was to obtain a particular phenotypic variant for genetic analysis. Furthermore, in many instances, only a small fraction of the presumptive variants possessed the desired stable phenotype. Due to the intrinsic characteristics of each gene mutation assay, factors required to quantify mutagenesis need to be established individually. For the CHO/HGPRT assay, the following need to be considered:

Cell physiology during mutagen treatment: Except in those experiments designed for cell cycle study, we treat cells during the exponential growth state because some mutagens may act preferentially on the proliferating cells.

Medium for cell growth and mutant selection: The growth medium used should not allow preferential growth of wild type over mutants or vice versa. Employing several independent TG^r mutants, we have found that, with either medium F12 or the hypoxanthine-free F12 medium supplemented with 5% or 10% whole serum or dialyzed serum, there is no differential growth between wild-type cells and TG^r mutants. During mutant selection with TG, the selective medium is devoid of hypoxanthine because TG competitively inhibits hypoxanthine transport across the cell membrane (26), and the reverse is likely true.

TG concentration: Optimum TG concentration should be used to select for phenotypic variants of mutational origin which are not either leaky or epigenetic in nature (13). As discussed below, our protocol, which calls for the use of dialyzed fetal calf serum with the optimum TG level, selects >98% of TG^r variants which have reduced HGPRT activity (26).

Cell density for selection: For full recovery of a small fraction of TG^r mutants from an excess of wild-type cells, the ratio of mutants to wild-type cells should be such that the mutants will escape the cytotoxic effects of the purine analog metabolites converted from TG through the HGPRT activity in the wild-type cells. In several reconstruction experiments in which a known number of TG^r mutants were cocultured with an increasing number of wild-type cells, we have found that 100 to 200 mutants can be fully recovered in the presence of as many as 0.3×10^6 wild-type cells in a 100-mm culture dish (26). Exponential loss of mutants occurs when the number of wild-type cells exceeds this number. The mechanism of the wild-type cell-mediated killing of the mutants is not fully understood, although it has been generally attributed to the effect of "metabolic cooperation" between cells in contact with each other.

Phenotypic expression time: Since the selection of the mutants is based on the loss of HGPRT activity, a period of delay for expression of the TG^r phenotype is expected to allow completion of mutation fixation and to permit diluting out of the preexisting enzyme and mRNA coded for HGPRT. We have found that maximum stable expression of the TG^r phenotype is reached 7 to 9 days after mutagenesis and remains constant thereafter irrespective of the nature and dose of the mutagen. Different doses of EMS, MNNG, ICR-191, and UV light give the same profiles of phenotypic delay (26–28,30–32) as described above.

Characteristics of the CHO/HGPRT System: Evidence of the Genetic Basis of Mutation at a Specific Locus

Conclusive direct proof of the genetic origin of mutations in somatic cells should theoretically rely on demonstration that the affected hereditary alteration has resulted in a modified nucleotide sequence of the specific gene, causing modified coding properties which result in the production of altered protein with changes in the amino acid sequence. In the absence of such proof, one must rely on indirect criteria which are consistent with the concept that the observed phenotypic variations are genetic in nature. Such criteria include stability of altered phenotype, mutagen-induced increase in occurrence of stable variants, biochemical and physiological identification of the variant phenotype, and chromosomal localization of the affected gene (4,13).

Over the past 5 years, we have used the assay protocol described and have found in approximately 600 experiments that the spontaneous mutation frequency lies in the range of 1 to 5×10^{-6} mutant/cell. Various physical and chemical agents are capable of inducing TG resistance. In all chemical mutagens examined, mutation induction occurs as a linear function of the concentration (6–9,13–19,26–32). For example, mutation frequency increases approximately linearly with EMS concentration in this near-diploid cell line, conforming to the expectation that mutation induction occurs in the gene localized at the functionally monosomic X chromosome. However, in the tetraploid CHO cells, EMS does not induce an appreciable number of mutations, even at very high concentrations, as predicted theoretically (13).

There is no measurable frequency of spontaneous reversion ($<10^{-7}$ revertant/cell) with 13 TG-resistant mutants, all of which contain low, yet detectable, HGPRT activity. More than 98% of the presumptive mutants isolated either from spontaneous mutation or as a result of mutation induction are sensitive to aminopterin, incorporate hypoxanthine at reduced rates, and have less than 5% HGPRT activity (26). Studies in progress have also shown that mutants containing temperature-sensitive HGPRT activity can be selected, suggesting that mutation resides in the HGPRT structural gene (O'Neill, J. P., and Hsie, A. W., *unpublished observations*).

The CHO/HGPRT system appears to fulfill the criteria for a specific gene locus mutational assay and should be valuable in studies of mechanisms of mammalian cell mutagenesis and in determination of the mutagenicity of various physical and chemical agents.

Interrelationships of Mutagen-mediated Cellular Lethality and Mutation Induction

When EMS is employed as a mutagen, mutation induction occurs over the entire survival curve, which includes both the shoulder region, where there is no appreciable loss of cell survival, and the exponential portion, where cell killing increases exponentially with increasing mutagen concentrations (13, 14,30,31). Apparently, there is no threshold effect of mutation induction with EMS. It appears that mutation induction is a more sensitive parameter than toxicity for agents such as EMS.

Later experiments show that many physical and chemical agents such as X-ray (27), UV light (15), ICR-191 (28), isopropyl methanesulfonate (iPMS) (7), and diethylsulfate (DES) (7) exhibit "EMS-type" curves of cell survival and mutagenesis. However, there are agents, typified by MNNG (8) and N-ethyl-N'-nitro-N-nitrosoguanidine (ENNG) (8), which do not exhibit an appreciable shoulder region in the survival curve, and for which mutation induction always occurs concomitantly with the loss of cell survival ("MNNG type").

As our studies on chemical mutagenesis have progressed, we have found other types of interrelationships of mutagenicity and cytotoxicity. Most promutagenic agents are neither toxic nor mutagenic to the cells in the absence of S_9-mediated metabolic activation. With S_9, benzo(a)pyrene [B(a)P] is mutagenic and cytotoxic with the EMS-type cell-survival curve (Brimer, P. A., O'Neill, J. P., San Sebastian, J. R., and Hsie, A. W., *unpublished data*). A direct-acting mutagen, 4-nitroquinoline-1-oxide (4-NQO), is highly cytotoxic and mutagenic to the CHO cells, and its cytotoxicity and mutagenicity decrease when it is treated with S_9 (San Sebastian, J. R., and Hsie, A. W., *unpublished data*). S_9 also greatly decreases the mutagenicity and cytotoxicity of ICR-191 (Fuscoe, J. C., and Hsie, A. W., *unpublished data*). As discussed below, the fact that a slight modification of the structure of the chemical may affect either the cytotoxicity or mutagenicity further demonstrates that the cytotoxicity and mutagenicity of chemical mutagens are separable.

A Study of EMS Exposure Dose: Differential Effects on Cellular Lethality and Mutagenesis

Earlier, we found that the frequency of EMS-induced mutation to TG resistance in cells treated for a fixed period of 16 hr is a linear function over a large range of mutagen concentrations, including both the shoulder region and the exponentially killing portion (13,14). To investigate whether EMS-induced mutagenesis can be quantified further, we treated cells with several concentrations of EMS for intervals of 2 to 24 hr. Mutation induction increased linearly with EMS concentrations of 0.05 to 0.4 mg/ml for incubation times of up to 12 to 14 hr. However, cell survival decreased exponentially with time over the entire 24-hr period. This difference in the time course of cellular lethality vs mutagenicity might be due to the formation of toxic nonmutagenic breakdown products in the medium with longer incubation times, or it might reflect a difference in the mode of action of EMS in these two biological effects. Further studies using varying concentrations (0.05 to 3.2 mg/ml) of EMS for 2 to 12 hr showed that the manifestation of cellular lethality and mutagenesis occurs as a function of EMS exposure dose: the biological effect is the same for different combinations of concentration multiplied by duration of treatment which yield the same product. From these studies, the mutagenic potential of EMS can be described as 310×10^{-6} mutant (cell mg ml^{-1} hr)$^{-1}$ (30). Thus, the CHO/HGPRT system appears to be suitable for dosimetry studies which are essential for our understanding of the molecular mechanisms involved in mammalian mutagenesis.

Structure–Activity Relationships of Chemical Mutagens

Since EMS mutagenicity is quantifiable, it appears that the CHO/HGPRT system should be useful for studies of the relationship between structural characteristics of the chemical mutagens and their mutagenicity.

The dose-response relationships of cell killing and mutation induction of two alkylsulfates [dimethylsulfate (DMS), DES] and three alkyl alkanesulfonates [methyl methanesulfonate (MMS), EMS, and iPMS] have been compared under identical experimental conditions. We observed that cytotoxicity decreased with the size of the alkyl group: DMS > DES; MMS > EMS > iPMS. All agents produced linear dose-response of mutation induction, in the order DMS > DES; MMS > EMS > iPMS, when comparison was made based on mutants induced per unit mutagen concentration. However, when comparisons were made at 10% survival, the relative mutagenic potency was: DES > DMS; EMS > MMS > iPMS (7).

Similar comparative studies were extended to two nitrosamidines (MNNG and ENNG) and three nitrosamides [N-methyl-N-nitrosourea (MNU), N-ethyl-N-nitrosourea (ENU), and N-butyl-N-nitrosourea (BNU)] differing in the nature of their alkylating group. Based on mutants induced per unit mutagen, the order of relative mutagenic potency was: MNNG > ENNG > MNU > ENU > BNU. The order was the same when comparisons were made at 10% cell survival (8). Nitrosation appears to be essential for nitrosamidines to exert mutagenicity, since N-methyl-N'-nitroguanidine is not mutagenic even at concentrations 50,000 higher than its nitroso analog, MNNG (San Sebastian, J. R., and Hsie, A. W., *unpublished data*).

Structure-activity relationships of antitumor agents would be of interest in screening for desirable chemotherapeutic drugs. Initially, we have chosen to investigate eight ICR compounds (ICR-191, -170, -292, -372, -191—OH, -170—OH, -292—OH, -372—OH). They contain structural similarities and differences which allow a study of the role of the heterocyclic nucleus and the alkylating side chain in the mutagenic activity. The first four contain a single 2-chloroethyl group (nitrogen mustard) on the side chain and are mutagenic, with the tertiary amine types (-170 and -292) three to five times more mutagenic than the secondary amine types (-191 and -372). All the hydroxy derivatives of the first four compounds are nonmutagenic although they are toxic. This indicates that the 2-chloroethyl group is required for mutation induction (28,29,

Fuscoe, J. C., O'Neill, J. P., and Hsie, A. W., *unpublished data*), and suggests that cytotoxicity is dissociable from mutagenicity.

Correlation of Mutagenicity in the CHO/HGPRT Assay with Reported Carcinogenicity in Animal Tests

To date, we have studies in different stages of completion of the mutagenesis of 108 chemicals. Most of these are various carcinogens and their structural analogs: 27 polycyclic hydrocarbons [e.g., B(*a*)P, benzo(*e*)pyrene, pyrene, and 7,12-dimethylbenz(*a*)anthracene], 16 nitrosamines (e.g., dimethylnitrosamine (DMN), diphenylnitrosamine, 2-methyl-1-nitrosopiperidine, and 2,5-dimethylnitrosopiperidine), 5 quinolines (e.g., quinoline and 4-NQO), 11 direct-acting alkylating agents (e.g., dimethylsulfate, methyl methanesulfonate, and MNU), 16 heterocyclic nitrogen mustards (e.g., ICR-191, -191—OH, -170, and -170—OH), 5 aromatic amines (e.g., 2-acetylaminofluorene and fluorene), 13 miscellaneous chemicals (e.g., captan, folpet, and saccharin), and 18 metallic compounds [e.g., $MnCl_2$ and *cis*-$Pt(NH_3)_2Cl_2$ (*cis*-DDP)] (16,19, Hsie, A. W., *unpublished data*).

In our studies of 108 chemicals, 83 have been reported to be either carcinogenic or noncarcinogenic in animal studies. Mutagenicity in the CHO/HGPRT assay of 76 of these agents correlated with the documented carcinogenicity in animals (20,38). The existence of a high correlation (76/83 = 92%) between mutagenicity and carcinogenicity recommends this assay for screening of chemical carcinogens. However, this result should be viewed with caution, since so far only limited classes of chemicals have been tested and some of the preliminary results remain to be confirmed.

Preliminary Development and Validation of the CHO Genetic Toxicity Assay for the Simultaneous Determination of Cytotoxicity, Mutagenicity, Chromosome Aberrations, and Sister-Chromatid Exchanges

We have so far shown that CHO cells are useful for studies of the cytotoxicity and mutagenicity of various chemicals. The CHO cells and other hamster cells in culture were also found to be suitable for studies of chemical-induced chromosome and chromatid aberrations and sister-chromatid exchanges. In our preliminary investigations, we have found that these assays are useful in evaluating the cytogenetic effects of B(*a*)P and DMN when CHO cells are coupled with the standard microsome preparation described earlier (San Sebastian, J. R., and Hsie, A. W., *unpublished data*). We expect that the successful development and validation of the multiplex CHO cell genetic toxicity system will be extremely valuable from both the scientific and economic points of view in genetic toxicology, because this system will allow the simultaneous determination of four distinct biological effects: *cytotoxicity* or *cloning efficiency* measures the reproductive capacity of a single cell to develop into a colony; *single gene mutagenesis* involves changes in the nucleotide sequence of DNA of a specific gene resulting in the acquisition of a novel or altered phenotype; *chromosome aberrations* involve microscopically identifiable changes in the number and/or structure of the chromosome; and *sister-chromatid exchange* measures the extent of double-strand exchange in the DNA duplex after breaks and rejoining of subunits of chromatids, each of which consists of one DNA duplex.

Metal Mutagenesis: A Preliminary Study of the Cytotoxicity and Mutagenicity of Pt, Mn, and Other Metallic Compounds

Platinum Compounds

cis-DDP is a widely used inorganic antitumor agent. It is particularly effective in combination chemotherapy against testicular and advanced bladder cancer (5). In addition to its antitumor activity, *cis*-DDP is known to enhance prophage induction, inactivate transforming DNA and viruses, induce filamentous growth and mutagenesis in bacteria, selectively inhibit DNA synthesis, produce chromosomal abnormalities, and kill various types of cells (5). The *trans* isomer is, biologically, either completely inactive or less effective than the *cis* form. Chemically similar molecules with such different biological effects offer an opportunity for investigation of the mechanisms of mutagenicity and toxicity. It is generally thought that *cis*- and *trans*-DDP produce different lesions in DNA which, in turn, result in different levels of expression of biological activity. Studies with *S. typhimurium* indicate that *cis*-DDP, the most potent of the Pt(II) chloroamines, is a base-substitution mutagen (22,24).

Employing the CHO/HGPRT assay, we found that *cis*-DDP is mutagenic and cytotoxic. *trans*-DDP shows highly reduced activity even over a concentration range ten times greater than the *cis* isomer. Also biologically relatively inactive are $K_2[PtCl_4]$ and $[Pt(NH_3)_4]Cl_2$. Based on the slope of mutation induction and cell survival, the relative mutagenicity and cytotoxicity of *cis*-DDP: *trans*-DDP: $K_2[PtCl_4]$: $[Pt(NH_3)_4]Cl_2$ are 100:0.7:1.5:0.02 and 100:0.6:0.7:0.1, respectively (Johnson, N. P., Hoeschele, J. D., Rahn, R. O., and Hsie, A. W., *unpublished data*). The relative mutagenicity of these compounds in CHO cells is similar to that found in earlier studies in the *Salmonella* system (22) and is parallel with the documented antitumor activity (5).

cis-DDP was found to bind to CHO DNA linearly, indicating that mutagenicity and cytotoxicity correlate with DNA binding for this compound. A preliminary experiment with *trans*-DDP shows that this compound also binds to DNA linearly, with a slope similar to that of *cis*-DDP (Johnson, N. P., Herschele, J. D., Rahn, R. O., and Hsie, A. W., *unpublished data*). Evidently, *cis*-DDP exhibits a much greater potency per DNA lesion than *trans*-DDP.

Covalent bonds are formed when platinum compounds react with DNA which are stable under conditions employed for isolation of platinum-bound DNA.

Most other metals bind weakly to DNA, therefore quantification of the extent of DNA binding in the cell is more difficult with them. Furthermore, with platinum the kinetics of binding are sufficiently slow to be monitored and the physical and chemical properties of the platinum compounds can be easily modified by altering the ligands. These features recommend the choice of platinum compounds for studies of the mechanism of chemical modification of cellular DNA by metals. The CHO/HGPRT system offers an additional advantage in that DNA modification can be related to cytotoxicity and mutagenicity. Since both *cis*- and *trans*-DDP bind to DNA but the *cis* isomer is much more mutagenic, it would be interesting to investigate the reasons for the different mutagenicity per DNA lesion of these two steric isomers.

Mn and 13 Other Metallic Compounds

Mn(II) has been shown to be mutagenic in the bacteria *E. coli* and *S. typhimurium,* bacteriophage T4, and yeast, and to be weakly carcinogenic in rats and mice (11).

Although $MnCl_2$ is neither mutagenic nor cytotoxic when assayed under the standard conditions for testing all direct-acting mutagens including *cis*-DDP, these biological effects are evident when the ionic composition of the medium is altered. For example, when CHO cells were treated in medium lacking $FeSO_4$, $CaCl_2$, $CuSO_4$, and $ZnSO_4$, but containing the usual amount of $MgCl_2$ (0.6 mM), $MnCl_2$ (90 μM) induced mutation to 100×10^{-6} mutant/cell (from the spontaneous mutation frequency of about 5×10^{-6} mutant/cell) and reduced cloning efficiency by 33%. The effect of $MnCl_2$ was dependent on the concentration of $MgCl_2$. Both the mutagenicity and toxicity produced by $MnCl_2$ (90 μM) were abolished by the addition of excess $MgCl_2$ (2.5 mM). Treatment with $MnCl_2$ in divalent-cation-deficient medium supplemented with $CaCl_2$ also produced mutation induction (60×10^{-6} mutant/cell at 14 μM $MnCl_2$) (9). The unusual environment required for demonstration of mutagenicity of $MnCl_2$ makes assessment of its biological hazard difficult.

The mutagenicities of 13 other metallic compounds were also determined, and the preliminary result shows that the carcinogenic metallic compounds $NiCl_2$, $BeSO_4$, $CdCl_2$, $FeSO_4$, $CuSO_4$, $AgNO_3$, $Pb(CH_3COO)_2$, and $ZnSO_4$ are mutagenic, whereas the noncarcinogenic metallic compounds RbCl, H_2SeO_3, $TiCl_4$, and $MgCl_2$ are not mutagenic (Couch, D. B., Forbes, N. L., and Hsie, A. W., *unpublished data*). The results of these studies remain to be confirmed in view of our recent findings that metal mutagenesis is sensitively modified by the physiological state of the cells during treatment. For example, the mutagenicity of mutagenic metals at high concentrations, which causes both severe growth inhibition and/or significant cellular lethality, is either not demonstrable or much less than expected from the linear dose-response at low dose-range. This observation suggests that mutation induction by metallic compounds requires active DNA synthesis during treatment with the cells. One possible mechanism of metal-induced mutagenesis could, thus, be that the metallic compound causes infidelity in DNA replicative synthesis, leading to the expression of an altered phenotype of a specific gene; this notion is consistent with the finding that carcinogenic and mutagenic metals enhance error of DNA synthesis in the cell-free system (34,35). Growth inhibition with inactive DNA synthesis during treatment would, therefore, disfavor mutation induction by mutagenic metals.

Since the ionic composition of the assay medium and cellular physiology apparently affect the manifestation of mutation induction of $MnCl_2$, it appears likely that quantitative metal mutagenesis requires further refinement of assay conditions, perhaps uniquely defined for each metallic compound. We have just begun to undertake such a demanding yet challenging task.

SUMMARY

We have defined a set of stringent conditions required for quantification of specific gene mutation in a mammalian cell system, CHO/HGPRT. More than 98% of the 6-thioguanine-resistant variants have been shown to be deficient in hypoxanthine-guanine phosphoribosyl transferase (HGPRT) activity in Chinese hamster ovary (CHO) cells. In all direct-acting chemical mutagens studied, mutation induction increases linearly as a function of the concentration, with no apparent threshold. Some chemicals induce mutation at noncytotoxic concentrations; others induce mutation only with a concomitant loss of cell survival. In one dosimetry study, ethyl methanesulfonate produced 310×10^{-6} mutant (cell mg ml^{-1} hr)$^{-1}$. The sensitive and quantitative nature of the CHO/HGPRT assay has been utilized for studies of the structure-activity (mutagenicity) relationships of various classes of chemicals, including alkylating agents and heterocyclic nitrogen mustards. When rat liver S_9-mediated metabolic activation is present, procarcinogens such as benzo(a)pyrene and 2-acetylaminofluorene are mutagenic, whereas their noncarcinogenic structural analogs pyrene and fluorene are not. Mutagenicity as determined in the CHO/HGPRT assay appears to correlate well (76/83 = 92%) with the reported carcinogenicity in animals of 108 chemicals being examined.

The system also appears to be suitable for studies of the mutagenicity and cytotoxicity of metallic compounds. We found that cis-dichlorodiammine Pt(II), [cis-Pt(NH$_3$)$_2$Cl$_2$], a widely used inorganic antitumor agent, is cytotoxic and mutagenic and that its mutagenicity correlates with its binding to DNA. However, trans-dichlorodiammine Pt(II), K$_2$[PtCl$_4$], and [Pt(NH$_3$)$_4$]Cl$_2$ exhibit greatly reduced biological activities. Among 14 other metals studied, we found that carcinogenic metallic compounds such as $MnCl_2$, $NiCl_2$, and $BeSO_4$ are mutagenic, whereas noncarcinogenic compounds such as $MgCl_2$ and H_2SeO_3 are not. Determination of metal mutagenicity is apparently complicated by the ionic composition of the medium. For example, the mutagenicity and cytotoxicity increase with increasing $MnCl_2$ concentration when cells are treated with divalent-cation-deficient medium containing normal levels of $MgCl_2$; however, both of these effects are diminished by increasing $MgCl_2$ concentration. The unusual environment required for demonstration of the mutagenicity of $MnCl_2$ complicates the assessment of its biological hazard. This may account in part for varying results in studies of the mutagenicity of other metallic compounds. Further refinement of the assay conditions, especially with respect to the ionic environment necessary for quantifying mutagenicity of each metallic agent, is in progress.

ACKNOWLEDGMENT

This research was supported jointly by the Environmental Protection Agency (IAG-D5-E681), the National Center for Toxicological Research, and the Office of Health and Environmental Research, U.S. Department of Energy, under contract W-7504-eng-26 with Union Carbide Corporation. NPJ and DBC are supported by Carcinogenesis Training Grant CA 05296 from the National Cancer Institute. JRSS is supported by Monsanto Toxicology Fund.

REFERENCES

1. Abe, S., and Sasaki, M. (1977): Chromosome aberrations and sister chromatid exchanges in Chinese hamster cells exposed to various chemicals. *J. Natl. Cancer Inst.,* 58:1635–1641.
2. Ames, B. N., McCann, J., and Yamasaki, E. (1975): Methods for detecting carcinogens and mutagens with the *Salmonella*/mammalian-microsome mutagenicity test. *Mutat. Res.,* 31:347–364.
3. Casto, B. D., Meyers, J., and DiPaolo, J. A. (1979): Enhancement of viral transformation for evaluation of carcinogenic or mutagenic potential of inorganic metal salts. *Cancer Res.,* 39:193–198.
4. Chu, E. H. Y., and Powell, S. S. (1976): Selective systems in somatic cell genetics. *Adv. Hum. Genet.,* 7:189–258.
5. Committee on Medical and Biologic Effects of Environmental Pollutants, National Academy of Sciences, USA (1977): Platinum-Group Metals. 232 pp. National Academy of Sciences, Washington, D.C.
6. Couch, D. B., Forbes, N. L., and Hsie, A. W. (1978): Comparative mutagenicity of alkylsulfate and alkanesulfonate derivatives in Chinese hamster ovary cells. *Mutat. Res.,* 57:217–224.
7. Couch, D. B., and Hsie, A. W. (1976): Dose-response relationships of cytotoxicity and mutagenicity of monofunctional alkylating agents in Chinese hamster ovary cells. *Mutat. Res.,* 38:399.
8. Couch, D. B., and Hsie, A. W. (1978): Mutagenicity and cytotoxicity of congeners of two classes of nitroso compounds in Chinese hamster ovary cells. *Mutat. Res.,* 57:209–216.
9. Couch, D. B., and Hsie, A. W. (1978): Metal mutagenesis: Studies of the mutagenicity of manganous chloride and 14 other metallic compounds in the CHO/HGPRT assay. In: *Program of the Annual Meeting of the Environmental Mutagen Society,* San Francisco, March 9–13, 1978, p. 74.
10. Ferm, V. H. (1972): The teratogenic effects of metals on mammalian embryos. *Adv. Teratol.,* 5:51–75.
11. Flessel, C. P. (1978): Metals as mutagens. In: *Inorganic and Nutritional Aspects of Cancer,* edited by G. N. Schrauzer, Chapter 9. Plenum Press, New York.
12. Fradkin, A., Janoff, A., Lane, B. P., and Kuschner, M. (1975): *In vitro* transformation of BHK12 cells grown in the presence of calcium chromate. *Cancer Res.,* 35:1058–1063.
13. Hsie, A. W., Brimer, P. A., Machanoff, R., and Hsie, M. H. (1977): Further evidence for the genetic origin of mutations in mammalian somatic cells: The effects of ploidy level and selection

stringency on dose-dependent chemical mutagenesis to purine analogue resistance in Chinese hamster ovary cells. *Mutat. Res.,* 45:271–282.
14. Hsie, A. W., Brimer, P. A., Mitchell, T. J., and Gosslee, D. G. (1975): The dose-response relationship for ethyl methanesulfonate-induced mutations at the hypoxanthine–guanine phosphoribosyl transferase locus in Chinese hamster ovary cells. *Somat. Cell. Genet.,* 1:247–261.
15. Hsie, A. W., Brimer, P. A., Mitchell, T. J., and Gosslee, D. G. (1975): The dose-response relationship for ultraviolet light-induced mutations at the hypoxanthine-guanine phosphoribosyl transferase locus in Chinese hamster ovary cells. *Somat. Cell Genet.,* 1:383–389.
16. Hsie, A. W., Couch, D. B., O'Neill, J. P., San Sebastian, J. R., Brimer, P. A., Machanoff, R., Riddle, J. C., Li, A. P., Fuscoe, J. C., Forbes, N. L., and Hsie, M. H. (1979): Utilization of a quantitative mammalian cell mutation system, CHO/HGPRT, in experimental mutagenesis and genetic toxicology. Presented at Chemical Industry Institute of Toxicology (CIIT) Workshop on "Strategies for Short-Term Testing for Mutagens/Carcinogens," Research Triangle Park, North Carolina, August 11–12, 1977. In: *CRC Uniscience Monographs in Toxicology,* edited by B. E. Butterworth, CRC Uniscience Monographs *(in press).*
17. Hsie, A. W., Li, A. P., and Machanoff, R. (1977): A fluence-response study of lethality and mutagenicity of white, black, and blue fluorescent lights, sunlamp, and sunlight irradiation in Chinese hamster ovary cells. *Mutat. Res.,* 45:333–342.
18. Hsie, A. W., Machanoff, R., Couch, D. B., and Holland, J. M. (1978): Mutagenicity of dimethyl nitrosamine and ethyl methanesulfonate as determined by a quantitative host-mediated CHO/HGPRT assay. *Mutat. Res.,* 51:77–84.
19. Hsie, A. W., O'Neill, J. P., Couch, D. B., San Sebastian, J. R., Brimer, P. A., Machanoff, R., Fuscoe, J. C., Riddle, J. C., Li, A. P., Forbes, N. L., and Hsie, M. H. (1978): Quantitative analyses of radiation- and chemical-induced lethality and mutagenesis in Chinese hamster ovary cells. *Radiat. Res.,* 76:471–492.
20. IARC, (1972–76): IARC Monograph on the Evaluation of Carcinogenic Risk of Chemicals to Man, Vols. 1–10. IARC, Lyons.
21. Ishidate, M., Jr., and Odashima, S. (1977): Chromosome tests with 134 compounds on Chinese hamster cells *in vitro:* A screening for chemical carcinogens. *Mutat. Res.,* 48:337–354.
22. Leconte, P., Macquet, J., Butour, J., and Paoletti, C. (1977): Relative efficiencies of a series of square-planar platinum (II) compounds on *Salmonella* mutagenesis. *Mutat. Res.,* 48:139–144.
23. McCann, J., Choi, E., Yamasaki, E., and Ames, B. N. (1975): Detection of carcinogens as mutagens in the *Salmonella*/microsome test: Assay of 300 chemicals. *Proc. Natl. Acad. Sci. USA,* 72:5135–5139.
24. Monti-Bragadin, C., Tamaro, M., and Banfi, E. (1975): Mutagenic activity of platinum and ruthenium complexes. *Chem. Biol. Interact.,* 11:469–472.
25. Murray, M. J., and Flessel, C. P. (1976): Metal-polynucleotide interactions: A comparison of carcinogenic and non-carcinogenic metals *in vitro. Biochim. Biophys. Acta,* 425:256–261.
26. O'Neill, J. P., Brimer, P. A., Machanoff, R., Hirsch, G. P., and Hsie, A. W., (1977): A quantitative assay of mutation induction at the hypoxanthine-guanine phosphoribosyl transferase locus in Chinese hamster ovary cells: Development and definition of the system. *Mutat. Res.,* 45:91–101.
27. O'Neill, J. P., Couch, D. B., Machanoff, R., San Sebastian, J. R., Brimer, P. A., and Hsie, A. W. (1977): A quantitative assay of mutation induction at the hypoxanthine-guanine phosphoribosyl transferase locus in Chinese hamster ovary cells (CHO/HGPRT system): Utilization with a variety of mutagenic agents. *Mutat. Res.,* 45:103–109.
28. O'Neill, J. P., Fuscoe, J. C., and Hsie, A. W. (1978): Mutagenicity of heterocyclic nitrogen mustards (ICR compounds) in cultured mammalian cells. *Cancer Res.,* 38:506–509.
29. O'Neill, J. P., Fuscoe, J. C., and Hsie, A. W. (1978): Structure–activity relationship of antitumor agents in the CHO/HGPRT system: Cytotoxicity and mutagenicity of 8 ICR compounds. Program of the Annual Meeting of the American Society for Cancer Research, Washington, D.C., April 5–8, 1978.
30. O'Neill, J. P., and Hsie, A. W. (1977): Chemical mutagenesis of mammalian cells can be quantified. *Nature,* 269:815–817.
31. O'Neill, J. P., and Hsie, A. W. (1978): Phenotypic expression time of mutagen-induced 6-thioguanine resistance in Chinese hamster ovary cells (CHO/HGPRT system). *Mutat. Res.,* 59:109–118.

32. Riddle, J. C., and Hsie, A. W. (1978): An effect of cell cycle position on ultraviolet light-induced mutagenesis in Chinese hamster ovary cells. *Mutat. Res.*, 52:409–420.
33. Shin, Y., and Eichhorn, G. (1975): Different susceptibility of DNA and RNA to cleavage by metal ions. *Nature,* 254:358–359.
34. Sirover, M. A., and Loeb, L. A. (1976): Metal-induced infidelity during DNA synthesis. *Proc. Natl. Acad. Sci. USA,* 73:2331–2335.
35. Sirover, M. A., and Loeb, L. A. (1976): Infidelity of DNA synthesis *in vitro:* Screening for potential mutagens or carcinogens. *Science* 194:1434–1436.
36. Sugimura, T., Sato, S., Nagao, M., Yahagi, T., Matsushima, T., Seino, U., Takeuchi, M., and Kawachi, T. (1976): Overlapping of carcinogens and mutagens. In: *Fundamentals in Cancer Prevention,* edited by P. N. Magee, S. Takayama, T. Sugimura, and T. Matsushima, pp. 191–215. University of Tokyo Press/University Park Press, Tokyo/Baltimore.
37. Sunderman, F. W., Jr. (1978): Carcinogenic effects of metals. *Fed. Proc.,* 37:40–46.
38. USPHS, Survey of Compounds which Have Been Tested for Carcinogenic Activity. (1967–73): USPHS Publication No. 149.

023
In Vitro Clonal Growth Assay for Evaluating Toxicity of Metal Salts

M. E. Frazier and T. K. Andrews

Molecular Biology and Biophysics Section, Biology Department, Pacific Northwest Laboratory, Battelle Memorial Institute, Richland, Washington 99352

ABSTRACT

A clonal growth assay using VERO, a monkey cell line, has been employed in our laboratory to provide a rapid and inexpensive estimate of the toxicity of metallic compounds. High dilution factors of the clonal growth assay were used to ensure minimal cell-to-cell interaction before clone formation. As a result, these isolated cells are more stringent in their nutritional requirements than high-density, monolayered cultures of the same cells.

The compounds under test were added (at varying concentrations) to isolated cells, and the degree of colony formation relative to control cultures was used as a measure of toxicity. Using the relative plating efficiency (RPE) concept, we have generated concentration response curves for several metallic compounds. When the RPE_{50} (the concentration where the plating efficiency equals 50% of control cultures) was selected as a point of comparison, the ranking of metal toxicity in VERO cells was V > Cr > Ru > Cd > Hg > Ni > Co > Zn > Cu > Mn > Pb > Fe > Mg > Ca. This ranking is consistent with the relative hazard ratings and toxicity ratings of these same compounds for humans. Further, there is a positive correlation between the concentrations of metallic compounds causing toxicity in VERO cells and the concentrations of these same compounds resulting in lethality for experimental animals.

The results of the VERO clonal growth assay were compared with results from 20-hr viability tests using either WI38 cells or alveolar macrophages; the sensitivity of the VERO system ranged from 4,400 times greater (chromium) to 4 times greater (copper).

Our results indicate that it would be reasonable to use cell cultures to screen metallic compounds and estimate their relative toxicity to higher organisms. This information would also aid in setting priorities and in designing further investigations using animal models.

Toxic and carcinogenic determinations measuring effects of metals on laboratory animals are slow and expensive. Further, the results obtained may vary, depending on a variety of factors (e.g., the chemical form of the metal being evaluated, the animal species being used, and the age of the animals). Therefore, when it becomes necessary to examine a large number of metallic compounds and the numerous combinations of these materials present in the environment, the number of variable parameters may be large enough to preclude a definitive analysis in animals. For such studies, *in vitro* assays would be useful in providing estimates of relative toxicity. These values could then be used to set priorities for testing compounds, alone or in combination, in animals. However, there is a paucity of reports on such assays in the literature and, in general, a lack of information on the toxic effects of metallic compounds in cell cultures. This is surprising in light of our ability to manipulate the cultured cell's "environment," and in view of the fact that cultured cells represent a less complex model system than whole animals.

The most significant early work in this area was by Verne (17), who for more than 20 years examined the toxicity of metal chlorides and drugs in cultured cells. Verne first addressed the following questions: (a) How can toxic effects upon cells be evaluated in culture?; (b) What will be the sensitivity of different cell strains?; and (c) Can cell cultures be used to provide qualitative and quantitative assays for metal toxicity? Unfortunately, these early studies were done using plasma clot and hanging drop cultures. Mauersberger and Zorn (9) were among the first investigators to attempt to answer these same questions using modern cell culture techniques. In their studies, the toxicity of $CuSO_4$ varied considerably, depending on the cell line used and the conditions of cultivation (e.g., percent serum and cell numbers). Several subsequent studies by other investigators dealt primarily with chromosomal effects of metals on cultured cells (11,12,16).

Only since the advent of concern about the consequences of metals in the environment has much consideration been given to examining the toxicity of a variety of metals using *in vitro* systems. Until our study, the alveolar macrophage 20-hr viability test (8,19,20) represented the most complete effort to study and compare the toxicity of several metal compounds using one assay system.

A clonal growth assay using VERO, a monkey kidney cell, has been employed in our laboratory to provide rapid and inexpensive estimates of the toxicity of metallic compounds, chelating agents, and complex organic mixtures. VERO cells were chosen as indicators because of their high plating efficiency, because they have a predictable and reproducible growth rate, and cell morphology, as well as excellent colony-forming qualities.

In the clonal growth assay, the objective is to observe development of several colonies, each of which originates from a single cell derived from the same parent population. For the purpose of this assay, the cells are considered to be identical.

The clonal growth assay has many applications. In fact, it can be used to study any factor that affects cell growth, survival, morphology, or any other functional property that can be readily scored without removing the colonies from the culture vessel.

Several features make this assay a potentially valuable tool for toxicity testing. First, there is little danger of carryover of essential nutrients with the cell inoculum because, in the culture vessel, the volume of the medium is several million times greater than the cell volume. For this same reason, modification or "conditioning" of the medium by the cells is not a problem. Furthermore, due to the physical distance between colonies, cellular interaction among different colonies is minimal. Because of this high dilution factor, cloned cells are more sensitive to toxic agents and more stringent in their nutritional requirements than high-density cultures of the same cells. Finally, the techniques appear to be particularly useful when only small amounts of material are available for test.

Once the system is in place, the assay can be used to obtain several kinds of data. First, the plating efficiency (which is determined by dividing the number of colony-producing cells by the total number of cells in the inoculum) allows an objective numerical comparison between selected variables (e.g., different concentrations of a metal). Furthermore, the growth rate, and variations of this rate among individual colonies, can provide a clearer analysis for studying effects of toxic materials on cell populations than the growth rate of mass cultures. Finally, morphological changes in individual colonies can be observed. In this report, only the data obtained from plating efficiency studies will be discussed.

EXPERIMENTAL METHODS

Preparation of Cell Cultures

Stock cultures of VERO, a monkey kidney cell line (14), were grown as monolayer cultures in NCTC-135 medium (6) supplemented with 10% fetal bovine serum (FBS) and containing 50 μg/ml gentamycin. All cultures were incubated at 35°C in 95% air and 5% CO_2. In preparation for the assay procedure, cultures were trypsinized with 0.1% trypsin-phosphate-buffered saline (4,5) solution to obtain monodispersed cells. The resultant cells were centrifuged and washed two times with culture medium, and cell viability was determined using trypan blue. Cell counts were performed using a hemocytometer and the cells were then diluted in culture medium (containing 20% FBS) so that 100 μl of the cell culture medium mixture contained approximately 50 cells. Then, using a sterile Hamilton syringe, 100 μl of culture fluid was delivered to each well of a microtiter II plate (Falcon Plastics).

Preparation of Test Chemical

The test chemical was solubilized in 0.1 N HCl made with double-distilled water. Initial dilutions were then made in NCTC-135 medium without FBS and filter-sterilized using a 0.22-μm filter. Serial dilutions were made from this stock using growth medium (also without FBS) at two times the desired final concentration. One hundred μl of the compound to be assayed were added to each of 10 wells of 2–4 replicate microtiter plates containing approximately 50 VERO cells per well. The outside wells of the microtiter plates were filled with growth media only, in order to minimize evaporation of media from the test wells.

Following a 4-day incubation period, the culture medium was removed, the wells were rinsed with phosphate-buffered saline, and the cells were fixed with 70% methanol. The cells were stained with crystal violet, rinsed with distilled water, and air-dried.

An inverted microscope (20×) was used to visualize the individual microtiter well. Colonies containing eight or more cells were scored as a clone.

Calculations

The colonies in each well were counted, the numbers were recorded, and relative plating efficiencies were determined for each dilution of the tested compound. The relative plating efficiency was then calculated by dividing the total number of colony-producing cells in the 10 control wells—those without test compound—into the total number of colony-producing cells in each of the 10 experimental wells (those containing a given concentration of test compound). The relative plating efficiencies from each of the replicated plates from at least two experiments were then added together to give a mean relative plating efficiency for the test compound at a given chemical concentration.

Metal Analyses

In the preparation of samples for metal analysis, sheets of Mylar were placed in 35-mm film projection holders, and a 1-cm circle of no. 41 Whatman filter paper was mounted on each Mylar sheet, using silicone grease. The metal-containing culture fluids or standards were added to the filter paper using micropipettes. The samples were dried using using heat lamps, and the amounts of metal were determined.

Four standard solutions were prepared for each metal chloride assayed. Using these solutions, standard curves were generated for each of the metals. The actual amount of metal added to the cultured cells was then determined by assaying those samples directly and comparing them with the standard curves. The amount of metal ion is expressed in $\mu g/ml$ of metal.

The concentration of metal present in these samples was determined using X-ray fluorescence. A filtered zirconium secondary source was used for analysis of V, Mn, Fe, Co, Ni, Cu, Zn, Hg, and Pb, and a filtered ^{241}Am source (100 mCi) was used for analysis of Cd and Ru. The filtered zirconium secondary source was excited by a 1.6-kW tungsten X-ray tube. All samples required 15-min analyses. Both tube and isotope-excited systems were equipped with similar 80-mm^2 Si(Li) detectors with a resolution of 200 eV FWHM at 6.4 KeV (Kevex, Inc., Burlingame, California).

A multielement peak analysis method (10) was used to analyze the resulting X-ray spectra. Net peak areas were corrected for overlap and used directly to compare samples to standards. $K\alpha$ peaks were used for all elements except Hg and Pb, for which the $L\alpha$ and $L\beta$ peaks, respectively, were used.

RESULTS

Using mean RPE values and the metal analysis data, we were able to generate concentration response curves for each of the metals tested (six are shown in Fig. 1). Each point represents the mean value determined from replicated plates of at least two separate experiments. The standard deviation in these experiments ranges from ± 2 to ± 10%.

An RPE_{50} (the concentration of test material which reduces the cloning efficiency by 50%) for each of the test materials was then determined by interpolation from the dose response curves. Using these RPE_{50} values, we have listed (Table 1) the metals in order of toxicity (on a μg/ml of metal basis), proceeding from the most toxic, vanadium, to the least toxic, calcium.

There is little data in the literature relating to the toxicity of metals in cell cultures. However, when we compared our values to the few that are available, we found our results were consistent with other reported findings. For example, Paton and Allison (11) reported cytotoxic effects in WI38 cells after 24-hr exposures with cadmium, cobalt, and mercury levels essentially identical to those reported here. Likewise, a statistically significant reduction in ciliary activity (beat frequency) of cultured tracheal cells occurred at concentrations of cadmium as low as 0.7 μg/ml (1). In corroboration of our value for chromium, it has been reported by Tsuda and Kato (16) that 0.065 μg/ml of chromium causes a seven-fold increase in numbers of lethal chromosomal abberations in hamster embryo cells versus unexposed controls. Finally, in support of our nickel values,

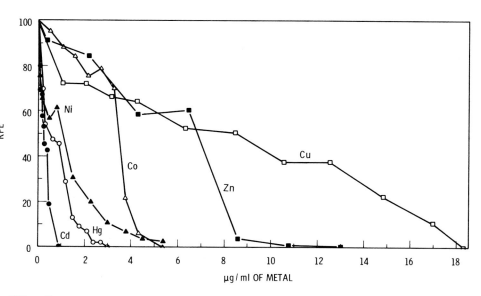

FIG. 1. Response curves showing the effects of various concentrations of metal ions on the relative plating efficiency (RPE) of VERO cells.

TABLE 1. *In vitro cytotoxicity of metals as measured with VERO cells in the clonal growth assay*

Test agent[a]	RPE_{50}[b] dose (μg/ml)[c]
V	0.03
Cr	0.06
Ru	0.1
Cd	0.3
Hg	0.5
Ni	1.1
Co	3.5
Zn	6.8
Cu	8.6
Mn	15
Pb	37
Fe	59
Mg	>2,000
Ca	>3,000

[a] All metals were tested as chloride salts except Cr, which was tested as chromate.
[b] RPE_{50} = the point at which the relative plating efficiency is 50% that of control cultures.
[c] Metal concentrations were determined using X-ray fluorescence (10).

a recent article by Christian and Nelson (3) reported a 50% reduction in growth of L929 cells at about 0.02 mM nickel, which is equivalent to our RPE_{50} value of 1.1 µg/ml with nickel. In an attempt to evaluate the sensitivity of our system, we have compared our values to the values obtained from alveolar macrophage (8,19,20) and WI38 (18) viability assays (Table 2), tests which are currently in use by the Environmental Protection Agency.

First, it should be noted that parameters and time periods measured by these systems differ from ours. The alveolar macrophage and WI38 cell systems measure viability, as determined by trypan blue exclusion after 20 hr of exposure. The VERO cell system, in contrast, measures a functional viability in which the individual cell must undergo several cell divisions in the presence of the test compound before it is scored as viable. Further, the VERO cells are a continuously dividing cell line, and the results of our studies with them relate to metal effects on a dynamic system. The alveolar macrophage, on the other hand, is considered to be an end cell, a static system. Nonetheless, since the alveolar macrophage and WI38 cell assays represent the only other *in vitro* toxicity assays with sufficient numbers of data points for comparison, we will compare the results from these three procedures. We have chosen the lowest published values for metal toxicity to the alveolar macrophage and WI38 cell systems. In every instance, the VERO system is the more sensitive. The increased sensitivity varies from a high of 4,400 times more sensitive with regard to chromium, to a low of about 4.5 times more sensitive with copper.

TABLE 2. *Comparison of data from alveolar macrophage, WI38 viability tests and the VERO clonal growth assay*

Metal ion	Viability—50% level at 20 hr			RPE_{50} of VERO cells, chloride (μg/ml)	Increased sensitivity of VERO assay
	WI38 Chloride (μg/ml)	Alveolar macrophage			
		Chloride (μg/ml)	Sulfates (μg/ml)		
V	14	5	8.7	3×10^{-2}	167 [b]
Cr	548.2	263	—	6×10^{-2} [a]	4.4×10^{3}
Ru	—	—	—	0.1	—
Cd	30.4	9	9	0.3	30
Hg	—	9.4	20.6	0.5	19
Ni	200.8	112.1	347	1.1	100
Co	—	—	—	3.5	—
Zn	—	195	53.1	6.8	8
Cu	—	40.4	38.7	8.6	4.5
Mn	430.7	261.5	241.4	15	16
Pb	—	—	—	37	—
Fe	—	—	—	59	—

[a] Tested as chromate.
[b] Value from the viability determination/value from VERO clonal growth assay.

In Fig. 2, we have attempted to correlate our RPE_{50} values with lethal dose (LD) values from the toxicology handbook (15). For this comparison, we selected the lowest concentration of the test material resulting in a lethal dose to a mammal. The correlation between our RPE_{50} and the LD was 0.985 (removing

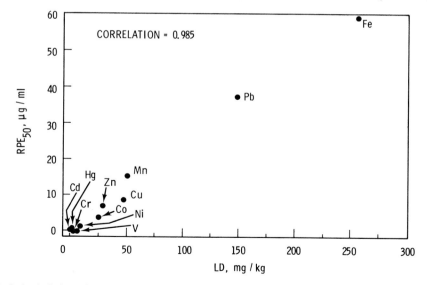

FIG. 2. Lethal dose (LD) values are plotted versus RPE_{50} values. When all 11 data points are included, the correlation is equal to 0.985.

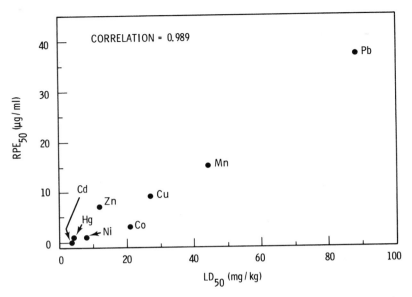

FIG. 3. The RPE$_{50}$ values obtained from the VERO clonal growth assay are plotted versus LD$_{50}$ values from Bienvenu et al. (2). The correlation between these two data sets is 0.989.

any two points does not appreciably alter the observed correlation). This correlation coefficient was derived from only 11 data points, but it suggests that the comparison of RPE$_{50}$ with LD values may be a reasonable means of estimating toxicity to higher organisms from cell culture data.

A similar comparison of RPE$_{50}$ with LD$_{50}$ values (Fig. 3) from Bienvenu et al. (2) yielded an equivalent correlation. We made certain assumptions in these two correlation comparisons, and there is some variability within the systems being compared; however, we do not intend to use these calculations to quantitatively prove our system, merely to show that the VERO clonal assay is qualitatively valid.

To further validate the clonal growth assay system as a means of estimating toxicity, we examined the relationships between our results and observed toxicity in humans (Table 3). We compared our RPE$_{50}$ values with the relative hazard ratings for these same compounds under industrial conditions. Sax's toxic hazard rating (13) is 0 = no toxicity, 1 = slight, 2 = moderate, and 3 = highly toxic. While this ranking system has obvious ambiguities and shortcomings (primarily the lack of a quantitative basis), our data are in agreement with Sax's ranking: i.e., the most toxic metals in our system (e.g., Cr, Hg, and Cd) have rankings of 3 in Sax's system, and our least toxic (Ca and Mg) have rankings of 0.

In the last column of Table 3, we have listed another system of toxicity rating or toxicity class (7). This system is based on the probably lethal oral dose (PLD) for a 70-kg human, using mortality rather than morbidity as an

TABLE 3. Comparison of in vitro cytotoxicity with toxicity and relative hazard ratings

Test agent	RPE$_{50}$ dose, (μg/ml)	Relative hazard[a] rating	Toxicity rating[b] or class
V	3×10^{-2}	variable	5
Cr	6×10^{-2}	3	4 or 5
Ru	0.1	unknown	not related
Cd	0.3	3	5 (?)
Hg	0.5	3	5
Ni	1.1	1–3	4 (?)
Co	3.5	1	4 (?)
Zn	6.8	variable	4
Cu	8.6	1–2	4
Mn	15	0–2	3 (?)
Pb	37	0–3	3 or 4
Fe	59	1	3
Mg	>2,000	0	3 (?)
Ca	>3,000	0	>1

[a] Reference 15.
[b] Reference 13.

endpoint. Significant clinical illness usually results at less than 10% of the probable lethal dose. The system has a quantitative basis: the values are for acute toxicity (a single dose)—a mean lethal dose, not a minimal fatal dose. The ranking is:

Class or Rating	PLD
6 = super toxic	Less than 5 mg/kg
5 = extremely toxic	5–50 mg/kg
4 = very toxic	50–500 mg/kg
3 = moderately toxic	0.5–5 g/kg
2 = slightly toxic	5–15 g/kg
1 = practically nontoxic	Above 15 g/kg

This toxicity rating is also in good agreement with our VERO cytotoxicity assay, supporting our contention that *in vitro* methods can be used to provide meaningful metal toxicity data.

In summary, a clonal growth assay using VERO cells has been modified to provide a rapid and inexpensive estimate of toxicity for metal compounds. The rankings of metal toxicity in our VERO cell assay are consistent with ranking of toxicity in man (7,13) and experimental animals (2,15). Further, the clonal growth assay appears to be a more sensitive measure of cytotoxicity than 20-hr viability tests (8,19,20).

Clonal growth assay techniques should be useful for studying metal–metal interactions, including analysis of synergistic, antagonistic, and additive effects of various metal combinations. Similar cell culture methods could also be used to investigate basic mechanisms of metal transport and to study the action of these elements at the molecular level (e.g., metabolic modifications, metal ion DNA interactions, etc.).

We are currently studying metal:chelon complexes in an attempt to relate the efficacy and toxicity of chelons to their chemical structures in order to predict which complexing agents would be most beneficial for use in chelation therapy.

ACKNOWLEDGMENTS

This work was supported by the U.S. Department of Energy under Contract EY-76-C-06-1830. We would like to thank our technicians, B. B. Thompson and M. J. Hooper, and our statistician, M. A. Wincek.

REFERENCES

1. Adalis, D., Gardner, D. E., Miller, F. J., and Coffin, D. L. (1977): Toxic effects of cadmium on ciliary activity using a tracheal ring model system. *Environ. Res.*, 13:111–120.
2. Bienvenu, P., Nofre, C., and Cier, A. (1963): The comparative general toxicity of metal ions. Relationship with periodic classification. *C. R. Acad. Sci., (Paris)* 256:1043–1044.
3. Christian, R. T., and Nelson, J. (1978): Coal: Response of cultured mammalian cells corresponds to the prevalence of coal worker pneumoconiosis. *Environ. Res.*, 15:232–241.
4. Dulbecco, R., and Vogt, M. (1954): Plaque formation and isolation of pure lines with poliomyelitis viruses. *J. Exp. Med.* 99:167–182.
5. Dulbecco, R., and Vogt, M. (1954): One step growth curve of Western equine encephalomyelitis virus on chicken embryo cells grown *in vitro* and analysis of virus yields from single cells. *J. Exp. Med.*, 99:183–199.
6. Evans, V. J., Bryant, J. C., Kerr, H. A., and Schilling, E. L. (1964): Chemically defined media for cultivation of long term cell strains from four mammalian species. *Exp. Cell. Res.*, 36:439–474.
7. Gosselin, R. E., Hodge, H. C., Smith, R. P., and Gleason, M. N., editors (1976): *Clinical Toxicology of Commercial Products: Acute Poisoning,* 4th edition. Williams and Wilkins, Baltimore.
8. Huisingh, J. L., Campbell, J. A., and Waters, M. D. (1977): Evaluation of trace-element interactions using cultured alveolar macrophages. In: *Pulmonary Macrophage and Epithelial Cells,* edited by C. L. Sanders, R. P. Schneider, G. E. Dagle, and H. A. Ragan, pp. 346–357. CONF-760927 NTIS, Springfield, Virginia.
9. Mauersberger, B., and Zorn, C. (1967): The Effect of $CuSO_4$ on the growth of mammalian cell cultures. *Exp. Cell. Res.*, 48:688–690.
10. Nielson, K. K., and Garcia, S. R. (1977): Use of X-ray scattering in absorption corrections for X-ray fluorescence analysis of aerosol loaded filters. In: *Advances in X-Ray Analysis,* Vol. 20, edited by H. F. McMurdie, C. S. Barrett, J. B. Newkirk, and C. O. Ruud, pp. 497–506. Plenum Press, New York.
11. Paton, G. R., and Allison, A. C. (1972): Chromosome damage in human cell cultures induced by metal salts. *Mutat. Res.*, 16:332–336.
12. Rohr, G., and Bauchinger, M. (1976): Chromosome analyses in cell cultures of the Chinese hamster after application of cadmium sulfate. *Mutat. Res.*, 40:125–130.
13. Sax, N. I., editor (1975): *Dangerous Properties of Industrial Materials,* 4th edition. Van Nostrand Reinhold, New York.
14. Simizu, B., Rhim, J. S., and Wiebenga, N. H. (1967): Characterization of the Tacaribe group

of arboviruses. I. Propagation and plaque assay of Tacaribe Virus in a line of African Green monkey kidney cells (VERO). *Proc. Soc. Exp. Biol. Med.,* 125:119–123.
15. Spector, W. S., editor (1956): *Handbook of Toxicology, Vol. I.* W. B. Saunders, Philadelphia.
16. Tsuda, H., and Kato, K. (1976): Potassium dichromate-induced chromosome aberrations and its control with sodium sulfite in hamster embryonic cells *in vitro. Gann,* 67:469–470.
17. Verne, J. (1954): Cellular sensitivity to drug action in short-term tissue cultures: *In vitro* correlations with sensitivity *in vivo. Ann. N.Y. Acad. Sci.,* 58:1195–1201.
18. Waters, M. D., Abernethy, D. R., Garland, H. R., and Coffin, D. L. (1974): Toxic effects of selected metallic salts on strain WI-38 Human Lung fibroblasts. *In Vitro,* 10:342.
19. Waters, M. D., Gardner, D. E., Aranyi, C., and Coffin, D. L. (1975): Metal toxicity for rabbit alveolar macrophages *in vitro. Environ. Res.,* 9:32–47.
20. Waters, M. D., Gardner, D. E., and Coffin, D. L. (1974): Cytotoxic effects of vanadium on rabbit alveolar macrophage *in vitro. Toxicol. Appl. Pharmacol.,* 28:253–263.

# Problems in Metal Carcinogenesis

Arthur Furst

Institute of Chemical Biology, University of San Francisco, San Francisco, California 94117

Metals and their compounds as primary carcinogens are a relatively new subset in the entire field of chemical carcinogenesis; in reality, inorganic substances have been suspect as human carcinogens for many years. For example, Hutchinson in 1888 (31) attributed skin cancer in humans as a result of arsenic exposure. Also, before 1900, it was noted that men who loaded chromate ores in gunny sacks developed ulcers of the nose as well as nasal pharynx cancer or lung cancer. Animal experiments using metals and inorganic salts were conducted many decades ago. As early as 1926, Michalowsky (40) injected zinc chloride in the testes of roosters and found after a time that testicular teratomas developed. Schinz and Uehlinger (52) reported a new principle of carcinogenesis when they reported that cobalt among other metals when implanted into rats also induced sarcomas. It is therefore surprising that until recently, most reviews on chemical carcinogenesis ignored inorganic ions as a class.

Metals, their compounds and alloys must be constantly evaluated as possible causes of cancer, for as new metallic materials are being manufactured, some occupational hazards may result (6,45). Also, metals, because they are not biodegradable, may be permanent contaminants in our biosphere. As we increase the burning of coal, extremely large amounts of fly ash and bottom ash containing heavy metal will result. Some sources of crude oil also contain metals. Thus, from the point of view of occupational and environmental exposure and hence carcinogenic potential, metals as a class must be studied more thoroughly and certainly more carefully.

A modern consideration relative to metals as primary carcinogenic agents is the need to critically evaluate the laboratory information on how the data are generated which suggests that an inorganic agent is a potential carcinogen. The Congress of the United States has given a number of regulatory agencies the mandate to eliminate carcinogens from the environment or work place, or if this is not possible, at least to minimize the exposure of the population in general or some populations specifically to those agents which can cause cancer in humans.

Standards can be promulgated logically and intelligently only when valid animal experiments are conducted. Extrapolations of the results of the animal experiments to what can be expected in humans can only be made if the animal experiments and the data generated from them are good. Unfortunately, insufficient numbers of experiments are conducted with the final objective of relating the results to humans. Thus, all who are interested in this aspect of the occupational and environmental problems must work with what is available, and often conclusions must be made, at the present time, even if the animal experiments were not designed for these extrapolations.

What information is available to government agencies from the published literature?

Perhaps the first line of inquiry is reviews. Recently, a number of reviews have appeared on metals as carcinogens; these have been written mainly by Sunderman (59,61,62,63) and by Furst (14–17,19). Luckey and Venugopal (38) also recently discussed this topic, as did Kanisawa (33) who wrote a review in Japanese; unfortunately this review is not readily available. All of these reviews are general in nature and do not consider an important question critically: how relevant are these animal experiments as related to humans? Even from the experiments conducted, at the present time we do not have adequate information even to address this question, let alone to answer it. Lacking are many important kinds of information. Perhaps this review can stimulate research projects so that the needed information will eventually be obtained.

Much more data must be accumulated on the relation of the route of administration of the inorganic agent to the endpoint cancer, so that a true assessment can be made of the human hazard. Exposure levels of carcinogenic metallic agents has been the topic of a review (13). While levels are important, they are not adequate by themselves. Humans are exposed to potential carcinogens and these chemicals impinge on people almost exclusively by the inhalation route, by dermal contact, or by ingestion.

An analysis of the experimental designs using animals reveals that in the vast majority of studies, the agent under investigation was administered by a parenteral route. Falin and Gromtseva (12) repeated Michalowsky's work of injecting zinc chloride in the testes of roosters resulting in teratomas; in a similar experiment Falin and Anissimowa (11) used copper sulfate with positive results. Reviere et al. (50) treated rats by this procedure and induced testicular tumors. In spite of the fact that these induced tumors are transplantable (1), the relevancy of these experiments to the human problem must be questioned. More recently, Damjanov et al. (8) found that nickel subsulfide injected intratesticularly into rats induced malignant testicular neoplasms after 20 months.

Subcutaneous implantation of solids was the route of choice for practically all investigators over two decades ago, but this even persists today. The publication which has been most frequently cited is by Oppenheimer et al. (47). This group placed a number of metallic foils under the skin of rats and obtained fibrosarcomas at the site. The elements studied included silver foil and stainless steel. Tin foil failed to produce tumors, but this preparation crumbled; the change in physical environment was not appreciated at that time. Much earlier, Schinz and Uehlinger (52) (which preceded the Oppenheimer experiments by 14 years) implanted cobalt, for example. Fibrosarcomas were also found.

Subsequently, Nothdurft made a comprehensive study of the subcutaneous implantation route and published a series of papers (42,43). He reported the following produced local sarcomas: platinum discs, silver discs, gold discs, as well as some nonmetallic implants such as ivory. Nothdurft also published an interesting short note (44); he demonstrated that the tumor yield and time of development of these growths are dependent upon which anatomical site the solid is implanted. Druckrey et al. (9) injected mercury liquid by the intraperitoneal route and after 23 months found peritoneal sarcomas. Bryson and Bischoff (5) reviewed the field of induction of sarcomas following the implants of silicates. And most recently, we now recognize a related field of carcinogenesis which can be called nonspecific solid state carcinogenesis (17) or better yet, the term (introduced by Nothdurft) to describe this field is *foreign body* carcinogenesis. Brand and co-workers have been investigating this phenomenon; he has recently reviewed it (4).

What clouds the field of metal carcinogenesis is that the physical form of an implanted material is an important factor in the determination of which solid will or will not induce a tumor. It now appears that a smooth surface is the important property for the foreign body carcinogenic action. An exception is found in at least one case, nickel subsulfide (21). Aluminum foil implanted subcutaneously was active, whereas rough surface iron was not (46). In spite of the literature which is accumulating on this aspect of carcinogenesis which seems to imply that this is a poor route to study, papers still appear where the investigator uses the subcutaneous route, or a route which is similar. For example, the implanting of a platinum disc in the eye of a rat resulted in a tumor (10).

A few experiments have been conducted using the intravenous route. Schmahl and Steinhoff (51) injected colloidal silver intravenously and obtained tumors; colloidal gold was inactive. Sunderman's group in their extensive experimentation with nickel compounds injected nickel carbonyl intravenously (37) and found a variety of tumors in rats, both malignant and benign; significantly, no squamous cell carcinoma of the lung developed.

Shimkin and co-workers (56) summarized their work on the intraperitoneal administration of soluble salts which influenced the appearance of the *spontaneous* pulmonary adenomas in Strain A mice (58).

At least one intrarenal study was conducted. Jasmin and Riopelle (32) injected nickel subsulfide (Ni_3S_2) into the renal pole of rats and $\frac{7}{16}$ of the animals developed renal carcinomas. Several other metallic compounds were tested similarly with negative results; pure nickel was one of the negative compounds.

An experiment was conducted wherein the agent, nickel oxide, was implanted in the brain and neoplastic growths (sarcoma and meningiomas) were found; the same agent implanted intramuscularly produced no tumors (57). Dermal absorption studies using inorganic or metal organic compounds are almost nonexistent. Yet experiments of this nature should be done. It is expected naturally that aqueous soluble compounds of nickel, cadmium, zinc, and others will not be absorbed through the skin. However, since nickel and chromium can sensitize the skin, and since these two compounds can cause a dermatitis, it is certainly possible that these (and perhaps other ions) can eventually penetrate through the dermal layer and get into the bloodstream. Organometallic compounds which are lipid-soluble would have a greater chance of penetrating the various layers of the skin and of being absorbed. Occupational problems may result when workers are exposed to organometallic agents. Compounds such as nickelocene, the pi-complex sandwich of nickel between two molecules of the cation of cyclopentadiene are definitely oil soluble. Skin painting experiments should be conducted for this route as a logical one to mimic the human experience. Compounds such as nickelocene should be applied either as a powder (which is a difficult procedure) or as a suspension in an inert vehicle. Solutions of metalocenes in various organic solvents should also be tried by skin painting but conclusions should not be based on this last experiment alone.

The most logical route to study the effects of particulates on experimental animals is by inhalation. Data derived from these experiments are the most applicable to humans. The inhalation route approaches reality. Much is now known about the filtration of particulates by the nose, and the deposition of inhaled particles in various parts of the respiratory tract; finally, the nature and rate of clearance of inhaled particles has also been extensively investigated. The characteristics of, and data of deposition of various particle sizes (aerodynamic median diameter in microns) for various animals with, perhaps, the rat being most investigated (49), are well known.

The data on air flow to lungs of humans, and rates of mixing of particulates and anatomical sites of respiratory tract where deposition takes place, are being accumulated (66). However, the clearance rates—especially the initial effects—may vary from species to species; thus, more information is needed on the mechanisms of clearance from the broncheatracheal area, the means by which the mucociliary transport takes place, and the mechanism of macrophage action and what kills macrophages. In addition, little has been done on the dissolving capacity and limitation of the fluids of the lungs.

In spite of the lack of knowledge of detailed mechanism, much more experimental work must be done on the inhalation aspect of metal carcinogenesis. As yet, not too many manuscripts detail inhalation studies. Some knowledge of pulmonary cell responses to metallic oxides exist (7). Hueper (30) did find pulmonary lesions in guinea pigs and rats when these test animals inhaled nickel powder over a long period of time. Later the doctors Sunderman (65) exposed rats to gaseous nickel carbonyl, the gas generated in the Mond processes for refining nickel. They found not only pulmonary carcinomas, but studied the subcellular distribution of nickel in these exposed rats. Wehner et al. (71) reported on some special exposure chambers to study inhalation experiments with rodents. He also exposed hamsters to nickel oxide and cobalt oxide (70) as well as nickel oxide and cigarette smoke (69). No cancer was found, but the hamsters exposed to nickel oxide were diagnosed as having pneumoconiosis. Ottolenghi et al. (48) reported that in an inhalation experiment with rats exposed to "nickel sulfide," lung tumors did develop. (The compound used in their experiment was *nickel subsulfide,* Ni_3S_2, but in the trade this is called "nickel sulfide." Thus, the title and even abstracts of this paper can be misleading). The lungs of the treated animals (both sexes combined) developed 15 adenomas compared to 1 in the control group; 10 adenocarcinomas compared to 1 and 3 squamous cell carcinomas compared to zero in the controls.

Short of inhalation—which is the best route—the intratracheal instillation method will undoubtedly give more useful information than that derived from the parenteral routes. The main disadvantage to this route is that the nasal pharynx region does not filter out the larger particles. The advantage is that the material under investigation is in direct contact with the lung surfaces.

Ho and Furst (28) demonstrated that the relatively inert ferric oxide is distributed well in the mouse lungs; but no evidence of tumor formation was found after one year. Kasprzak et al. (34) did not find any lung neoplasm when nickel subsulfide was administered intratracheally once; however, they did find premalignant pathological reactions when nickel subsulfide was combined with benzo(*a*)pyrene; this confirmed the additive effects of these two agents reported by Sunderman's group (39) who gave this combination intramuscularly. Other means of placing metals directly on lung tissue can be by an intrathoracic injection. By this route, Furst et al. (18) produced mesotheliomas with cadmium or nickel. This is a traumatic technique which is similar to the placing of chromate hooks in the lungs by Laskin et al. (36).

Metals and inorganic ions are usually not ingested in either occupational or environmental settings. The major exception is that during the bronchial clearance of inhaled fine particles, the mucus which contains these impurities may be swallowed rather than expectorated.

A variety of trace elements are present in normal food, and in addition a large number of people in this country take dietary supplements which include minerals: these may contain copper, zinc, calcium, magnesium, iron, manganese, etc. No carcinogenic hazard has ever been scientifically related to the ingestion of trace minerals.

The inadvertent exposure of people to heavy metals in drinking water (2,35) should stimulate more studies in this field.

Perhaps the most work on the administration of heavy metal ions in drinking water was conducted by the late Dr. Schroeder. He was scrupulously careful to keep all metals out of his animal laboratories. He administered a variety of compounds over their lifetime to both rats and mice.

The agents were dissolved in the drinking water at levels which would be similar to that to which humans would be exposed. In some mice studies, nickel, chromium, lead, cadmium, and titanium at 5 ppm failed to produce tumors; nickel appeared to decrease the incidence of tumors in the female mice (53). Chromium did not seem to produce more tumors than were found in the controls in similar mouse studies (54), but the rhodium- and palladium-treated animals had a greater incidence of tumors than did the controls. Schroeder and Mitchener (55) also studied selenium and tellurium in rats. They concluded that these elements were active in these studies; however, much work on selenium has gone on and will be covered by other reviews.

Walters and Roe (68) evaluated tin as Na_2SnCl_6 in the drinking water of mice for a prolonged period. No excess tumors were noted. Their parallel work with zinc oleate failed to reveal an increase of lymphoma and pulmonary adenomas over the controls. Zinc, as a possible carcinogen, is in a state of confusion. An entire review is necessary to cover this element.

The available literature information on the toxicology and especially the carcinogenicity of a specific agent becomes the major basis for the decision by various government agencies as to what the exposure level of that agent should be in either the workplace or the environment. The National Institute for Occupational Safety and Health has issued a series of criteria documents; these discuss the background for setting standards to help "meet the need for preventing impairment of health from occupational exposure to. . . ."

Nickel and its compounds have been the most extensively studied of all metals as to their potential carcinogenicity. Excellent reviews exist on this subject (60,61,64).

In spite of these reviews, the criteria document in Occupational Exposure to Inorganic Nickel (41) recognized that further research is needed, especially in the area of epidemiology as well as "animal studies . . . to characterize the acute and chronic toxicities of the many nickel compounds for which insufficient information is available."

We recommend that in future research, multiple doses should be tested; and the time of appearance of the tumors should be noted if possible. The doses used should include at least one level which is comparable to that to which humans are exposed. If possible, more inhalation studies should be conducted; short of this, both mice and rats should be exposed by the intratracheal technique to get lung effects. Hamsters and rats have been used mainly for intratracheal studies. Ho (27) has summarized the reasons for using mice as the model for pulmonary exposure. Grasso and Crampton (22) also believe that mice are valuable in carcinogenic testing.

What are finally needed to make intelligent evaluations are the acute toxicity, such as an LD_{50} or LC_{50}, long-term studies at high, medium, and low doses, the time of appearance of the first tumors, the time chart of the percentage of animals with tumors (here serial sacrifice may be necessary), and definite pathological diagnosis of the histological slides. If possible, the routes of administration should bear some relationship to that to which humans are exposed.

Thus, for metallic orthopedic implants, which have been found to produce sarcoma in dogs (24), an intramuscular study of solids in rats makes sense. Gaechter et al. (20) studied a series of metallic rods of seven materials which can be used in humans, and found no significant increase in tumors in Sprague Dawley rats. Heath et al. (25) found activity in wear particles but . . .

> those who cannot remember the past
> are condemned to repeat it.
>
> George Santayanna

How can intraperitoneal implants of solids be interpreted from a practical point of view (3)? This is not to negate the pure theoretical aspect of a study, but these studies continue to appear (23).

However, we still have the problem of usual cancer incidence among occupational workers exposed to metals. This has been reviewed by Hernberg (26) and Houten et al. (29). Copper smelter workers have been studied in Japan and found to have an increase of lung and colon cancer over their cohorts (67). What animal experiments conducted to date will help any regulatory body decide how to deal with exposure levels of copper in the work place? The answer at the present time is: none.

ACKNOWLEDGMENT

The co-trustees of the Carrie Baum Browning Fund are gratefully acknowledged.

REFERENCES

1. Anissimova, V. (1939): Experimental zinc teratomas of the testis and their transplantation. Preliminary communication. *Am. J. Cancer,* 36:229–232.
2. Berg, J. W., and Burbank, F. (1972): Correlations between carcinogenic trace metals in water supplies and cancer mortality. *Ann. N.Y. Acad. Sci.,* 199:249–264.
3. Bischoff, F., and Bryson, G. (1977): Intraperitoneal foreign body reaction in rodents. *Res. Commun. Chem. Pathol. Pharmacol.,* 18(2):201–214.
4. Brand, K. G. (1976): Solid state or foreign body carcinogenesis. In: *Scientific Foundations of Oncology,* edited by T. Symington and R. L. Carter, pp. 490–495. Wm. Heinemann, London.
5. Bryson, G., and Bischoff, F. (1967): Silicate-induced neoplasms. *Prog. Exp. Tumor Res,* 9:77–164.
6. Carnow, B. W. (1976): Discussion of Part V; carcinogenesis in the metal industry. *Ann. N.Y. Acad. Sci.,* 271:496–504.
7. Casarett, L. J., Casarett, M. G., and Whalen, S. A. (1971): Pulmonary cell responses to metallic oxides. *Arch. Int. Med.,* 127:1090–1098.
8. Damjanov, I., Sunderman, F. W., Jr. Mitchell, J. M., and Allpass, P. R. (1978): Induction of testicular sarcomas in Fischer rats by intratesticular injection of nickel subsulfide. *Cancer Res.,* 38(2):268–276.
9. Druckrey, H., Hamperl, H., and Schmahl, D. (1957): Carcinogenic action of metallic mercury after intraperitoneal administration to rats. *Z. Krebsforsch.,* 61:511–519.
10. Evgen'eva, T. P. (1972): Pigmented tumors in rats induced by introduction of platinum and cellophane films into the chamber of the eye. *Bull. Exp. Biol. Med.,* 74:1296–1298.
11. Falin, L. I., and Anissimowa, W. W. (1940): Pathogenese der Experimentellen Teratoiden Geschwüllate der Geschlechtadrülsen. Teratoid e Hodengechwulst Beim Hahn, Erseugt Durch Einfuhrung von $CuSO_4$-Losung. *Z. Krebsforsch.,* 50(3/5):339–351.
12. Falin, L. I., and Gromtseva, K. E. (1939): The pathogenesis of experimental teratoid tumors of the genital glands. I. Experimental zinc sulfate teratoma of the testicle of roosters. *Arch. Sci. Biol. U.S.S.R.,* 5(3): 101–11.
13. Fishbein, L. (1976): Environmental metallic carcinogens: an overview of exposure levels. *J. Toxicol. Environ. Health,* 2(1):77–109.

14. Furst, A. (1963): *The Chemistry of Chelation in Cancer*, pp. 17–18. Charles C Thomas, Springfield, Illinois.
15. Furst, A. (1977): Inorganic agents as carcinogenesis. In: *Advances in Modern Toxicology*. 3:209–229.
16. Furst, A. (1978): An overview of metal carcinogenesis. In: *Inorganic and Nutritional Aspects of Cancer*, pp. 1–12. Plenum Press, New York.
17. Furst, A. (1971): Trace elements related to specific chronic diseases: cancer. In: *Environmental Geochemistry in Health and Disease*, edited by H. L. Cannon and H. C. Hopps, pp. 109–130. Geological Society of America, Inc. Boulder, Colorado.
18. Furst, A., Cassetta, D. M., and Sasmore, D. P. (1973): Rapid Induction of Pleural Mesotheliomas in the Rat. *Proc. West. Pharmacol Soc.*, 16:150–153.
19. Furst, A., and Haro, R. T. (1969): A Survey of Metal Carcinogenesis. *Prog. Exp. Tumor Res.*, 12:102–133.
20. Gaechter, A., Alroy, J., Andersson, G. BJ., Galante, J., Rostoker, M., and Schajowicz, F. (1977): Metal carcinogenesis: a study of the carcinogenic activity of solid metal alloys in rats. *J. Bone JT Surg. AM*, 59(5):622–624.
21. Gilman, J. P. W., and Herchen, H. The effect of physical form of implant on nickel sulphide tumourigenesis in the rat. *Acta Unio Int. Cancrum*, 19(3/4):615–619.
22. Grasso, P., and Crampton, R. F. (1972): The value of the mouse in carcinogenicity testing. *Food Cosmet. Toxicol.*, 10(3):418–26.
23. Griss, P., Warner, E., Buchinger, R., Busing, C. M., and Heimke, G. (1977): Experimental investigation on non-specific foreign body sarcoma induction of AL203-ceramic implants. *Scand. J. Haematol.*, 32:87–98.
24. Harrison, J. W., McLain, D. L., Hohn, R. B., Wilson, G. P., Chalman, J. A., and McGowan, K. N. (1976): Osteosarcoma associated with metallic implants—Report of two cases in dogs. *Clin. Orthop.*, 116:253–257.
25. Heath, J. C., Freeman, M. A., and Swanson, S. A. (1971): Carcinogenic properties of wear particles from prostheses made in cobalt–chromium alloy. *Lancet*, 1:564–6.
26. Hernberg, S. (1977): Incidence of cancer population with exceptional exposure to metals. In: *Origins of Human Cancer*, edited by A. H. H. Hiatt, J. D. Watson, and J. A. Winsten, pp. 147–157. Cold Spring Harbor Laboratory, Cold Spring Harbor, New York.
27. Ho, W. (1975): The mouse as a potential model for testing the effects of pulmonary exposure to combustion products. *J. Fire and Flame/Combustion Toxicology*, 2:226–243.
28. Ho, W., and Furst, A. (1973): Intratracheal instillation method for mouse lungs. *Oncology*, 27:385–393.
29. Houten, L., Brass, I. D., Viadana, E., and Sonnosso, G. (1977): Occupational cancer in men exposed to metals. *Adv. Exp. Med. Biol.*, 91:93–102.
30. Hueper, W. C. (1958): Experimental studies in metal carcinogenesis. IX. Pulmonary lesions in guinea pigs and rats exposed to prolonged inhalation of powdered metallic nickel. *Arch. Path.*, 65:600–607.
31. Hutchinson, J. (1888): On some examples of arsenic-keratoses of the skin and of arsenic cancer. *Trans. Path. Soc. London*, 39:352–363.
32. Jasmin, G., and Riopelle, J. L. (1976): Renal carcinomas and erythrocytosis in rats following intrarenal injection of nickel subsulfide. *Lab. Invest.*, 35(1):71–78.
33. Kanisawa, M. (1971): Aspects of metal carcinogenesis. *Ann. Rept. Inst. Food Microbiol. (Chiba Univ.)*, 24:1–35.
34. Kasprzak, K. S., Marchow, L., and Breborowicz, J. (1973): Pathological reactions in rat lungs following intratracheal injection of nickel subsulfide and 3,4-benzpyrene. *Res. Commun. Chem. Pathol. Pharmacol.*, 6(1):237–245.
35. Kraybill, H. F. (1976): Distribution of chemical carcinogens in aquatic environments. *Prog. Exp. Tumor Res.*, 20:3–34.
36. Laskin, S., Kuschner, M., and Drew, R. T. (1970): Studies in pulmonary carcinogenesis. In: *Inhalation Carcinogenesis*, edited by E. Nettlesheim. *AEC Symposium Series*, 18:321–350.
37. Lau, T. J., Hackett, R. L., and Sunderman, F. W., Jr. (1972): The carcinogenicity of intravenous nickel carbonyl in rats. *Cancer Res.*, 32(10):2253–2258.
38. Luckey, T. D., and Venugopal, B. (1977): *Metal Toxicity in Mammals, Vol. 1*, pp. 129–260. Plenum Press, New York.

39. Maenza, R. M., Pradham, A. M., and Sunderman, F. W. (1971): Rapid induction of sarcomas in rats by combination of nickel sulfide and 3,4-benzpyrene. *Cancer Res.*, 31:2067–2071.
40. Michalowsky, I. (1926): Die experimentelle Erzeugung einer teratoiden Neubildung der Hoden beim Hahn. *Zentbl. Allg. Path. Path. Anat.*, 38:585.
41. NIOSH (1977): Criteria for a recommended standard occupational exposure to inorganic nickel. U.S.D.H.E.W., National Institute for Occupational Safety and Health *(DHEW (NIOSH) Pub.* 77–164, May.
42. Nothdurft, H. (1955): Experimental production of sarcomas in rats and mice by implantation of round disks of gold, platinum, silver, or ivory. *Naturwissenschaften*, 42:75–76.
43. Nothdurft, H. (1960): Production of tumors by the implantation of foreign bodies. *Abhandl. Deut. Akad. Wiss. Berlin, Klasse Med.*, 3:80–89.
44. Nothdurft, H. (1962): Unterschiedliche ausbeutenan subcutanen Fremdkorpersarkomen der Ratte in Abhangigkeit von Korperregion. *Naturwissenschaften*, 49:18–19.
45. Norseth, T. (1977): *Industrial Viewpoints on Cancer Caused by Metals as an Occupational Disease in Origins of Human Cancer*, edited by A. H. H. Hiatt, J. D. Watson, and J. A. Winsten, pp. 159–167. Cold Spring Harbor Laboratory, Cold Spring Harbor, New York.
46. O'Gara, R. W., and Brown, J. M. (1967): Comparison of the carcinogenic actions of subcutaneous implants of iron and aluminum in rodents. *J. Natl. Cancer Inst.*, 38(6):947–952.
47. Oppenheimer, B. S., Oppenheimer, E. T., Danishefsky, I., and Stout, A. P. (1956): Carcinogenic effect of metals in rodents. *Cancer Res.*, 16:439–441.
48. Ottolenghi, A. D., Haseman, J. K., Payne, W. W., Falk, H. L., and Macfarland, H. N. (1975): Inhalation studies of nickel sulfide in pulmonary carcinogenesis of rats. *J. Natl. Cancer Inst.*, 54(5):1165–1172.
49. Raabe, O. G., Yeh, H. C., Newton, G. J., Phelan, R. F., and Velasquez, D. J. (1977): Deposition of inhaled particles on monodispersed aerosols in small rodents. In: *Inhaled Particles IV*, edited by W. N. Walton. Pergamon Press, Oxford.
50. Rivière, M. R., Chouroulinkov, I., and Guerin, M. (1959): Testicular tumors in the rat after injection of zinc chloride. *Compt. Rend.*, 249:2649–2651.
51. Schmahl, D., and Steinhoff, D. (1960): Experimental carcinogenesis in rats with colloidal silver and gold solutions. *Z. Krebsforsch.*, 63:586–591.
52. Schinz, H. R., and Uehlinger, E. (1942): Der Mettallkrebs; ein neues Prinzip der Krebserzeugung. *Z. Krebsforsch.*, 52:425–437.
53. Schroeder, H. A., Balassa, J. J., and Vinton, W. H. (1964): Chromium, lead, cadmium, nickel and titanium in mice. *J. Nutr.*, 83(3):239–250.
54. Schroeder, H. A., and Mitchener, M. (1971): Scandium, chromium (VI), gallium, yttrium, rhodium, palladium, indium in mice: Effects on growth and life-span. *J. Nutr.*, 101:1431–1438.
55. Schroeder, H. A., and Mitchener, M. (1971): Selenium and tellurium in rats: Effect on growth, survival and tumors. *J. Nutr.*, 101:1531–1540.
56. Shimkin, M. B., Stoner, G. D., and Theiss, J. C. (1977): Lung tumor response in mice to metals and metal salts. *Adv. Exp. Med. Biol.*, 91:85–91.
57. Sosinski, E. (1975): Morphological changes in rat brain and skeletal muscle in the region of nickel oxide implantation. *Neuropatol. Pol.*, 13(3–4):479–483.
58. Stoner, G. D., Shimkin, M. B., Troxell, M. C., Thompson, T. L., and Terry, L. S. (1976): Test for carcinogenicity of metallic compound by the pulmonary tumor response in strain A. mice. *Cancer Res.*, 36:1744–1747.
59. Sunderman, F. W., Jr. (1978): Carcinogenic effects of metals. *Fed. Proc.*, 37(1):40–46.
60. Sunderman, F. W., Jr. (1973): The current status of nickel carcinogenesis. *Ann. Clin. Lab. Sci.*, 3:156–180.
61. Sunderman, F. W. (1977): Metal carcinogenesis. *Advances in modern toxicology.* (2):257–295.
62. Sunderman, F. W. (1971): Metal carcinogenesis in experimental animals. *Food Cosmet. Toxicol.*, 9(1):105–120.
63. Sunderman, F. W., Jr. (1976): A review of the carcinogenicities of nickel, chromium and arsenic compounds in man and animals. *Prev. Med.*, 5(2):279–294.
64. Sunderman, F. W., Jr. (1977): A review of the metabolism and toxicology of nickel. *Ann. Clin. Lab. Sci.*, 7:377–393.
65. Sunderman, F. W., and Sunderman, F. W., Jr. (1963): Studies of nickel carcinogenesis. *Am. J. Clin. Pathol.*, 40(6):563–575.

66. Thomas, R. G. (1972): An interspecies model for retention of inhaled particles. In: *Assessment of Airborne Particles,* edited by T. T. Mercer, P. E. Morrow, and W. Stöber, pp. 405–420. Charles C Thomas, Springfield, Illinois.
67. Tokudome, S., and Kuratsume, M. (1976): A cohort study on mortality from cancer and other causes among workers at a metal refinery. *Int. J. Cancer,* 17:310–317.
68. Walters, M., and Roe, F. J. C. (1965): The effects of zinc and tin administered orally to mice over a prolonged period. *Food Cosmet. Toxicol.,* 3(2):271–276.
69. Wehner, A. P., Busch, R. H., Olson, R. J., and Craig, D. K. (1975): Chronic inhalation of nickel oxide and cigaret smoke by hamsters. *Am. Ind. Hyg. Assoc. J.,* 36(11):801–810.
70. Wehner, A. P., and Craig, D. K. (1972): Toxicology of inhaled NiO and CoO in Syrian golden hamsters. *Am. Indus. Hyg. Assoc. J.,* 33:146–155.
71. Wehner, A. P., Craig, D. K., and Stuart, B. O. (1972): An aerosol exposure system for chronic inhalation studies with rodents. *Am. Ind. Hyg. Assoc. J.,* 33:483–487.

Trace Element Interactions in Carcinogenesis*

Gerald L. Fisher

Radiobiology Laboratory, University of California, Davis, California 95616

Recently, a great emphasis has been placed on the study of the role of metals in carcinogenesis. While most efforts have been directed to the evaluation of metals as carcinogens, this chapter emphasizes another aspect of metal carcinogenesis, namely, trace element interactions. Three areas will be reviewed: (a) trace elements as antagonists or synergists of metal carcinogenesis; (b) trace elements as antagonists or synergists of organic chemical carcinogenesis, and (c) alterations, associated with carcinogenesis, of homeostatic levels of endogenous essential trace elements.

Much of the work on metals as antagonists or synergists in chemical carcinogenesis is extremely preliminary. However, careful study of metal interactions will provide insight into the mechanisms of chemical carcinogenesis. In this chapter, the effect of metals on the immune response will not be discussed. Although this is an important area of research, it is not within the scope of this chapter. Also, the effect of metals on bacterial and mammalian cell cultures is considered to be beyond our present scope. These areas are discussed in Chapters 4 and 8.

In summarizing combination effects in chemical carcinogenesis, Schmahl (58) has presented a useful summary of parameters important from a pathological standpoint (without concern for route of exposure). As presented in Table 1, antagonistic or synergistic responses from coexposure to trace elements and carcinogens may lead to a difference in incidence of tumor. Also, a difference in latency—that is, time to first tumor appearance—may be observed. Further, the minimum dosage required to produce the biological endpoint may also be changed. A very important aspect of interaction as discussed elsewhere in this volume by Furst is the location and the origin of the tumor. Data will be presented to indicate that metals may change the site of the origin of the tumor without affecting the latency, incidence, or minimum dosage for tumor development. Detailed histological evaluation is most important. The incidence of malignant and nonmalignant tumors may be affected by metal interactions. In many of the animal systems that have been studied, a premalignant lesion can be identified; for example, adenomas and papillomas are well-demonstrated prema-

---

* This chapter is dedicated to Dr. Perry R. Stout whose enthusiasm and genius will always be inspirational to scientists studying the biological role of trace elements.

TABLE 1. *Parameters for evaluation of interactions in chemical carcinogenesis*

1. Incidence of tumors
2. Latency period
3. Minimum dosage
4. Location and origin of tumor
5. Histology (nonmalignant, premalignant, malignant)
6. Incidence of metastases

Modified from Schmahl (58).

lignant lesions in murine, lung, and skin models, respectively. And finally, there is a suggestion that the incidence of tumor metastases may be uniquely altered without affecting the first five parameters presented in Table 1.

METAL INTERACTIONS IN METAL CARCINOGENESIS

There have been few studies directed toward the evaluation of trace-element interactions in metal-induced chemical carcinogenesis. Gunn et al. (26) reported that subcutaneous (s.c.) coinjection of zinc acetate protected against cadmium-chloride-induced testicular tumors in rats and mice. Coinjection of zinc with cadmium reduced the incidence from 68% to 12%, at 11 months, of interstitial cell tumors in rats. More dramatic results were observed in a similar study in mice; 14-month incidience of interstitial cell tumors was reduced from 77% to 0%. A subsequent study (27) indicated that coinjection of Cd and Zn also resulted in inhibition of Cd-induced sarcoma at the site of injection. Although mechanistic evaluation has not been performed, the authors suggest that Cd may interfere with Zn regulation of cell division. Interestingly, when cadmium powder was injected intramuscularly (i.m.) into the thigh of rats and zinc powder injected into the opposite leg, no antagonism of cadmium-induced fibrosarcoma was observed (24). These results suggest that the chemical form and intimacy of chemical mixture may play an important role in metal–metal interactions.

More recently, Sunderman and co-workers (66) clearly showed that manganese metal, when coinjected with nickel subsulfide into the thigh muscles of rats, afforded protection against nickel subsulfide carcinogenesis. Aluminum oxide, copper powder, or chromium powder did not affect the sarcoma incidence at the site of injection. The tumor incidence of the $Mn-Ni_3S_2$ (equimolar) exposed group was 63% compared to 98% in the rats receiving Ni_3S_2 alone. In a subsequent study in which equimolar ratios of manganese and nickel subsulfide were used, a reduction in sarcoma incidence at the site of injection from 77% to 7% was observed. This experiment was performed with a single i.m. injection of 1.2 mg Ni_3S_2, approximately half the mass used in the previous report (66). Detailed laboratory studies using $^{63}Ni_3S_2$ indicated that mixture with manganese diminished the *in vitro* solubility of the carcinogen in rat serum (65). However, manganese did not affect the *in vivo* excretion or mobilization from the site of injection. The subcellular concentration of nickel was diminished by mixture with manganese, possibly by manganese antagonism of the nickel inhibition of RNA polymerase activity.

METAL INTERACTIONS IN ORGANIC CHEMICAL CARCINOGENESIS: ANTAGONISM

Much of the literature in the area of trace element antagonism of organic chemical carcinogenesis is preliminary and oftentimes conflicting. Kobayashi et al. (40) have reported that inhalation or s.c. injection of aluminum inhibits the development of lung adenomas and adenocarcinomas in nitroquinoline-oxide-treated (s.c.) mice. Inhaled aluminum was effective as $AlCl_3$ solution or particulate Al_2O_3; injected aluminum was in the form of dissolved $AlCl_3$. The results of this study should be interpreted carefully since a relatively large number of mice died during the course of the study, no body weight data are presented, and it appears that the mice were sacrificed 7 months after initiation of the study.

A number of studies have investigated zinc as an inhibitor of organic chemical carcinogenesis. Early work by Poswillo and Cohen (52), using zinc with dimethylbenzanthracene (DMBA) in skin-painting experiments, suggested that dietary zinc afforded protection against carcinoma induction in hamsters. After 6 months, only 1 of 9 animals receiving 100 ppm dietary zinc in combination with DMBA exposure to the cheek pouch manifested carcinoma compared to 13 of 15 animals receiving a normal zinc diet (22 ppm) and DMBA exposure. A subsequent study of salivary gland tumors in rats by Ciapparelli et al. (7) also suggested that dietary zinc supplementation may retard DMBA-induced tumor growth. In this preliminary study, zinc supplementation resulted in a decreased rate of tumor growth and an apparent change in the tumor histology, manifested as a decrease in the carcinomatous epithelium and an increase in inflammatory response. Thus, these authors hypothesized that zinc supplementation may result in stimulation of the immune system and subsequent host-mediated tumor rejection. However, in a more detailed study, Edwards (12) repeated the experiment of Poswillo and Cohen (52) and found no protection of dietary zinc against tumor incidence resulting from skin painting with DMBA in Syrian golden hamsters. A suggestion of extended survival time was observed with zinc supplementation, in agreement with the observations of Ciapparelli et al. (7).

An interesting report, again rather preliminary, by Duncan and Dreosti (11) suggested that either low or high levels of dietary zinc may protect against tumor induction. In this work, the 10-week incidence of skin papillomas resulting from painting with 3-methylcholanthrene was evaluated.

Insight into a possible mechanism of the effect of zinc on chemical carcinogenesis may be gleaned by evaluation of a series of studies (3,9,46) using transplanted tumor models in zinc-deficient animals and pair-fed or weight-matched controls.

Dietary zinc deficiency resulted in a striking decrease in the tumor growth rate and subsequent increased survival of rats injected with Walker 256 carcinosarcoma (9,46). Tumor transplantation was successful in 68% of the zinc-deficient animals compared to 100% of the controls. Similar results have been reported by Barr and Harris (3) using injected leukemic cells as an ascites tumor in mice. These studies clearly indicate that diets deficient in zinc can reduce the rate of tumor growth in laboratory animals. Although most of these studies are preliminary and the data often conflicting, it does appear that alteration of the dietary levels of zinc may affect the course of chemically induced and transplanted tumors. Because zinc is required for DNA-synthesis and subsequent cell division, it is possible that the decreased rate of tumor development in zinc-deficient animals is associated with decreased cell turnover. On the other hand, decreased tumorigenesis with increased dietary zinc may be due to stimulation of the immune system.

Increased levels of dietary copper have been shown in a number of studies to be protective against hepatocarcinogens. Fare and Howell (14) presented data indicating that increased dietary copper (0.5% cupric oxyacetate hexahydrate) protected against dietary methoxy-dye-induced liver tumors. However, no protection was afforded against ear duct tumors or, in a subsequent study (15), skin tumors induced by painting. It was suggested that the protection by copper feeding was due to "competitive binding with the carcinogen for available protein sites in the liver." Subsequent studies by Yamane and co-workers (69) indicated that the effect of copper in protecting against liver carcinogenesis associated with the methoyoxy-dyes probably results from enhanced activity of azo-reduction in the liver. Copper was found to stimulate azoreductase without affecting the activity of either N-demethylation or aromatic hydroxylation. Thus, it is suggested that the protection of the liver by copper is uniquely associated with more rapid catabolism of the carcinogen. An apparently similar situation, but very different biochemical mechanism, has been described for ethionine-induced liver tumors. Kamamoto et al. (36) have studied hepatoma induction by ethionine in rats receiving copper administered in the diet for varying times up to 20 weeks. Cofeeding of copper with ethionine for a period varying from 12 weeks to the total experimental time of 20 weeks completely protected against hepatoma development in the mice. Subsequent work by Brada et al. (5) suggested that the copper-ethionine complex can be slowly solubilized and absorbed from the intestinal lumen. The complex is more slowly catabolized than ethionine alone, and results in increased hepatotoxicity. Hence, there is a suggestion of decreased carcinogenicity due to increased cytotoxicity. In summary, it is believed that the copper protection against methyoxy-dye hepatoma results from enhanced azo-reduction—i.e., enhanced catabolism. Copper protection against ethionine-induced hepatoma is thought to be due to decreased catabolism and increased hepatotoxicity.

Although selenium relationships to cancer are discussed in detail elsewhere in this volume, studies of selenium as an antagonist of chemical carcinogenesis will be briefly overviewed in this chapter. Many laboratory animal studies have documented a protective effect of dietary selenium against a variety of organic chemical carcinogens. Research by Shamberger (61) indicated that skin papillomas induced by DMBA-croton oil skin painting were markedly reduced in incidence when the mice were dietarily supplemented with selenium as selenide or selenite. At 20 weeks after DMBA exposure, 90% versus 40% incidence of papilloma was observed in control versus selenium-supplemented animals. Carcinomas were also depressed in about the same ratio with cofeeding of selenium. Using the same system, selenium again afforded protection against skin papillomas associated with benzo(a)pyrene (BaP). A 94% tumor incidence was observed in those animals painted with BaP alone compared to 44% in animals receiving BaP and diets supplemented with either selenite or selenide. More recently, Jacobs (34) reported that dietary supplementation of selenium antagonized colon carcinogenesis associated with injection of dimethylhydrazine (DMH). Eighty-seven percent of the rats receiving DMH alone manifested tumors whereas only 40% of those supplemented with dietary selenium as the selenite showed tumors. Selenium, however, did not affect the incidence of tumors in rats receiving methylazoxymethanol (MAM). It should be pointed out, however, that for animals receiving MAM and selenium, the total number of tumors was decreased nearly 50%. Thus, although the average number of MAM-induced tumors per animal was depressed, the tumor incidence was not affected by selenium. Recently, Jacobs and co-workers (35) reported protection against the mutagenic activity of 2-acetylaminofluorene (AAF) and its hydroxyderivatives in the Ames Salmonella assay system. Selenium supplementation of the culture system resulted in approximately 60% of the mutagenic activity of AAF alone. This is important work in that it suggests that short-term bioassays, such as a bacterial assay, may provide useful information on potential interactions of metal and organic carcinogens.

Although somewhat aside from organic chemical carcinogenesis, dietary selenium also has been shown to decrease the incidence of spontaneous tumors in inbred C3H mice (59). Dietary supplementation of drinking water with SeO_2 resulted in a 10% incidence of spontaneous mammary tumors at 15 months compared to an 82% incidence in untreated controls. Thus, it appears that dietary selenium supplementation may decrease incidence of chemically induced or spontaneous tumors in laboratory animals. The implications of these studies and correlation with epidemiological studies are discussed elsewhere in these proceedings.

METAL INTERACTIONS IN ORGANIC CHEMICAL CARCINOGENESIS: SYNERGISM

A number of preliminary studies have indicated a potential synergism for coexposure to trace elements and organic chemical carcinogens. Using a modification of the tracheal implantation system (9,49), Lane and Mass (43) demonstrated that exposure to a mixture of chromium carbonyl and BaP resulted in no difference in tumor incidence compared to BaP exposure alone. However, there was a suggestion of an increased rate of metastasis associated with the coexposure to chromium carbonyl and BaP. Three out of 7 coexposed tumor-bearing animals had lung metastases compared to 0 out of 6 receiving the BaP alone. Exposure to chromium carbonyl alone resulted in 2 of 10 grafts developing malignancy.

Finogenova (17) reported that cobalt may synergistically interact with 20-methylcholanthrene (20-MC) in skin tumor induction. The study indicates a decrease in the latency of papilloma induction with no effect on tumor incidence when animals were injected with $CoCl_2$ and painted with 20-MC compared to those receiving only 20-MC. Early work by Fare (13) indicated that admixture of copper acetate with DMBA resulted in a decreased incidence of skin tumors if the vehicle for skin painting was acetone. However, when the vehicle for skin-painting was olive oil, there was no synergism of copper with DMBA. Most recently, Ishinishi et al. (33) have presented preliminary data on tumor incidence resulting from coexposure of rats to As_2O_3 and BaP by intratracheal instillation. Three out of 7 rats receiving the mixture manifested tumors compared to 1 out of 7 animals receiving BaP alone. These preliminary data, which suggest that arsenic may be a cocarcinogen, are in conflict with previous studies which failed to demonstrate the cocarcinogenicity of arsenic exposure with methylcholanthrene (47) or with diethylnitrosamine (42).

The studies described thus far on the synergistic interaction of trace elements with chemical carcinogens are preliminary. Certainly, additional well-designed experiments are required to extend these initial efforts. However, one model system of metal interactions with chemical carcinogens has been studied in detail. Saffiotti and co-workers (54–57) have developed an experimental method using ferric oxide and BaP for efficient induction of bronchogenic carcinoma in Syrian golden hamsters. This method involves intratracheal intubation of an intimate mixture of Fe_2O_3 and BaP. Using this technique up to an 100% incidence of tracheobronchial tumors may be reproducibly achieved. Early techniques using BaP dispersed in gelatin resulted in low and variable tumor yield.

The enhanced carcinogenicity of the Fe_2O_3-BaP system was thought to be due to slow release of the carcinogen from the "inert" carrier dust, thus providing increased penetration of the carcinogen to the bronchial epithelium. It should be emphasized that tumor induction was almost solely associated with the tracheobronchial region and very few tumors developed in the peripheral lung. A subsequent study by Kennedy and Little (39) indicated that intubation with the BaP-Fe_2O_3 mixtures produced a high incidence of tracheobronchial tumors, while intubation with ^{210}Po-Fe_2O_3 predominantly induced peripheral lung tumors. It was hypothesized that the difference between the ^{210}Po-Fe_2O_3 and BaP-Fe_2O_3 systems related to release rate of the agent. The BaP was eluted from the iron oxide carrier during clearance of the particles on the mucociliary escalator, resulting in exposure of upper airway cells. In contrast, the ^{210}Po-

Fe$_2$O$_3$ mixture was firmly bound, resulting in the greatest delivered dose to the unciliated peripheral lung. In a subsequent study, however, Schreiber et al. (60) demonstrated a species difference in the effect of BaP-Fe$_2$O$_3$ exposure on lung tumors in rats and hamsters. Again, the Syrian golden hamsters had tumors predominantly in the tracheobronchial region and very few in the peripheral lung. However, using rats and identical instillation techniques, tumors predominated in the peripheral lung. Although different sites of tumor origin were observed in the hamster and the rat, no difference in BaP clearance times was detected. This important finding indicates that clearance rates alone cannot explain the site of tumor origin or carcinogenicity of the BaP-Fe$_2$O$_3$ mixture.

Henry et al. (28) demonstrated that the method of preparation of the BaP-Fe$_2$O$_3$ mixture markedly affected the carcinogenic potential. Four preparative techniques were evaluated (Table 2). In the first treatment group, the BaP-Fe$_2$O$_3$ mixture was prepared by precipitating BaP onto Fe$_2$O$_3$, resulting in an intimate association between the carrier and the carcinogen. Second, BaP was ground with Fe$_2$O$_3$ as described by Saffiotti et al. (54). Thirdly, Fe$_2$O$_3$ was added to a gelatin suspension of BaP just prior to instillation. A gelatin suspension of BaP alone was used in the fourth group. Dramatic differences in the calculated tumor incidence at 40 weeks were observed (Table 2). Animals receiving either intimately mixed preparation (Groups 1 or 2) had greater than 80% tumor incidence compared with no tumors in Groups 3 and 4. In the lifetime study, 80% of the animals receiving intimate BaP-Fe$_2$O$_3$ mixtures were tumor-bearing compared with approximately 15% incidence for animals receiving BaP in gelatin alone or in combination with Fe$_2$O$_3$. Animals receiving BaP intimately mixed with Fe$_2$O$_3$ showed slower clearance rates for BaP than those receiving the less intimately associated mixtures (Groups 3 and 4). However, it should be pointed out that the particle sizes used in the treatment groups were different. The most carcinogenic treatments were associated with the large particles, i.e., 50–80% of the total mass was associated with particles greater than 15 µm, whereas finer particles were employed in the other two treatment groups. Most

TABLE 2. *Effect of the physical properties of intratracheally intubated hematite mixtures on the incidence of tumors in syrian golden hamsters*

Group 1—B(a)P precipitated from acetone onto Fe$_2$O$_3$
Group 2—grinding B(a)P with Fe$_2$O$_3$
Group 3—B(a)P-gelatin suspension + Fe$_2$O$_3$
Group 4—B(a)P-gelatin suspension

	Group:			
	1	2	3	4
Tumor Incidence (%) at 40 weeks	100	90	0	0
Tumor incidence (%) at 100 weeks	100	100	45	45
Respiratory clearance halftime (hr)	4	11	1.5	1
Particle size (% of mass > µm)	53% > 15	77% > 15	90% < 10	95% < 5

Modified from Henry et al. (28).

recently, Stenback and Rowland (63) compared the relative carcinogenicity of two BaP-Fe$_2$O$_3$ mixtures prepared using either large (64% by mass > 10 μm) or smaller-sized particles (98% < 10 μm). Intratracheal instillation of the large particles in hamsters resulted in 65% incidence of tumor compared to only 10% incidence when relatively small particles were used. Clearance studies indicated that the large particles were more slowly removed from the respiratory tract than the smaller particles. Thus, increased carcinogenicity was associated with increased lung retention. Feron et al. (16) have compared in hamsters the rate of respiratory tract clearance of an Fe$_2$O$_3$-BaP mixture to that of saline suspensions of BaP alone. No significant difference in clearance rates was observed. The authors concluded that the use of BaP with particle sizes equivalent to the BaP-Fe$_2$O$_3$ mixture resulted in similar retention times and efficacy of BaP alone as a respiratory tract carcinogen.

Stenback et al. (64) compared TiO$_2$, Al$_2$O$_3$, and C to Fe$_2$O$_3$ as BaP-carrier dusts for respiratory tract carcinogenesis induced by intubation in hamsters. As a BaP carrier, TiO$_2$ was as effective as Fe$_2$O$_3$ in producing respiratory tract tumors; in contrast, C and Al$_2$O$_3$ were only slightly more carcinogenic than BaP alone. The differences in carcinogenicity could not be ascribed to a particle size effect. Similarly, Nettesheim et al. (50) compared the tumorigenicity of injected diethylnitrosamine (DEN) in hamsters concurrently inhaling either synthetic smog, Fe$_2$O$_3$ dust, or both. Animals receiving Fe$_2$O$_3$-DEN treatment demonstrated a higher incidence of tumors in the bronchiolar-alveolar region compared to DEN alone, while no difference in tumor incidence in the tracheobronchial region was observed. It was hypothesized that Fe$_2$O$_3$ acted as a cocarcinogen at the site of maximum accumulation. Inhalation of synthetic smog did not enhance, and possibly diminished, the carcinogenicity of injected DEN.

In summary, these studies indicate that Fe$_2$O$_3$ may act as a cocarcinogen. Particle size has clearly been demonstrated to be an important feature of the Fe$_2$O$_3$-BaP system, in that after intratracheal intubation, larger particles are more slowly cleared from the respiratory tract than smaller particles. However, particle size alone cannot explain species differences in site of tumor origin nor the efficacy of Fe$_2$O$_3$ as a BaP-carrier compared to other particles.

ALTERATION OF HOMEOSTATIC LEVELS OF ENDOGENOUS, ESSENTIAL TRACE ELEMENTS

This section will be limited to a discussion of alteration of copper, zinc, and selenium levels in the serum or plasma of cancer patients. These elements were chosen because of the increasing body of literature which indicates potential utility of serum analysis in the differential diagnosis and, most importantly, prognosis of human cancer.

Serum copper levels (SCL) have been studied in detail and have been demonstrated to be elevated in patients with a variety of tumors (Table 3). Furthermore, the degree of elevation has been correlated with disease activity for a variety

TABLE 3. *Serum copper levels in humans with malignant neoplasia*

Patients	Serum copper levels (μg/100 ml)
Adult controls	100
Male	104
Female	113
Mammary carcinoma	229
Lung carcinoma	192
Gastrointestinal carcinoma	157
Hepatic carcinoma	209
Acute leukemias	334
Lymphoma	298
Melanoma	164

Modified from Fisher and Shifrine (22).

of tumor types (Table 4). In this regard, the most detailed studies of SCL have been performed at the M. D. Anderson Hospital and Tumor Institute. In their study of 236 patients with malignant lymphoma, elevation in SCL was demonstrated to correlate with disease activity (31). Patients responding to treatment generally displayed a decline in SCL to normal levels; nonresponders

TABLE 4. *Association of elevated serum copper levels with clinical disease stage in human cancer patients*

Clinical disease stage	Serum copper levels (μg/100 ml)
Cervical carcinoma	
Stage I	162
Stage II	184
Stage III	202
Stage IV	219
Bladder carcinoma	
Stage I	119
Stage II	144
Stage III	175
Stage IV	217
Hodgkin's disease	
(Active)	165
(Inactive)	121
Stage I	133
Stage II	164
Stage III	167
Stage IV	190
Osteosarcoma	180
Primary; no metastases	162
Metastatic	195

Modified from Fisher and Shifrine (22).

did not display a change in their elevated SCL. Patients responding to therapy, who subsequently have relapse of the disease, were generally observed to display increased SCL prior to clinical signs. Similarly, these researchers have demonstrated elevated SCL in children (67) and adults (30) with Hodgkin's disease. Elevated SCL in the active disease were found to decrease in patients responding to therapy and to increase again in relapse. These authors have pointed out that analysis of SCL is an independent laboratory test, not correlative with other clinical chemical assays. Warren et al. (68) have also reported close correlation between SCL and the prognosis of Hodgkin's disease. The M. D. Anderson group has reported similar findings in acute leukemia patients (29). Furthermore, in these patients the degree of elevation of SCL was correlative with percentage of blast cells in bone marrow. Fisher and Shifrine (21) have also reported a similar correlation of SCL with peripheral blood blast cells in radiation-induced leukemia in the dog. Similarly, Delves et al. (10) and Illicin (32) have reported elevated SCL in acute leukemia and a decline in SCL associated with therapeutic response.

Elevations of SCL have also been described for carcinoma patients. O'Leary and Feldman (51) have demonstrated that the degree of elevation of SCL in women with cervical carcinoma increases as the stage of the disease increases, and that those patients responding to treatment have nearly normal SCL. Albert et al. (2) have also reported that patients with bladder carcinoma display elevated SCL which correlate with stages of the disease. Elevated SCL have also been reported in patients with mammary carcinoma (8), bronchial carcinoma (41), and gastric carcinoma (38).

Fisher and Shifrine (21) have described SCL elevations in dogs with radiation-induced and "spontaneous" osteosarcoma. Upon amputation of tumors of the appendicular skeleton and no clinical signs of metastasis, SCL were observed to return to near normal levels. Interestingly, four dogs with radiographically suspicious lesions, thought to possibly be osteosarcoma, were biopsied. Histological examination of the biopsy specimens indicated that the lesions were not malignant. Similarly, SCL were not elevated. These observations prompted further research on SCL in human patients with osteosarcoma. Fisher et al. (20) reported that the degree of elevation of SCL in osteosarcoma correlated with the extent and activity of disease; the highest copper levels were associated with metastatic disease and the poorest prognoses.

A variety of other clinical and physiological conditions such as pregnancy, oral contraception, inflammation, and infection are also associated with markedly elevated serum copper levels. Thus, interpretation of serum copper data in cancer diagnosis and prognosis should be done with careful consideration of factors affecting serum copper homeostasis (18).

Although the mechanisms of elevation of SCL in neoplasia have not been demonstrated, Shifrine and Fisher (62) and Warren et al. (68) have reported that the elevation in serum copper appears to be due to elevation in the serum glycoprotein, ceruloplasmin. Work by the NIH group (37,48,53) demonstrated that glycoprotein homeostasis is controlled by the number of sialic acid residues on the molecule. Upon desialylation or cleavage of two sialic acid residues from sialylated ceruloplasm, the asialo-ceruloplasmin is rapidly catabolized by the liver. Recent work by Bernacki and Kim (4) has indicated that metastasizing mammary tumors in rats release sialyltransferase into peripheral blood. Also it has been reported that tumor cells have increased surface concentrations of sialyltransferase as well as sialic acid (22). Thus, Fisher and Shifrine (22) have hypothesized that the increased concentration of ceruloplasmin associated with cancer development in humans is due to resialylation of asialo-ceruloplasmin by sialyl transferase associated with tumor cell development. This is a working hypothesis which has not been evaluated. In summary, it has been demonstrated that the increased serum copper levels observed in human cancer patients indicate increased concentration of ceruloplasmin which is hypothesized to result from decreased catabolism rather than increased anabolism of the glycoprotein.

Changes in serum zinc levels (SZL) have also been described in cancer patients (Table 5). In 1959, Addink and Frank (1) reported that blood and serum from cancer patients generally shows subnormal zinc levels unless the tumor is associated with tissues rich in zinc (e.g., bone, lung). Return to normal levels was usually associated with favorable prognosis. These observations have generally been supported in later studies. Delves et al. (10) have described subnormal plasma zinc levels in leukemic children. Response to therapy was associated with increases in plasma zinc to normal levels. McBean et al. (44) reported subnormal SZL in patients with prostatic carcinoma; however, they indicated that the decreased SZL may be due to the advanced age (67–79 years) of the patients. Fisher et al. (20) reported that patients with primary osteosarcoma had elevated SZL, while those with metastases had depressed SZL. Amputated patients who were clinically tumor-free had nearly normal SZL. Although SZL may be useful in neoplastic disease prognosis, serum zinc homeostasis appears

TABLE 5. *Serum or plasma zinc levels in humans with malignant neoplasia*

Patients	Serum or plasma zinc levels (μg/100 ml)
Prostatic carcinoma	93
Leukemia	79
Multiple myeloma	91
Various malignancies	105
Downs's syndrome	64
Osteosarcoma (primary)	207
Osteosarcoma (secondary)	88
Bronchial carcinoma	72
Normal	113

Modified from Fisher (18).

to be less tightly regulated than copper homeostasis. This may be explained, in part, by the observation that anxiety and stress may dramatically alter SZL in humans (23). In dogs, however, although stress has been demonstrated to markedly alter SZL, little effect on SCL has been observed (19). Because copper levels generally are increased and zinc levels generally decreased in cancer patients, the copper:zinc ratio has been described as a useful indicator of disease activity (10,20).

Recently, data have been presented to indicate that serum selenium levels are decreased in human patients with carcinomas but not with reticuloendothelial tumors (6,45). Increase or return to near normal selenium levels is associated with a decreased incidence of multiple primary tumors in carcinoma patients as well as a decrease in disease recurrence. Further studies are required to evaluate the utility of serum selenium analyses in cancer diagnosis and prognosis. One drawback, however, is the extremely low serum selenium levels and the difficulty of selenium analysis. While copper and zinc in serum are easily and accurately analyzed using readily available atomic absorption spectrophotometry, selenium is presently analyzed by the relatively expensive and unavailable technique of neutron activation analysis.

SUMMARY

Most of the research on metal interactions in metal- and organic-chemical carcinogenesis is preliminary and oftentimes conflicting. With regard to metal carcinogenesis, there is a paucity of reports on metal interactions. The studies of manganese antagonism of nickel carcinogenesis are exemplary in this area. With regard to metal antagonism of organic chemical carcinogenesis, two systems appear to be understood: copper with methoxy dye and copper with ethionine. Selenium appears to be an effective inhibitor of organic chemical carcinogenesis. There is extreme disagreement in the literature with regard to zinc's role as an anticarcinogen. With regard to synergisms, the hematite-benzo(a)pyrene system has been studied in detail. It appears that particle size and chemistry are important in this interaction. With regard to alteration of homeostatic levels, analysis of serum copper levels may be useful in the differential diagnosis and prognosis of cancer development in humans. The mechanism for copper increase appears to be increased levels of circulating ceruloplasmin which may be due to decreased catabolism rather than increased anabolism. Copper-to-zinc ratios also appear to be useful in the prognosis of leukemia and osteosarcoma. Further work is required to evaluate the utility of serum selenium analyses.

ACKNOWLEDGMENTS

This work was supported by the U.S. Department of Energy. The author gratefully acknowledges the clerical assistance of Charles Baty and review of the manuscript by B. A. Prentice, K. L. McNeill, and C. E. Chrisp.

REFERENCES

1. Addink, N. W. H., and Frank, L. J. P. (1959): Remarks apropos of analysis of trace elements in human tissues. *Cancer,* 12:544–551.
2. Albert, L., Hienzsch, E., Arndt, J., and Kriester, A. (1972): Bedeutung und Veranderungen des Serum-kupferspiegels während und nach der Bestrahlung von Harnblasmkarzinomen. *J. Urol.,* 8:561–566.
3. Barr, D. H., and Harris, J. W. (1973): Growth of the P388 leukemia as an ascites tumor in zinc-deficient mice. *Proc. Soc. Exp. Bio. Med.,* 144:284–287.
4. Bernacki, A. J., and Kim, U. (1977): Concomitant elevations in serum sialyltransferase activity and sialic acid content in rats with metastasizing mammary tumors. *Science,* 195:577–580.
5. Brada, Z., Altman, N. H., and Bulba, S. (1975): The effect of cupric acetate on ethionine metabolism. *Cancer Res.,* 35:3172–3180.
6. Broghamer, W. L., McConnell, K. P., and Blotcky, A. L. (1976): Relationship between serum selenium levels and patients with carcinoma. *Cancer,* 37:1384–1388.
7. Ciapparelli, L., Retief, D. H., and Fatti, L. P. (1972): The effect of zinc on 9,10-dimethyl-1,2-benzanthracene (DMBA) induced salivary gland tumours in the albino rat—A preliminary study. *So. Afr. J. Med. Sci.,* 37:85–90.
8. DeJorge, F. B., Goes, J. S., Guedes, A. B., and De Ulhoa Cintra, A. B. (1965): Biochemical studies on copper, copper oxidase, magnesium, sulfur, calcium, and phosphorus in cancer of the breast. *Clin. Chim. Acta,* 12:403–406.
9. DeWys, W., Pories, W. J., Richter, M. C., and Strain, W. H. (1970): Inhibition of Walker 256 carcinosarcoma growth by dietary zinc deficiency. *Proc. Soc. Exp. Bio. Med.,* 135:17–22.
10. Delves, H. T., Alexander, F. W., and Lay, H. (1973): Copper and zinc concentration in the plasma of leukaemic children. *Brit. J. Haematology,* 24:525–531.
11. Duncan, J. R., and Dreosti, I. E. (1975): Zinc intake, neoplastic DNA synthesis, and chemical carcinogenesis in rats and mice. *J. Natl. Cancer Inst.,* 55:195–196.
12. Edwards, M. B. (1976): Chemical carcinogenesis in the cheek pouch of Syrian hamsters receiving supplementary zinc. *Arch. Oral Biology,* 21:133–135.
13. Fare, G. (1964): Protein binding during mouse skin carcinogenesis by 9,10-dimethyl-1,2-benzanthracene. The effect of copper acetate and the non-random distribution of induction times among mice given identical treatment. *Brit. J. Cancer,* 18:768–776.
14. Fare, G., and Howell, J. S. (1964): The effect of dietary copper on rat carcinogenesis by 3-methoxy dyes. I. Tumors induced at various sites by feeding 3-methoxy-5-aminoazobenzene and its n-methyl derivative. *Cancer Res.,* 24:1279–1283.
15. Fare, G., and Orr, J. W. (1965): The effect of dietary copper on rat carcinogenesis by 3-methoxy dyes. II. Multiple skin tumors by painting with 3-methoxy-4-dimethylaminobenzene. *Cancer Res.,* 25:1784–1791.
16. Feron, V. J., De Jong, D., and Rijk, M. A. H. (1976): Clearance of benzo(a)pyrene from the respiratory tract of hamsters following its intratracheal instillation with or without ferric oxide. *Zbl. Bakt. Hyg.,* 163:441–447.
17. Finogenova, M. A. (1973): Effect of cobalt on induced carcinogenesis of the skin. Translated from: *Byulleten' Eksperimental' noi Biologii i Meditsini,* 75:73–75.
18. Fisher, G. L. (1975): Function and homeostasis of copper and zinc in mammals. *Sci. Total Environ.,* 4:373–412.
19. Fisher, G. L. (1977): Effects of disease on serum copper and zinc values in the beagle. *Am. J. Vet. Res.,* 38:935–940.
20. Fisher, G. L., Byers, V. S., Shifrine, M., and Levin, A. S. (1976): Copper and zinc levels in serum from human patients with sarcomas. *Cancer,* 37:356–363.
21. Fisher, G. L., and Shifrine, M. (1977): Serum-copper and serum-zinc levels in dogs and humans with neoplasia. In: *Biological Implications of Metals in the Environment.* 15th Annual Hanford Life Sciences Symposium, Richland, Washington, September, 1975, pp. 507–522.
22. Fisher, G. L., and Shifrine, M. (1978): Hypothesis for the mechanism of elevated serum copper levels in cancer patients. *Oncology,* 35:22–25.
23. Flynn, A., Strain, W. H., Pories, W. J., and Hill, O. A. (1973): Blood serum zinc as an indicator of acute stress. In: *Trace Substances in Environmental Health VII,* edited by D. D. Hemphill, pp. 271–276. University of Missouri Press, Columbia, Missouri.
24. Furst, A., and Cassetta, D. (1972): Failure of zinc to negate cadmium carcinogenesis. *Proc. Am. Assoc. Cancer Res.* (abstracts), p. 62.

25. Griesemer, R. A., Nettesheim, P., and Marchok, A. C. (1976): Fate of early carcinogen-induced lesions in tracheal epithelium. *Cancer Res.*, 36:2959–2664.
26. Gunn, S. A., Gould, T. C., and Anderson, W. A. D. (1963): Cadmium-induced interstitial cell tumors in rats and mice and their prevention by zinc. *J. Natl. Cancer Inst.*, 31:745–759.
27. Gunn, S. A., Gould, T. C., and Anderson, W. A. D. (1964): Effect of zinc on cancerogenesis by cadmium. *Proc. Soc. Exp. Biol. Med.*, 115:653–657.
28. Henry, M. C., Port, C. D., and Kaufman, D. G. (1975): Importance of physical properties of benzo(a)pyrene-ferric oxide mixtures in lung tumor induction. *Cancer Res.*, 35:207–217.
29. Hrgovcic, M., Tessmer, C. F., Brown, B. W., Wilbur, J. R., Mumford, D. M., Thomas, F. B., Shullenberger, C. C., and Taylor, G. (1973): Serum copper studies in the lymphomas and acute leukemias. *Prog. Clin. Cancer*, 5:121–152.
30. Hrgovcic, M., Tessmer, C. F., Thomas, F. B., Fuller, L. M., Gamble, J. F., and Shullenberger, C. C. (1973): Significance of serum copper levels in adult patients with Hodgkin's disease. *Cancer*, 31:1337–1345.
31. Hrgovcic, M., Tessmer, C. F., Thomas, F. B., Ong, P. S., Gamble, J. F., and Shullenberger, C. C. (1973): Serum copper observations in patients with malignant lymphoma. *Cancer*, 32:1512–1524.
32. Illicin, G. (1971): Serum copper and magnesium levels in leukaemia and malignant lymphoma. *Lancet*, 2:1036.
33. Ishinishi, N., Kodama, Y., Nobutomo, K., and Hisanaga, A. (1977): Preliminary experimental study on carcinogenicity of arsenic trioxide in rat lung. *Environmental Health Perspectives*, 19:191–196.
34. Jacobs, M. M. (1977): Inhibitory effects of selenium on 1,2-dimethylhydrazine and methylazoxymethanol colon carcinogenesis. *Cancer*, 40:2557–2564.
35. Jacobs, M. M., Matney, T. S., and Griffin, A. C. (1977): Inhibitory effects of selenium on the mutagenicity of 2-acetylaminofluorene (AAF) and AAF derivatives. *Cancer Lett.*, 2:319–322.
36. Kamamoto, Y., Makiura, S., Sugihara, S., Hiasa, Y., Arai, M., and Ito, N. (1973): The inhibitory effect of copper on Dl-ethionine carcinogenesis in rats. *Cancer Res.*, 33:1129–1135.
37. Kawasaki, T., and Ashwell, G. (1976): Chemical and physical properties of an hepatic membrane protein that specifically binds asialoglycoproteins. *J. Biol. Chem.*, 251:1296–1302.
38. Keiderling, W., and Scharpf, H. (1954): Uber die Klinische Bedeutung der Serumkupferung Serumeisenbestimmung bei neoplastichen Krankheitszustanden. *Munsch. Med. Wschr.*, 95:437–439.
39. Kennedy, A. R., and Little, J. B. (1974): The transport and localization of benzo(a)pyrene-hematite and hematite-^{210}Po in the hamster lung following intratracheal instillation. *Cancer Res.*, 34:1344–1352.
40. Kobayashi, N., Katsuki, H., and Yamane, Y. (1970): Inhibitory effect of aluminum on the development of experimental lung tumor in mice induced by 4-nitroquinoline 1-oxide. *Gann*, 61:239–244.
41. Kolaric, K., Roguljic, A., and Fuss, V. (1975): Serum copper levels in patients with solid tumors. *Tumori*, 61:173–177.
42. Kroes, R., van Logten, M. J., Berkvens, J. M., deVries, T., and van Esch, G. J. (1974): Study on the carcinogenicity of lead arsenate and sodium arsenate and on the possible synergistic effect of diethylnitrosamine. *Fd. Cosmet. Toxicol.*, 12:671–679.
43. Lane, B. P., and Mass, M. J. (1977): Carcinogenicity and cocarcinogenicity of chromium carbonyl in heterotopic tracheal grafts. *Cancer Res.*, 37:1476–1479.
44. McBean, L. D., Smith, J. C., Berne, B. H., and Halsted, J. A. (1974): Serum zinc and alpha$_2$-macroglobulin concentration in myocardial infarction, decubitus ulcer, multiple myeloma, prostatic carcinoma, Down's syndrome, and nephrotic syndrome. *Clin. Chim. Acta*, 50:43–51.
45. McConnell, K. P., Broghamer, W. L., Blotcky, A. J., and Hurt, O. J. (1975): Selenium levels in human blood and tissues in health and in disease. *J. Nutr.*, 105:1026–1031.
46. McQuitty, T., DeWys, W. D., Strain, W. H., Robb, C. G., Apgar, J., and Pories, W. J. (1970): Inhibition of tumor growth by dietary zinc deficiency. *Cancer Res.*, 30:1387–1390.
47. Milner, J. E. (1969): The effects of ingested arsenic on methylcholanthrene-induced skin tumors in mice. *Arch. Environmental Health*, 18:7–11.
48. Morell, A. G., Gregoriadis, G., Scheinberg, I. H., Hickman, J., and Ashwell, G. (1971): The role of sialic acid in determining the survival of glycoproteins in the circulation. *J. Biol. Chem.*, 246:1461–1467.

49. Nettesheim, P. (1976): Precursor lesions of bronchogenic carcinoma. *Cancer Res.*, 36:2654–2658.
50. Nettesheim, P., Creasia, D. A., and Mitchell, T. J. (1975): Carcinogenic and cocarcinogenic effects of inhaled synthetic smog and ferric oxide particles. *J. Natl. Cancer Inst.*, 55:159–169.
51. O'Leary, J. A., and Feldman, M. (1970): Serum copper alterations in genital cancer. *Surgical Forum*, 21:411–412.
52. Poswillo, D. E., and Cohen, B. (1971): Inhibition of carcinogenesis by dietary zinc. *Nature*, 231:447–448.
53. Pricer, W. E., and Ashwell, G. (1971): The binding of desialylated glycoproteins by plasma membranes of rat liver. *J. Biol. Chem.*, 2456:4825–4833.
54. Saffiotti, U., Cefis, F., and Kolb, L. H. (1968): A method for the experimental induction of bronchogenic carcinomas. *Cancer Res.*, 28:104–124.
55. Saffiotti, U., Cefis, F., Kolb, L. H., and Shubik, P. (1965): Experimental studies of the conditions of exposure to carcinogens for lung cancer induction. *J. Air Pollution Control Association*, 15:23–25.
56. Saffiotti, U., and Kaufman, D. G. (1974): Carcinogenesis of laryngeal carcinoma. *Laryngoscope*, 84:454–466.
57. Saffiotti, U., Montesano, R., Sellakumar, A. R., and Kaufman, D. G. (1972): Respiratory tract carcinogenesis induced in hamsters by different dose levels of benzo(a)pyrene and ferric oxide. *J. Natl. Cancer Inst.*, 49:1199–1204.
58. Schmahl, D. (1976): Combination effects in chemical carcinogenesis (experimental results). *Oncology*, 2:73–76.
59. Schrauzer, G. N., and Ishmael, D. (1974): Effects of selenium and of arsenic on the genesis of spontaneous mammary tumors in inbred C_3H mice. *Ann. Clin. Lab. Science*, 4:441–447.
60. Schreiber, H., Martin, D. H., and Pazmino, N. (1975): Species differences in the effect of benzo(a)pyrene-ferric oxide on the respiratory tract of rats and hamsters. *Cancer Res.*, 35:1654–1661.
61. Shamberger, R. J. (1970): Relationship of selenium to cancer. I. Inhibitory effect of selenium on carcinogenesis. *J. Natl. Cancer Inst.*, 44:931–936.
62. Shifrine, M., and Fisher, G. L. (1976): Ceruloplasmin levels in sera from human patients with osteosarcoma. *Cancer*, 38:244–248.
63. Stenback, F., and Rowland, J. (1978): Role of particle size in the formation of respiratory tract tumors induced by benzo(a)pyrene. *Europ. J. Cancer*, 14:321–326.
64. Stenback, F., Rowland, J., and Sellakumar, A. (1976): Carcinogenicity of benzo(a)pyrene and dusts in the hamster lung (instilled intratracheally with titanium oxide, aluminum oxide, carbon and ferric oxide). *Oncology*, 33:29–34.
65. Sunderman, F. W., Jr., Kasprzak, K. S., Lau, T. J., Minghetti, P. P., Maenza, R. M., Becker, N., Onkelinx, C., and Goldblatt, P. J. (1976): Effects of manganese on carcinogenicity and metabolism of nickel subsulfide. *Cancer Res.*, 36:1790–1800.
66. Sunderman, F. W., Jr., Lau, T. J., and Cralley, L. J. (1974): Inhibitory effect of manganese upon muscle tumorigenesis by nickel subsulfide. *Cancer Res.*, 34:92–95.
67. Tessmer, C. F., Hrgovcic, M., and Wilbur, J. (1973): Serum copper in Hodgkin's disease in children. *Cancer*, 31:303–315.
68. Warren, R. L., Jelliffe, A. M., Watson, J. V., and Hobbs, C. B. (1969): Prolonged observations on variations in the serum copper in Hodgkin's disease. *Clin. Radiol.*, 20:247–256.
69. Yamane, Y., Sakai, K., Uchiyama, I., Tabata, M., Taga, N., and Hanaki, A. (1969): Effect of basic cupric acetate on the biochemical changes in the liver of the rat fed carcinogenic aminoazo dye. I. Changes in the activities of DAB metabolism by liver homogenate. *Chem. Pharm. Bull.*, 17:2488–2493.

Trace Metals in Health and Disease, edited by
N. Kharasch. Raven Press, New York © 1979.

Metals as Mutagenic Initiators of Cancer

C. Peter Flessel

Air and Industrial Hygiene Laboratory Section, California Department of Health Services, Berkeley, California 94704

ABSTRACT

Compounds of certain metals produce cancer in humans and animals. Analysis of organic chemical carcinogenesis indicates that many carcinogens are mutagens. This chapter explores the possible role of mutagenesis in the initiation of cancer by metals, with emphasis on chromium.

Chromium exhibits properties most consistent with a mutagenic initiation mechanism. Chromium (VI) is mutagenic in the Ames *Salmonella* assay and preferentially inhibits bacterial strains deficient in DNA repair. It induces chromosomal aberrations, promotes transformation of animal cells in culture, and increases copying errors by DNA polymerases *in vitro*. In addition, the mutagenic activity of chromium (VI) is much greater than chromium (III) and is strikingly decreased by microsomal metabolism. These results parallel those from animal cancer studies which also implicate Cr(VI) as the biologically active species in carcinogenesis. Cell-free investigations suggest that the ultimate intracellular mutagen is probably Cr(III).

Other metal carcinogens exhibiting genetic toxicity are arsenic, beryllium, cobalt, iron, lead, nickel, and zinc. Both As(V) and As(III) preferentially inhibit growth in DNA repair-deficient bacteria and produce chromosome damage in cell culture. Bacterial mutagenesis and cell transformation have been reported for As(III). Epidemiological evidence suggests that trivalent arsenic is carcinogenic in humans. Salts of divalent Be, Cd, Co, Fe, Pb, and Ni induce cell transformation and all but Fe also decrease the fidelity of DNA synthesis *in vitro*. In bacteria, Be and Fe are mutagenic and Cd preferentially inhibits DNA repair-deficient strains. Salts of Cd, Pb, Ni, and Zn also induce chromosomal abnormalities in animal or plant cells. Titanium, carcinogenic only in organometallic form, shows no genetic effects.

Among other metals, compounds of manganese (II), a reported weak carcinogen, have produced mutations in bacteria, phage, yeast, and animal cells, facilitated transformation of cells in culture, and increased DNA miscopying *in vitro*. Of particular interest are two carcinostatic metals, platinum and selenium. Compounds of both platinum [Pt(II)] and selenate [Se(VI)] proved mutagenic in bacterial and animal cells. Selenite [Se(IV)], nonmutagenic in bacteria, was found to inhibit the mutagenic effects of the potent mutagen-carcinogen acetylaminofluorene in bacteria and also proved mutagenic in animal cells. DNA-binding mechanisms are apparently involved in platinum effects while redox mechanisms may account for the actions of selenium.

In the case of chromium, evidence from genetic assays consistently suggests that carcinogenesis may involve mutagenic initiation. Intracellular reduction and binding to DNA best explain chromium's genetic toxicity.

It remains to be seen whether this correlation between mutagenesis and carcinogenesis is causal, especially since compounds of a number of metals not generally recognized as carcinogenic (antimony, copper, mercury, molybdenum, tellurium) also produce genetic damage.

INTRODUCTION

Numerous studies, summarized in the excellent reviews of Sunderman and Furst (47,48), have shown that compounds of certain metals produce cancer in humans and animals. While detailed molecular mechanisms remain obscure, it is natural to suppose that DNA mutations, clearly involved in the initiation of cancer by many organic carcinogens (27), may also be important in metal carcinogenesis. This chapter extends an earlier review of metal mutagenesis (13), and explores the possible relationship between metal mutagenesis and carcinogenesis, with particular emphasis on chromium.

Metal Carcinogens

Based on current evidence (47,48), at least 10 metals can be considered carcinogenic. Some compounds of four metals (As, Cd, Cr, and Ni) have been implicated as human carcinogens and compounds of nine metals (Be, Cd, Co, Cr, Fe, Ni, Pb, Ti, and Zn) have been shown to induce cancers in animals (Table 1).

Multistage Models of Cancer

According to the multistage model of carcinogenesis, cancer arises from a single cell which passes through a series of stages, the last of which leads to a clone of cells which grows irreversibly into a tumor (38). At the molecular level, the model distinguishes between two distinct phases: initiation and promotion. Initiation typically requires a single exposure and is irreversible, while promotion requires prior initiation, repeated exposure, and is reversible, up to a point. Generally, initiators are electrophilic agents which bind to cellular macromolecules to produce genetic damage. Most initiators also require metabolic activation (29). Promoters are a diverse collection of chemical substances, which include estrogenic hormones, cigarette smoke, and phorbal esters, the active constituents of croton oil. Functionally, promoters are related by their common propensity for encouraging the expression of precancerous mutagenic lesions (38).

GENETIC TOXICITY OF METALS

Measuring Genetic Toxicity

Among the metals, 9 of 10 carcinogens (47,48) exhibit some form of genetic toxicity—i.e., the capacity to damage DNA *in vivo* or *in vitro*. Genetic toxicities of metals have been compared by several methodologies (Table 1). Mutagenic effects, in bacterial (25,26,31) and animal (18) cells, and chromosomal aberrations in plant (20,21) and animal (34,42) cells, have been observed. Metal effects on bacterial DNA-repair (31) and *in vitro* DNA miscoding (46) have provided essential comparisons too. In addition, *in vitro* cell transformation has been

TABLE 1. *Genetic toxicity of metal carcinogens*

Metal	Form oxidation state	Carcinogenicity (47,48)	Mutagenicity Prokaryotes	Mutagenicity Eukaryotes (18)	DNA-repair (31)	Cell free DNA miscoding (46)	Chromosome effects	Animal cell transformation (5)
As	V	Humans −						
	III	+						
Be	II	Animals +	+(31)		+		+(34)	+
Cd	II	Both +	+(17)	++	+	++	+(34)	++
Co	II	Animals		++	−	++	+(16,42)	++
Cr	VI	Both +	+(25,26,36)		+		+(16)	+
	III	−	±(26,37)		−			
Fe	Dextrans	Animals +			+	±	+(16,51)	+
	II ⇌ III	−			−	+	−(51)	
Ni	Subsulfide	Both +	+(4)			−	+(16,20)	+
	II	+?		+			+(47)	
Pb	II	Animals		++	−	++	++(16,20)	++
Ti	Organometallic	Animals			−	+	+(21)	
	IV							−
Zn	II	Animals +/−? (co- or anti-carcinogen)		+		−	+(16)	+

References following column headings apply to all column entries unless otherwise indicated.

observed. Cell transformation assays do not measure genetic toxicity per se, having been designed to detect cellular changes which correlate with cancer (5). Many—but not all—mutagens induce cell transformation, while some potent promoters are nonmutagenic. At the molecular level, cell transformation apparently reflects perturbations in both DNA content and expression. Studies of metals indicate that compounds of nine metal carcinogens, all of which showed genetic toxicity, also induced transformation of animal cells in culture (Table 1).

An Overview of Metals

The genetic toxicity of metals has recently been reviewed (13). A number of metal compounds are mutagenic in bacteria or phage. These include compounds of arsenic (31), beryllium (T. Hollocher, *personal communication*), chromium (25,35), copper (53), iron (4), manganese (33), molybdenum (31), platinum (1) and selenium (25). Genetic evidence suggests that some arsenic, chromium, and molybdenum compounds may influence the accuracy of DNA repair processes in microorganisms (31,55). Elsewhere in this volume, Hsie and co-workers (18) showed that compounds containing beryllium, cadmium, copper, iron, lead, manganese, nickel, platinum, silver, and zinc induced mutations in animal cells in culture. Some chromium compounds have also been shown to produce "genetic effects" in eukaryotic cells (yeast) (3). Metals which gave positive results in DNA-repair tests included arsenic, cadmium, chromium, mercury, manganese, molybdenum, selenium, and tellurium (23,31,55). Some soluble salts of nine metals (silver, beryllium, cadmium, cobalt, chromium, copper, manganese, nickel, and lead) also increased errors in cell-free DNA synthesis (46). Aluminum, antimony, arsenic, chromium, cadmium, copper, lead, mercury, platinum, and tellurium compounds induced chromosomal aberrations or abnormal cell divisions in animal or plant cells (13,28,47). Transformation of hamster embryo cells in culture has been reported for certain salts of 15 metals: antimony, arsenic, beryllium, cadmium, cobalt, chromium, copper, iron, lead, manganese, nickel, platinum, silver, thallium, and zinc (5,51).

Animal and Human Carcinogens

Examining further the metal carcinogens (47,48), chromium exhibits properties most nearly consistent with the mutagenic initiation model (Table 1). Chromium (VI) is mutagenic in *Salmonella* (25,26,35), *E. coli* (31), and yeast (3), and preferentially inhibits bacterial strains deficient in DNA-repair (31,54). It induces chromosomal aberrations (16,51), promotes transformation of animal cells in culture (5,14) and increases copying errors by DNA polymerases *in vitro* (46). In addition, the mutagenic activity of chromium (VI) is much greater than chromium (III) and is strikingly decreased by microsomal metabolism (26,36). Cancer studies indicate that Cr(VI) is also a more potent carcinogen than Cr(III) (2).

Another metal carcinogen for which oxidation states and genetic toxicity have been compared is arsenic. Both As(V) and As(III) preferentially inhibit strains of bacteria deficient in DNA repair (31,55) and produce chromosome damage in cells in culture (34). Bacterial mutagenesis (31) and cell transformation (5) have been reported for As(III). Arsenic has not been tested in the cell-free DNA polymerase miscoding assay (L. Loeb, *personal communication*). Epidemiological evidence suggests that trivalent arsenic is carcinogenic in humans (47,48).

Among other carcinogens, common salts of divalent Be, Cd, Co, Fe, Pb, Ni, and Zn induce cell transformation (5) and all but Fe also decrease the fidelity of DNA synthesis *in vitro* (41). Among the active compounds in cell transformation are sodium arsenite, beryllium sulfate, cadmium acetate or chloride, lead acetate or monoxide, and nickel sulfate. Insoluble titanium dioxide does not cause transformation or enhance viral transformation (5). In bacteria, Be (T. Hollocher, *personal communication*) and Fe (4) are mutagenic and Cd preferentially inhibits strains deficient in DNA repair (31). Salts of Cd (42), Pb (21), Ni (20), and Zn (16) induce chromosomal abnormalities in animals or plant cells. Only titanium, among the metal carcinogens, fails to exhibit some form of genetic toxicity, while carcinogenic effects are observed only with a single organometallic species, titanocene (47,48).

Other Metals Relevant to Cancer

Among other metals relevant to cancer are manganese and silver, both special cases (Table 2). Soluble and organometallic compounds of Mn(II) are reported to be weakly carcinogenic (15). Manganese salts produced mutations in bacteria (8), phage (33), and yeast (40), and facilitated transformation of cells in culture (5). *In vitro*, manganese increased DNA miscopying (54) and polynucleotide mispairing (30). Since Mn(II) is partially oxidized to Mn(III) under most assay aerobic conditions, effects reported for divalent manganese may also involve the trivalent ion. Metallic silver is a reported solid state carcinogen (32) while soluble compounds of Ag(I) increased DNA miscopying *in vitro* (46).

Of particular interest, from the point of view of mechanisms, are two metals reported to inhibit carcinogenesis, selenium (44) and platinum (43). Observations on selenium are compounded by redox effects. Se(VI) is easily reduced to Se(IV)—e.g., by ascorbic acid. *In vivo*, Se(IV) is readily converted to Se(-II). Selenium's genetic effects have been variable depending on experimental test circumstances. Selenate [Se(VI)] proved mutagenic in the Ames *Salmonella* assay (25), while selenite [Se(IV)] was negative (25). Selenite was found to inhibit the mutagenic effects of the potent mutagen-carcinogen acetylaminofluorene (19) and also was active in an *E. coli* DNA-repair test, in which selenate was inactive (55). In animal cell assays, selenite induced DNA fragmentation, unscheduled DNA-repair synthesis, chromosome aberrations, and mitotic inhibition in cultured human fibroblasts, while selenate induced only unscheduled

TABLE 2. Genetic toxicity of other cancer-related metals

Metal	Form oxidation state	Carcinogenicity	Mutagenicity		DNA repair		Cell free DNA miscoding (46)	Chromosome effects	Animal cell transformation (5)
			Prokaryotes	Eukaryotes	Prokaryotes	Eukaryotes			
Ag	0	Animal, solid state (47,48)		+(18)			+		
Al	(I)	−(47,48)						+(13)	
Cu	(III)	Animal (47,48)	+(53)	+	−(31)		+	+(16)	+
	(II)	−/+? (co- or anti-carcinogen)							
Hg	(organohalides)		+(8,33)		+(31)			+(13,47)	
Mn	(II ⇌ III)	Animal (15)	+(31)	+(40)	+(31)			+(16)	
Mo	(VI)		+(1)	+(18)	+(31)		+	+(13)	+
Pt	(II) cis PDD[a]	Human, active anti-tumor agent (1,44)						+(28)	
	(II) trans PDD[a]	Human, inactive anti-tumor agent (1,44)	−(1)						
Sb	(V)	−(47,48)						+(34)	
Se	(VI)	Both, anti-carcinogen (19,44)	+(25)		−(55)	+(23)		−(23)	
	(IV)		−(25)						
Te	(IV)	−(47,48)			+(55)	+(23)		+(23,41) +(34)	

References following column headings apply to all column entries unless otherwise indicated.
[a] PDD, platinum (II) diaminodichloride.

DNA-repair synthesis (23). Selenite also induced sister-chromatid exchanges in human whole blood cultures (41).

An elegant example of "relatedness" between chemical structure and biological function—in this case between mutagenesis and anti-tumor activity—involves platinum. *Cis*-platinum (II) diamminodichloride (PDD)—an active human anti-tumor agent—induced auxotrophic mutations in *E. coli*, but isomeric *trans*-platinum (II) diamminodichloride—which lacks anti-tumor activity—was an ineffective mutagen (1). *Cis*-platinum (II)-diamminodichloride has also been observed to produce chromosomal aberrations, primarily chromatid breaks, in human lymphocytes in culture (28). At the molecular level, Eichhorn and coworkers (9) demonstrated that *cis*-PDD binds more rapidly than the *trans*-isomer to DNA and much more strongly inhibits *E. coli* RNA polymerase, decreasing the average chain length of the RNA produced.

Compounds of the following metals also exhibit genotoxic properties: aluminum (III) (13,47), antimony (V) (34), copper (II) (18,46–48,53), mercury organohalides (13,47,48), molybdenum (VI) (31,34), and tellurium (IV) (34).

CHROMIUM: A POSSIBLE MUTAGENIC INITIATOR

Among the metals surveyed here, chromium behaves the most like a mutagenic initiator.

Mutagenesis

Bacteria

Salmonella typhimurium

Chromium is the carcinogenic metal most active in the standard Ames *Salmonella*/mammalian microsome assay (26,35). The mutagenic activity of Cr(VI) was 3 to 5 revertants per nanomole (10^{-9}M) (26), several orders of magnitude less than found for the most active organic mutagen-carcinogens (27) but several orders of magnitude greater than for Cr(III) (25,35).

Salts of chromate and dichromate are frameshift mutagens in *Salmonella*, reverting strains with the his C3076 and his D3052 mutations, as well as strains with the his G46 base pair substitution which also carry the resistance transfer factor (pKM 101). Strains carrying the frameshift mutation his D3052 have a repetitive -C-G-C-G-C-G-C-G sequence near the site of the histidine mutation, while strains with the his C3076 mutation appear to have a run of C's at the site of the mutation. If direct binding of metal ions to DNA is involved in mutagenesis, then chromate and dichromate apparently show propensities for GC-rich regions of the bacterial chromosome. The fact that Cr(VI) can also revert point mutations in strains carrying the resistance (R) factor could indicate that metal ions potentiate error-prone DNA repair processes, which are known to be mediated by genes carried by the R factor (13).

The mutagenic activity of Cr(VI) in the *Salmonella*/microsome test is almost completely eliminated by microsomal metabolism (26,36). Incubation of chromate with rat liver microsomes in the presence of the enzyme cofactor NADPH resulted in reduction to trivalent chromium (17) as well as the loss of mutagenic activity. This is consistent with the finding that mutagenicity can be restored by oxidizing the inactive trivalent species to the hexavalent form with potassium permanganate (37).

Escherichia coli

The mutagenicity of chromium in *E. coli* has also been extensively examined. Venitt and Levy (52) demonstrated that Cr(VI), as Na_2CrO_4, K_2CrO_4, or $CaCrO_4$, induced reversions in a tryptophan auxotroph, *E. coli* WP2 (trp⁻). Positive results were obtained by spotting chromate solutions on plates containing lawns of trp⁻ bacteria and scoring for trp⁺ revertants. Bacteria were also treated in suspension after which the chromate was removed and the mutagenicity and lethality determined. Under conditions of 10–30% cell survival, the mutation frequency of trp⁺ revertants was several hundred times the background level. This study also suggested that chromate facilitated mutations in GC-rich regions of the chromosome. *E. coli* WP2 (trp⁻) carries an *ochre* mutation in the trpE gene. In gene expression, the *ochre* mutation is transcribed into messenger RNA as a "nonsense" codon, UAA, which (usually) is not translated by the genetic dictionary. Occasionally, second-site mutations occur (e.g., in the translation machinery) which permit reading of nonsense codons and lead to suppression of the *ochre*-type mutations. When 100 chromium-induced trp⁺ revertants were checked for the presence of such suppressors, 98 were discovered to contain them. In contrast, only about half of the spontaneous trp⁺ revertants tested carried suppressors. Since so few of the metal-induced revertants were altered at the primary DNA site, containing only AT-pairs, it was suggested that chromate preferentially induced mutations in GC-rich regions of DNA (52).

Nishioka (31) also studied chromium mutagenesis in *E. coli* WP2 (trp⁻). Suspensions of bacteria were treated briefly with $K_2Cr_2O_7$, at millimolar concentrations. This increased the frequency of trp⁺ revertants. However, when tested in a derivative of WP2 lacking a functional recA (DNA recombination-repair) gene, hexavalent chromium was inactive. This result again suggests that the recA gene product may enhance chromium mutagenesis, as well as the actions of many other chemical and physical agents.

Yeast

Genetic effects of potassium dichromate in yeast *(Schizosaccharomyces pombe)* have also been observed. Induction of forward mutations in a haploid strain and mitotic gene conversion in heteroallelic diploid strains were reported (3).

DNA Repair Assays

Cells deficient in DNA-repair (rec⁻) are much more sensitive to DNA-damaging agents than wild type (rec⁺) strains. When a chemical is more inhibitory for rec⁻ than for rec⁺ cells, it is reasonable to suspect "DNA-damaging capacity" (31). Chromium compounds giving strong positive DNA-repair "rec" effects in *B. subtilis* (31) and *E. coli* (54) were $K_2Cr_2O_7$ and K_2CrO_4. By contrast, Cr(III) in several forms was negative. Pretreatment of potassium dichromate with Na_2SO_3 also eliminated the positive effect (31).

Cell-free DNA Polymerase Miscoding

Direct analysis of metal effects on DNA replication without regard to factors involved in cellular uptake can be made in a cell-free DNA synthesizing system. The studies of Loeb and co-workers (46) have shown that compounds of Ag, Be, Cd, Co, Cr, Cu, Mn, Ni, and Pb increased misincorporation of bases into DNA in such a system. Arsenic, platinum, and selenium were not tested and iron and zinc proved negative. Recent experiments (54; L. Loeb, *personal communication*) suggest that the DNA may be the critical target site for "miscoding," at least in the case of Mn(II).

Among chromium compounds, chlorides of both Cr(II) and Cr(III) were reported to have increased the incorporation of an incorrect substrate, dGTP, into DNA using poly d(A–T) as the template and Avian Myeloblastosis Virus (AMV) as the source of the DNA polymerase (46). Since Cr(II) is readily oxidized to Cr(III) under aerobic conditions, results obtained using compounds of divalent chromium are presumably due to the trivalent species. Results with Cr(VI) compounds were mixed. CrO_3 was positive and $Na_2Cr_2O_7$ negative (46). The maximum increase in error frequency was observed for $CrCl_2$ at 0.65 mM and for CrO_3 at 16 mM. These results suggest that under cell-free conditions, trivalent chromium is a more potent inducer of genetic mistakes than hexavalent chromium.

Chromosome Damage in Eukaryotes

Numerous studies indicate that certain heavy metal salts, including those of chromium, produce chromosome aberrations or abnormal cell divisions in higher plants and animals (16,21,47). Early investigations on the effects of heavy metal salts on root-tip mitosis in bean plants (16) showed that nitrates of Fe, Co, Ni, Cu, Cr, Cd, Mn, and Zn induced fragmentation and chromosome loss in growing plant tissue. Co, Ni, Cr, Fe, and Zn were considered strong mutagens and Cu, Cd, and Mn weak mutagens.

Tsuda and Kato (51) studied chromium-induced chromosomal aberrations in hamster embryo cells exposed in culture. "The addition of $K_2Cr_2O_7$ . . . resulted in . . . chromosomal aberrations including gaps, breaks, and exchanges." These effects were reduced by the addition of a reducing agent,

Na$_2$SO$_3$. Among other chromium compounds examined, divalent and trivalent chromium salts were ineffective, ". . . whereas . . . hexavalent . . . CrO$_3$ was highly effective" (51). These results suggest consideration of pH, as well as redox mechanisms, since hydrolysis of Cr$_2$O$_7^=$ and CrO$_3$ can lower the pH.

In Vitro Transformation of Animal Cells

Cell transformation involves some qualitatively different cellular changes than those involved in mutagenesis. Many mutagens induce cell transformation. However, a number of potent promoters, such as the phorbal esters, are nonmutagenic. It is likely, therefore, that transformation measures changes related to both initiation and promotion.

Among the heavy metals shown to transform Syrian hamster fetal cells and to enhance the frequency of adenovirus transformed foci was calcium chromate (5).

In a detailed study of Cr(VI), Fradkin et al. (14) showed that calcium chromate-treated hamster kidney (BHK 21) cells were shortened and grew in random orientation. Normal cells were elongated and grew in parallel orientation. Transformed cells lost "anchorage dependence" (i.e., the requirement for a solid surface from which to grow) and grew into large clusters of cells in soft agar, whereas normal cells underwent only one or two divisions under these conditions. Chromium-transformed cells contained enlarged nuclei and grew at rates that surpassed the controls. Once they appeared, all of these changes in cell morphology and growth properties were irreversible. However, tumorigenesis *in vivo* using chromium-transformed cells was not reported (14).

Metabolism: Cellular Uptake and Reduction

In contrast to Cr(III), Cr(VI) can be readily taken up by cells where it may be reduced (25). In biological materials analyzed, the fraction of chromium as the trivalent species varied between ~ 40% and ~ 90%, the highest levels being found in chicken eggs (45). Chromium is unique among human trace elements in that very high concentrations are present at birth but these fall to low levels later. The high concentrations at birth suggest a selective mechanism for concentrating chromium from maternal sources by the developing embryo (45).

With regard to antigenicity, Treagan (50) has described the sensitivity to chromium as more frequently due to hexavalent chromium compounds, which penetrate the skin better than tri- or divalent chromium salts. However, since hexavalent chromium salts have poor protein-binding capacities, a conversion of hexavalent to trivalent chromium compounds has been postulated (39).

Cr(VI) mutagenicity is eliminated by incubation with microsomal preparations from liver or erythrocytes but not with those from lung, muscle, serum, or plasma (36), suggesting that tissue-specific metabolism is involved in reduction.

The mutagenic potential also can be eliminated by incubating Cr(VI) with reducing agents such as ascorbic acid or sodium sulfite (31,37). These findings could be relevant to the preferential localization of chromium-induced tumors in the lung, a tissue in which the pool of reducing power is low.

Effects on Macromolecular Synthesis

Chromium preferentially inhibits DNA synthesis (22). When administered to (BHK) fibroblasts, potassium dichromate almost immediately inhibited DNA synthesis, while inhibition of RNA and protein synthesis was much more gradual.

Binding to Macromolecules

Chromium in its several oxidation states forms a variety of ligand complexes. Favored ligand binding sites are nitrogen and oxygen. Trivalent chromium, especially, forms coordination complexes with biological ligands, including nucleic acids. Direct binding of Cr to DNA was suggested by the observation that DNA isolated from cells treated with $K_2Cr_2O_7$ contained as much as 1% Cr (22).

Danchin (7) examined the binding of Cr(III) ion to both nucleic acids and proteins. Chromium (III) formed highly insoluble ligand complexes with nucleic acids but soluble complexes with proteins.

Related studies of Eisinger et al. (10) on the DNA binding of metal ions in solution allowed grouping according to similar behavior. Mn(II), Cu(II), and Cr(III) each appeared tightly bound to phosphate groups on the outside of the DNA. Binding reached saturation at mole ratios of ~ 2.5 metal ions per phosphate. Addition of Cu(II) or Cr(III), in excess of this saturation density, resulted in the precipitation of DNA from solution. This behavior was unlike that seen with Mn(II) where no precipitation occurred even at the highest concentrations. Fe(III) was bound to sites inside the DNA, presumably to purine and pyrimidine bases, and also had the capacity to precipitate DNA at high concentrations. Ni(II), Fe(II), and Co(II) showed no interior base binding and only weak nonlocalized exterior phosphate binding. X-ray diffraction measurements of chromium-treated DNA fibers indicated that Cr(III) binding to DNA was not regular.

A Possible Mechanism of Chromium Mutagenesis/Carcinogenesis

These observations suggest a possible mechanism for chromium mutagenesis which may also be relevant to carcinogenesis. Chromium (VI) enters cells readily. Once inside, a portion of the chromium is reduced to chromium (III), in the process of which Cr(III) ions may become bound to the sugar-phosphate backbone of DNA. Binding may induce configurational changes in DNA, such as chain distortion or base puckering, which lead to errors during gene replication, recombination, or repair.

SUMMARY

It is likely that several mechanisms operate concurrently in the induction of cancer by metals. In addition to mutation effects emphasized in this chapter, metal ion promoter effects (12), involving messenger RNA transcription (9,11) and translation (49), should be examined.

Of the metal carcinogens, all except titanium are genetic toxins. Chromium is the most active in a variety of short-term tests. Genetic assays indicate that hexavalent chromium is a much more potent mutagen than trivalent chromium, presumably because it is more readily transported across membranes into cells. However, cell-free biochemical and physicochemical studies suggest that the ultimate mutagen–carcinogen is Cr(III), which may act by forming strong ligand complexes with DNA, thereby altering the coding properties.

It remains to be seen whether the apparent correlation between mutagenesis and carcinogenesis/anti-carcinogenesis reflects a general mechanism, especially since compounds of a number of metals not generally recognized as carcinogenic (aluminum, antimony, copper, mercury, molybdenum, tellurium) also produce genetic damage (13).

Compounding this uncertainty are wide variations in the experimental conditions used (e.g., in pH, redox conditions and the presence or absence of competing ions and complexing agents) in various test systems. Definitive studies on the relationship between metal mutagenesis and carcinogenesis remain to be done.

ACKNOWLEDGMENT

The author thanks Dr. E. Kothny for reviewing the manuscript.

REFERENCES

1. Beck, D., and Brubaker, R. (1975): Mutagenic properties of *cis*-platinum (II) diammino chloride in *Escherichia coli. Mutat. Res.*, 27:181–189.
2. Bidstrup, P., and Case, R. (1956): Carcinoma of the lung in workmen in the bichromates-producing industry in Great Britain. *J. Ind. Med.*, 13:260–264.
3. Bonatti, S., Meini, M., and Abbondandolo, A. (1976): Genetic effects of potassium dichromate in *Schizosaccharomyces pombe. Mutat. Res.*, 38:147–150.
4. Brusick, D., Gletten, F., Japannath, D., and Weeker, U. (1976): The mutagenic activity of ferrous sulfate for *Salmonella typhimurium. Mutat. Res.*, 38:386–387.
5. Casto, B., Meyers, J., and Di Paulo, J. (1979): Enhancement of viral transformation for evaluation of the carcinogenic or mutagenic potential of inorganic metal salts. *Cancer Res.*, 39:193–198.
6. Church, M., and Flessel, P. (1978): Manganese mutagenesis in *Salmonella typhimurium. Abs. Environmental Mutagen Society*, Ac-9.
7. Danchin, A. (1975): Labelling of biological macromolecules with covalent analogs of magnesium. II—Features of the Chromic Cr(III) ion. *Biochimie*, 57:875–880.
8. Demerec, M., and Hanson, J. (1951): Mutagenic action of manganous chloride. *Cold Spring Harbor Symp. Quant. Biol.*, 16:215–228.

9. Eichhorn, G., Shin, Y., Clark, P., Rikkond, J., Pitha, J., Tarien, E., Froehlich, J., and Rao, G. (1978): Essential and deleterious effects in the interaction of metal ions with nucleic acids. *(This volume.)*
10. Eisinger, J., Shulman, R., and Szymanski, R. (1962): Transition metal binding in DNA solutions. *J. Chem. Phys.,* 36:1721–1729.
11. Falchuk, K. (1979): The role of zinc in the biochemistry of the *E. gracilis* cell cycle. *(This volume.)*
12. Fisher, G. (1979): Trace element interaction in carcinogenesis. *(This volume.)*
13. Flessel, C. (1978): Metals as mutagens. In: *Inorganic and Nutritional Aspects of Cancer,* edited by G. Shrauzer, pp. 117–128. Plenum Press, New York.
14. Fradkin, A., Janoff, A., Lane, B., and Kuschner, M. (1975): *In vitro* transformation of BHK-21 cells grown in the presence of calcium chromate. *Cancer Res.,* 35:1058–1063.
15. Furst, A. (1978): Tumorigenic effect of an organomanganese compound on F 344 rats and Swiss albino mice: Brief communication. *J. Natl. Cancer Inst.,* 60:1171–1173.
16. Glass, E. (1956): Untersuchungen über die Einwirkung von Schwermettsalzen auf die Wurzelspitzenmitose von Vicia faba. *Zeitschrift fur Botanik,* 44:1–58.
17. Gruber, J., and Jennette, K. (1978): Metabolism of the carcinogen chromate by rat liver microsomes. *Biochem. Biophys. Res. Commun.,* 82:700–706.
18. Hsie, A., Johnson, N., Couch, B., O'Neil, P., and Forbes, N. (1979): Quantitative mammalian cell mutagenesis and a preliminary study of the mutagenic potential of metals. *(This volume.)*
19. Jansson, B., Jacobs, M., and Griffin, A. (1978): Gastrointestinal cancer: epidemiology and experimental studies. In: *Inorganic and Nutritional Aspects of Cancer,* edited by G. Schrauzer, pp. 305–322. Plenum Press, New York.
20. Komczynski, L., Nowak, H., and Rejniak, L. (1963): Effect of cobalt, nickel and iron on mitosis in the roots of the broad beam *(Vicia faba). Nature,* 198:1016–1017.
21. Levan, A. (1945): Cytological reactions induced by inorganic salt solutions. *Nature,* 156:751–752.
22. Levis, A., Buttignol, M., Bianchi, V., and Sponza, G. (1978): Effects of potassium dichromate on nucleic acid and protein synthesis and on precursor uptake in BHK fibroblasts. *Cancer Res.,* 38:110–116.
23. Lo, L., Koropatnick, J., and Stich, H. (1978): The mutagenicity and cytotoxicity of selenite, "activated" selenite and selenate for normal and DNA repair-deficient human fibroblasts. *Mutat. Res.,* 49:305–312.
24. Loeb, L., Zakour, R., Kurkel, T., and Koplitz, M. (1979): Metals and genetic miscoding. *(This volume.)*
25. Lofroth, G., and Ames, B. (1978): Mutagenicity of inorganic compounds in *Salmonella typhimurium*: Arsenic, chromium and selenium. *Mutat. Res.,* 53:65–66.
26. Lofroth, G. (1978): The mutagenicity of hexavalent chromium is decreased by microsomal metabolism. *Naturwissenschaften,* 65:207–208.
27. McCann, J., Choi, E., Yamasaki, E., and Ames, B. (1975): Detection of carcinogens as mutagens in the *Salmonella*/microsome test: Assay of 300 chemicals. *Proc. Natl. Acad. Sci. USA,* 72:5135–5139.
28. Meyne, J., and Lockhart, L. (1978): Cytogenetic effects of *cis*-platinum (II) diaminedichloride on human lymphocyte cultures. *Mutat. Res.,* 58:87–97.
29. Miller, J., and Miller, E. (1977): Ultimate chemical carcinogens as reactive mutagenic electrophiles. In: *Origins of Human Cancer,* edited by H. Hiatt, J. Watson, and J. Winsten, pp. 603–627. Cold Spring Harbor Press, Cold Spring Harbor, New York.
30. Murray, M., and Flessel, C. (1976): A comparison of carcinogenic and non-carcinogenic metals *in vitro. Biochim. Biophys. Acta,* 425:256–261.
31. Nishioka, H. (1975): Mutagenic activities of metal compounds in bacteria. *Mutat. Res.,* 31:185–190.
32. Oppenheimer, B., Oppenheimer, E., Danishefsky, I., and Stout, A. (1956): Carcinogenic effect of metals in rodents. *Cancer Res.,* 16:439–441.
33. Orgel, A., and Orgel, L. (1965): Induction of mutations in bacteriophage T4 with divalent manganese. *J. Mol. Biol.,* 14:453–457.
34. Paton, G., and Allison, A. (1972): Chromosome damage in human cell cultures induced by metal salts. *Mutat. Res.,* 16:332–336.
35. Petrilli, F., and DeFlora, S. (1977): Toxicity and mutagenicity of hexavalent chromium on *Salmonella typhimurium. Applied and Environmental Microbiology,* 33:805–809.

36. Petrilli, F., and DeFlora, S. (1978): Metabolic deactivation of hexavalent chromium mutagenicity. *Mutat. Res.,* 54:139–147.
37. Petrilli, F., and DeFlora, S. (1978): Oxidation of inactive trivalent chromium to the mutagenic hexavalent form. *Mutat. Res.,* 58:167–173.
38. Petro, R. (1977): Epidemiology, multistage models, and short-term mutagenicity tests. In: *Origins of Human Cancer,* edited by H. Hiatt, J. Watson, and J. Winsten, pp. 1403–1430. Cold Spring Harbor Press, Cold Spring Harbor, New York.
39. Polak, L., Turk, J., and Frey, J. (1973): Studies on contact hypersensitivity to chromium compounds. *Prog. Allergy,* 17:145–226.
40. Putrament, A., Baranowake, H., Ejchart, A., and Prazmo, W. (1975): Manganese mutagenesis in yeast. A practical application of manganese for the induction of mitochondrial antibiotic resistant mutations. *J. Gen. Microbiol.,* 62:265–270.
41. Ray, J., and Altenberg, L. (1978): Sister-chromated exchange induction by sodium selenite: Dependence on the presence of red blood cells or red blood cell lysate. *Mutat. Res.,* 54:343–354.
42. Rohr, G., and Bauchinger, M. (1976): Chromosome analyses in cell cultures of the Chinese hamster after application of cadmium sulphate. *Mutat. Res.,* 40:125–130.
43. Rosenberg, B. (1978): Noble metal complexes in cancer chemotherapy. In: *Inorganic and Nutritional Aspects of Cancer,* edited by G. Schrauzer, pp. 129–150. Plenum Press, New York.
44. Schrauzer, G. (1978): Trace elements, nutrition and cancer: Perspectives of prevention. In: *Inorganic and Nutritional Aspects of Cancer,* edited by G. Schrauzer, pp. 323–344. Plenum Press, New York.
45. Schroeder, H., Balassa, J., and Tipton, I. (1962): Abnormal trace metals in man-chromium. *J. Chron. Dis.,* 15:941–964.
46. Sirover, M., and Loeb, L. (1976): Infidelity of DNA synthesis *in vitro:* Screening for potential metal mutagens or carcinogens. *Science,* 194:1434–1436.
47. Sunderman, F. (1978): Carcinogenic effects of metals. *Fed. Proc.,* 37:76–79.
48. Furst, A. (1977): Inorganic agents as carcinogens. In: *Advances in Modern Toxicology,* edited by H. Kraybill and M. Mehlman, 3:1–12. Wiley, New York.
49. Szer, W., and Ochoa, S. (1964): Complexing ability and coding properties of synthetic polynucleotides. *J. Mol. Biol.,* 8:823–834.
50. Tregan, L. (1975): Metals and the immune response. *Research Communications in Chemical Pathology and Pharmacology,* 12:189–220.
51. Tsuda, H., and Kato, K. (1978): Chromosomal aberrations and morphological transformation in hamster embryonic cells treated with potassium dichromate *in vitro. Mutat. Res.,* 46:87–94.
52. Venitt, S., and Levy, L. (1974): Mutagenicity of chromates in bacteria and its relevance to chromate carcinogenesis. *Nature,* 250:493–495.
53. Weed, L. (1963): Effects of copper on *Bacillus subtilis. J. Bacteriol.,* 85:1003–1010.
54. Weymouth, L., and Loeb, L. (1978): Mutagenesis during *in vitro* DNA synthesis. *Proc. Natl. Acad. Sci. USA,* 75:1924–1928.
55. Yagi, T., and Nishioka, H. (1977): DNA damage and its degradation by metal compounds. *The Science and Engineering Review of Doshisha University,* 18:1–8.

Trace Metals in Health and Disease, edited by
N. Kharasch. Raven Press, New York © 1979.

Essential and Deleterious Effects in the Interaction of Metal Ions with Nucleic Acids

G. L. Eichhorn, Y. A. Shin, P. Clark, J. Rifkind, J. Pitha, E. Tarien, G. Rao, *S. J. Karlik, and *D. R. Crapper

*Laboratory of Cellular and Molecular Biology, Gerontology Research Center, National Institute on Aging, National Institutes of Health, Baltimore City Hospitals, Baltimore, Maryland 21124, and *Department of Physiology, Medical Sciences Building, University of Toronto, Toronto, Ontario M5S 1A8, Canada*

Virtually all biological processes in which nucleic acids are engaged involve the participation of metal ions. To understand the function of metal ions in such processes it is necessary, first of all, to understand how metal ions react with nucleic acids.

THE EFFECTS OF METAL IONS ON NUCLEIC ACID STRUCTURE

There are three principal binding sites for metals on the nucleic acid molecules: (a) the phosphate groups, (b) the heterocyclic bases, and (c) the 2 –OH groups in the case of RNA. We shall concern ourselves here with only the first two binding sites (10).

In the DNA double helix, the sugar phosphate backbone is on the surface of the molecule and the bases are on the inside and perpendicular to the helical axis. Metals binding to the phosphates or to the bases therefore have very different effects upon the structure of such a molecule. Metals binding to the phosphate stabilize the structure, whereas metals binding to the bases destabilize the structure because their binding competes with the hydrogen-bonding of the DNA bases. This dichotomy between phosphate binding stabilization and base binding destabilization is dramatically illustrated in Fig. 1A, which shows the "melting" of DNA—i.e., the transition from double helix to single strands (9). Since the double helix has a relatively low absorbance and the single strands a relatively high absorbance, a plot of absorbance as a function of temperature pinpoints the transition from double helix to single strands; the "melting temperature," or T_m, is the midpoint of that transition. Phosphate binding metal ions—e.g., magnesium—increase the melting temperature, (Fig. 1B); thus they increase the stability of the double helix. Base binding metal ions—e.g., copper ions— decrease the melting temperature, (Fig. 1C); they therefore decrease the stability of the double helix.

When double helical DNA is heated in the absence of divalent metal ions,

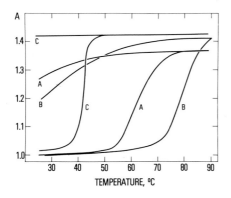

FIG. 1. Effect of divalent ions on DNA melting: (A) no metal, (B) Mg^{2+}, a phosphate-binding metal, (C) Cu^{2+}, a base binding metal. (From ref. 9.)

(Fig. 1A), random coils are produced irreversibly. The regeneration of the thermodynamic conditions for the stability of hydrogen-bonding does not reform the double helix because the complementary bases are no longer in register. There is some hydrogen bond reformation but this occurs primarily within the single strands of the DNA. On the other hand, when base binding divalent metal ions are present during the heating (Fig. 1C), they form crosslinks between the DNA chains in the denatured state and, as a result, the complementary bases remain in register so that, when the thermodynamic conditions for the stability of the double helix are regenerated, the bases can find each other again, and the double helix is reformed. Therefore, base binding metal ions can lead to the reversible unwinding and rewinding of DNA molecules or other polynucleotide helices.

Phosphate binding metal ions stabilize the DNA structure (Fig. 1B). This is because negatively charged phosphate groups on the surface of this structure repel each other, so that there is a tendency for them to get as far apart as possible, as by the unwinding of the double helix. If solid DNA is dissolved in distilled water, spontaneous unwinding therefore occurs. This unwinding can be prevented by neutralizing the negative charges on the phosphate by counter ions.

THE EFFECT OF METAL IONS ON DNA-POLYPEPTIDE BINDING

Within the cell, the stabilization of DNA molecules is not accomplished primarily by metal ions but by positively charged proteins, the histones, which contain large amounts of lysine. The polylysine molecule contains only positively charged lysines which form electrostatic bonds with DNA; it is therefore a model of the histones. The negative charges on the DNA phosphate groups interact with the positive charges on the polylysine. Recently, we have been interested in the effects of divalent metal ions on this interaction between DNA and polylysine. Since the metal ions bind to the same negative charges on the DNA phosphate groups as polylysine, they will compete with some of these positive charges on the lysines and displace them from their bonds to the DNA molecules. Thus the reaction of divalent metal ions can loosen the bonds between the DNA and the polylysine. This phenomenon is illustrated in Fig. 2, which contains melting curves of DNA polylysine complexes (containing an excess of DNA; the 1:1 DNA polylysine complex is not soluble). There is a biphasic transition: Step 1 involves the melting of the free DNA, and step 2 the melting of the DNA polylysine complex. (Polylysine, of course, stabilizes the DNA and increases the T_m.)

The effect of divalent metal ions is quite different in steps 1 and 2. In step 1, it is the same as in free DNA: The phosphate binding metals stabilize DNA and therefore increase T_m, while the base binding metals destabilize DNA and decrease T_m. But in step 2, all of the metal ions, whether they are phosphate binders or base binders, decrease the T_m. This is, of course, what we would expect from the displacement of the stabilizing polylysine from the DNA molecules by competition with metal ions which do not stabilize the DNA quite as strongly as the polylysine.

The interaction of DNA with proteins or polypeptides is quite complicated. Again, we use the DNA-polylysine complex as a model. This complex is highly associated, i.e., the DNA molecules interact with each other in such a way that they are packed anisotropically and produce a so-called ψ-complex, which is defined by a very intense circular dichroism or optical rotatory dispersion spectrum. There are two kinds of ψ spectra. We call the one that produces negative circular dichroism a $\psi(-)$ effect and the one that produces a positive circular dichroism a $\psi(+)$ effect (24). Some of the histones that are associated in the cell with DNA produce $\psi(-)$ structures and some of them produce $\psi(+)$ structures (1,14,18,22,23). Phosphate binding metal ions enhance either effect. Thus the addition of magnesium ions will enhance the $\psi(-)$ characteristics of a $\psi(-)$ structure and the $\psi(+)$ characteristics of a $\psi(+)$ structure. Base

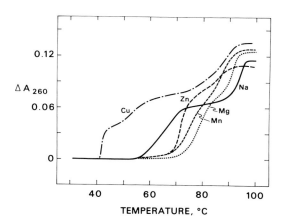

FIG. 2. Effect of divalent metals on melting of DNA-polylysine complex.

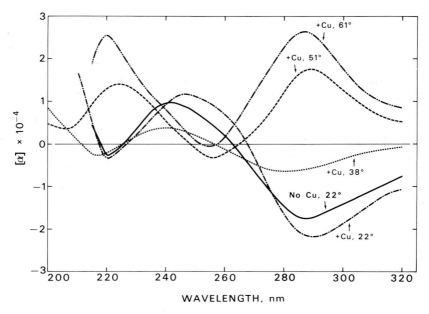

FIG. 3. Effects of Cu^{2+} on the optical rotatory dispersion of DNA-polylysine complex. (From ref. 24.)

binding metal ions produce a very different and more dramatic effect (24) as illustrated in Fig. 3. Copper ions bind to phosphate at room temperature and therefore copper ions produce an enhancement of the $\psi(-)$ effect at room temperature. However, when DNA is heated in the presence of copper, the copper binds to the bases, and as the temperature is increased, there is a gradual transition from the $\psi(-)$ structure to the $\psi(+)$ structure. Thus, the addition of a base binding metal ion significantly affects the way in which the DNA molecules in the DNA polylysine complex are arranged with respect to each other.

Figure 4 shows the effect of the addition of *cis*-dichlorodiammine platinum (II) to DNA-polylysine (24). (This complex is an important anti-tumor agent.) Platinum binds to the DNA bases, but this reaction is very slow. Therefore, there is a very slow transition from a $\psi(-)$ structure to a $\psi(+)$ structure in the presence of this platinum complex. Thus, there are two examples which indicate that metal to base binding is responsible for such transitions. Copper requires heat for base binding, and the platinum complex requires time for base binding, and in both cases the conditions required for base binding produce the transition.

This transition is reversible (Fig. 5). If copper is added to DNA, heating induces conversion to the $\psi(+)$ structure. The subsequent addition of either EDTA or a high concentration of sodium chloride will displace the copper from interaction with the DNA, resulting in a reversion to $\psi(-)$. Thus, metal

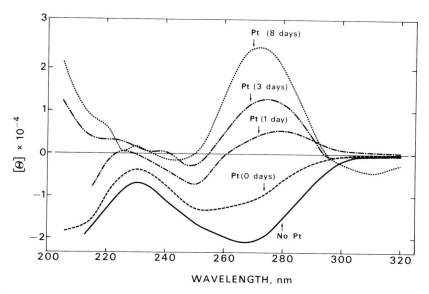

FIG. 4. Effect of *cis* [Pt(NH$_3$)$_2$Cl$_2$] on the circular dichroism of DNA-polylysine complex. (From ref. 24.)

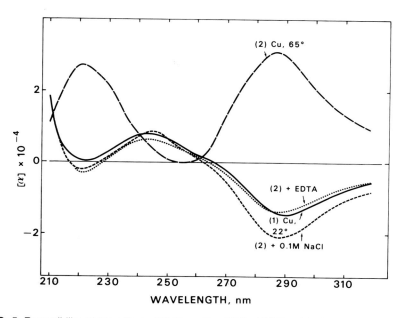

FIG. 5. Reversibility of the effect of Cu^{2+} on the ORD of DNA-polylysine. (From ref. 24.)

ions not only can change the way in which the DNA molecules are packed together, but they do so reversibly.

Not all base binding ions favor the ψ(+) structure. Base binding metal ions have tendencies to support either ψ(+) or ψ(−). If one starts with ψ(−) and adds a ψ(−)-favoring metal, the ψ(−) structure is enhanced and if one adds a ψ(+)-favoring metal, ψ(−) is converted to ψ(+). If one starts with a ψ(+) structure, the reverse phenomena occur. Therefore, metal ions are extremely important in the organization of DNA molecules into a compact structure.

METAL IONS IN THE CELL NUCLEUS AND CELLULAR AGING

Metal ions are very important to the structure of the cell nucleus. Electron micrographs of the cell nucleus by Monneron and Moulé (17) show the chromatin packed into granules in the presence of a relatively high concentration of magnesium (5.0 mM); the chromatin granules become dispersed at a lower concentration of magnesium (0.1 mM). This process is also reversible—e.g., the addition of magnesium to a nucleus with dispersed chromatin results in the granulation of the chromatin.

Some time ago, Dr. Holger von Hahn showed in our laboratory that there may be some changes in chromatin as a function of age, and that these changes involve metal ions (27,28). He obtained a Sephadex chromatogram of histones isolated from young and old nuclei in the presence of high concentrations of magnesium ions. Some of the chromatographic peaks are much lower in the old than in the young nuclei. If the isolation of nuclei is carried out in the absence of magnesium ions, the histone chromatographs from the young nuclei look similar to those from the old. In other words, isolation of the nuclei in the absence of magnesium ions produces the same effect as aging. We don't understand the reason for this phenomenon; since metal ions are involved in chromatin structure, perhaps the nature of this involvement changes with age.

It is generally believed that the aging phenomenon is genetically determined; it therefore stands to reason that there must be age changes in genetic information transfer—i.e., in chromatin structure, replication, transcription, and protein synthesis. Many laboratories have looked at age changes in various aspects of genetic information transfer. A study carried out in our laboratory (6) showed that the hydrolysis of chromatin with micrococcal nuclease is very different when the chromatin is obtained from young and old rats. Therefore, the accessibility of the DNA must be very different in old and young chromatin.

Perhaps one should ask what we could do if we discover the basic phenomenon that determines the aging process. The possibility of genetic engineering arises; obviously many technical difficulties are involved in genetic engineering, but the possibility exists that genetic engineering could become feasible in the future. But even so, there would remain the moral problems involved in genetic engineering, and these are indeed formidable. The question then arises: is it possible that there is an environmental impact upon genetic information transfer? If there is, it might be easier to control such environmental impact and the moral difficulties would be considerably less (11).

METAL IONS AND GENETIC INFORMATION TRANSFER

Possibly the most important substances which come into the cell from the environment are metal ions. Many metal ions are essential for biological processes; but whether essential or nonessential, all metal ions come into the cell from the environment. Metal ions are required for every step in genetic information transfer. Nevertheless, the presence of metal ions can cause errors in genetic information transfer; e.g., if the wrong metal ion is involved in any particular information transfer step, errors can result. This phenomenon is illustrated by an effect of metal ions on RNA synthesis. This synthesis requires that *ribo*nucleotides, which contain a 2'-hydroxyl group, but not *deoxy*nucleotides, which do not contain this group, be incorporated into the nucleic acid chain. Therefore, the enzyme must be able to recognize the difference between a ribonucleotide and a deoxynucleotide. RNA synthesis is achieved through the action of the enzyme RNA polymerase, which requires activation by divalent metal ions. The only metal ions capable of fostering the correct incorporation of ribonucleotides into RNA are manganese, magnesium, and cobalt. But manganese, which accomplishes this job most effectively, does not distinguish readily between ribonucleotide and deoxynucleotide incorporation (15). Magnesium and cobalt, on the other hand—which are somewhat less effective for ribonucleotide incorporation—discriminate reasonably well (19). The manganese acts like a fast typist who makes mistakes, and magnesium acts like a slower but accurate typist. Thus, RNA synthesis can be used to illustrate a way in which the wrong metal ion participating in a genetic information transfer process can cause mistakes.

Even an essential metal ion, when present in the wrong concentration, can lead to error. Some time ago, Szer and Ochoa (26) used poly(U)—which contains the UUU codons for phenylalanine—as a message in protein synthesis and measured the amount of correct phenylalanine incorporation as a function of increasing magnesium ion concentration. No incorporation occurred in the absence of magnesium. The incorporation was optimal at about 10 mM [Mg^{2+}]. They also measured the incorrect incorporation of leucine and found that leucine incorporation is maximal at about 20 mM [Mg^{2+}]. Thus, errors can result not only from the wrong metal ion, but even from the right metal ion in the wrong concentration.

Metal ions can lead to a variety of other deleterious effects in nucleic acid structure and function. We have already discussed the formation of crosslinks between DNA strands; if these occur in the cell, they could indeed be harmful. The reaction of phosphate binding metal ions with RNA results in the cleavage of the internucleotide linkages so that small oligomers and mononucleotides result. This degradation occurs only with RNA and not with DNA, since it requires the presence of the 2'-hydroxyl group (2–4). We have speculated that the ease of degradation of RNA compared to DNA could account for the selection of DNA rather than RNA as the primary genetic molecule (4).

Another potential deleterious effect is the change in specificity induced by metal ions in enzymes which act on nucleic acids—e.g., pancreatic deoxyribonuclease. It occurred to us that, since a nuclease enzyme must recognize a certain structure characteristic of a nucleotide, the structural change produced by metal binding may interfere with this recognition process (5). Since copper ions preferentially bind to guanines on the DNA chains, we supposed that in a DNA molecule that had reacted with copper, the enzyme would preferentially cleave sites adjacent to all bases other than guanine. End group analyses of the fragments produced by DNase digestion of copper-bound DNA shows much less cleavage at guanine than in the absence of copper. Thus, enzyme specificity is markedly changed by metal binding.

Finally, a potentially very deleterious phenomenon induced by metal ions is the mispairing of bases. The correct propagation of the genetic code, of course, requires the correct recognition of complementary bases through proper Watson-Crick hydrogen-bonding. If other kinds of hydrogen-bonding occur, then, of course, errors will occur. We postulated that in the presence of low concentrations of metal ions, the interaction between nucleic acid strands is relatively weak and therefore only the most stable hydrogen bonding would be allowed. At high concentrations of metal ions, however, this interaction may become so strong that not only the most stable hydrogen bonds but also less stable hydrogen bonds are produced. Then at high metal ion concentration, there would be less discrimination between strong base pairs—i.e., the complementary base pairs—and weaker base pairs. To test this hypothesis, we studied the interaction between synthetic polymers like, e.g., poly(A) and poly(I,U) (12). The possibilities in this reaction are that one can get complementary base pairing (U and A) and noncomplementary base pairing (I and A). In complementary base pairing, the I's will not hydrogen bond with the A's and loop out of the helix. In noncomplementary base pairing—or mispairing, as we call it—all possible base pairs react to produce a perfect helix. If the (I,U) co-polymer contains 50% I and 50% U, then the predicted stoichiometry for only complementary base pair interaction is 2:1. If, on the other hand, ambiguity, or mispairing, occurs, then the predicted stoichiometry is 1:1. The interaction of poly (I,U) with poly(A) was studied as a function of magnesium ion concentration. At low [Mg^{2+}], the stoichiometry is 2:1, characteristic of complementary base pairing, and at high [Mg^{2+}], it is 1:1, characteristic of ambiguity in base pairing, or mispairing. Thus, high metal ion concentration does induce mispairing of bases.

The transition between complementary base pairing and mispairing occurs between about 10 and 20 mM magnesium, the same concentration range as for the transition from the incorporation of the correct amino acid to incorrect amino acid in protein synthesis. This correlation can be readily explained, if one considers that the recognition of message by transfer RNA is dependent on the correct recognition of the Watson-Crick base pairs. At low magnesium ion concentration, only the strongest base pairs are formed, and these are the complementary base pairs. If the message is CUU, it recognizes the transfer RNA for phenylalanine which has the complementary GAA anticodon (U can replace C in the message, so that UUU in poly(U) also codes for phenylalanine). We have postulated that at high magnesium ion concentrations this correct recognition fails. Binding of leucyl instead of phenylalanyl transfer RNA to the message indicates recognition failure in only one base pair, since the codon for leucine is AUU or GUU. It appears that at high metal ion concentration, mispairing is permissible so that some leucyl RNA binds to the message and leads to the incorporation of leucine instead of phenylalanine. Although this correlation makes it appear quite reasonable that misincorporation of amino acids at high metal ion concentration is due to mispairing, that point is not proved because metal ions have many other effects on the protein-synthesizing system, such as the effect on the structure of ribosomes. But mispairing can surely have deleterious effects in many stages of information transfer.

We have indicated numerous deleterious effects that metal ions have on nucleic acid structure and on various aspects of genetic information transfer. Let us now consider some age changes in the concentrations of these metal ions in cells. Many studies of such age changes have revealed that they are quite significant (11). We then compare the following results: (a) there are dramatic age changes in the metal ion concentrations of cells; (b) metal ions are absolutely essential for every aspect of genetic information transfer; and (c) metal ions produce many deleterious effects. It seems therefore reasonable to suppose that the accumulation of the changes in metal ion concentrations with age do have an impact on genetic information transfer which, in turn, may influence the aging process. We have no proof for this assumption, but the hypothesis seems reasonable and worthy of considerable attention for the reasons cited earlier.

There is one phenomenon in which we know that metal ions are involved in a process that is very relevant to aging; it was discovered by Crapper and his group at the University of Toronto. They have shown that there is a high localized accumulation of aluminum in brain (7), and they have also found that the aluminum binds to the chromatin of the cell (8). We are presently engaged in collaboration with Crapper and Karlik at the University of Toronto to determine how aluminum ions react with DNA. Melting studies with DNA aluminum complexes show an increase in T_m at low concentrations of aluminum, and a decrease in T_m at high aluminum concentrations. At intermediate aluminum concentrations, there is biphasic melting (16). The reason for the two effects may be the existence of two important species in aqueous solutions of Al; namely, Al^{3+} and $[Al OH]^{2+}$; the latter ion increases the melting temperature and the former decreases it. Figure 6 shows plots of the melting temperatures of both types of aluminum DNA complexes as a function of increasing aluminum concentration. When the Al^{3+} plot is compared to similar plots of melting temperature versus metal ion per DNA phosphate (13), the aluminum profile looks very similar to the copper profile. We have thoroughly studied the reaction of copper ions with DNA and RNA (9,20). The major result of that reaction is the formation of both intermolecular and intramolecular crosslinks; by analogy,

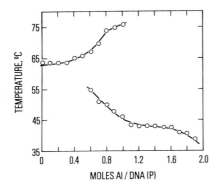

FIG. 6. Effect of aluminum concentration on melting temperature of aluminum complexes of DNA.

it appears that aluminum forms crosslinks between DNA chains. We also have more direct evidence for these crosslinks, which could be responsible for Alzheimer's disease phenomena, but of course this is purely speculative. Studies by Sincock (25) show that it is possible to influence longevity of low life forms—rotifers—by the use of complexing agents (25). Several different complexing agents were supplied during the growth of these rotifers and survival curves showed that all of the agents greatly increased the longevity of these animals. Thus, it is demonstrated that it is possible to affect longevity by metal complexing, although we do not have any idea whether this phenomenon has anything to do with genetic information transfer.

On the other hand, the *cis*-dichlorodiammine platinum complex has been shown to be an effective anti-tumor agent, and this activity is apparently due to the interference of this molecule with genetic information transfer (21). Thus longevity can be affected by metal complexing agents, and metal complexes can affect genetic information transfer. Perhaps it is possible that metal ions acting on genetic information transfer systems may have some influence on life-span also.

In conclusion, we have postulated that the many deleterious effects of metal ions have an impact on the aging process. These effects do not constitute a basic aging phenomenon, but they may have secondary effects which could be important because they may be more easily controlled than the primary phenomena.

REFERENCES

1. Adler, A. J., Moran, E. C., and Fasman, G. D. (1975): Complexes of DNA with histones f2a2 and f3. Circular dichroism studies. *Biochemistry,* 14:4179–4185.
2. Butzow, J. J., and Eichhorn, G. L. (1965): Degradation of polyribonucleotides by zinc and other divalent metal ions. *Biopolymers,* 3:95–107.
3. Butzow, J. J., and Eichhorn, G. L. (1971): On the mechanism of degradation of polyribonucleotides and oligoribonucleotides by zinc(II) ions. *Biochemistry,* 10:2019–2027.
4. Butzow, J. J., and Eichhorn, G. L. (1975): Different susceptibility of DNA and RNA to cleavage by metal ions. *Nature,* 254:358–359.

5. Clark, P., and Eichhorn, G. L. (1974): A predictable modification of enzyme specificity. Selected alteration of DNA bases by metal ions to promote cleavage specificity by deoxyribonuclease. *Biochemistry,* 13:5098–5102.
6. Clark, P., and Eichhorn, G. L. (1977): Age Changes in Rat Liver Chromatin Structure. Abstract presented at the 30th Annual Scientific Meeting of the Gerontological Society, San Francisco, California.
7. Crapper, D. R., Krishnan, S. S., and Quittkat, S. (1976): Aluminum, neurofibrillary degeneration and Alzheimer's disease. *Brain,* 99:67–80.
8. DeBoni, U., Scott, J. W., and Crapper, D. R. (1974): Intracellular aluminum binding; a histochemical study. *Histochemistry,* 40:31–37.
9. Eichhorn, G. L. (1962): Metal ions as stabilizers or destabilizers of the deoxyribonucleic acid (DNA) structure. *Nature,* 194:474–475.
10. Eichhorn, G. L. (1973): Complexes of polynucleotides and nucleic acids. In: *Inorganic Biochemistry,* edited by G. L. Eichhorn, pp. 1210–1240. Elsevier, Amsterdam.
11. Eichhorn, G. L. (1978): Aging, genetics, and the environment: Potential of errors introduced into genetic information transfer by metal ions. *Mech. Ageing Develop. (in press).*
12. Eichhorn, G. L., Berger, N. A., Butzow, J. J., Clark, P., Heim, J., Pitha, J., Richardson, C., Rifkind, J. M., Shin, Y., and Tarien, E. (1973): Some effects of metal ions on the structure and function of nucleic acids. In: *Metal Ions in Biological Systems,* edited by S. K. Dhar, pp. 43–66. Plenum Press, New York.
13. Eichhorn, G. L., and Shin, Y. A. (1968): Interaction of metal ions with polynucleotides and related compounds. XII. The relative effect of various metal ions on DNA helicity. *J. Am. Chem. Soc.,* 90:7323–7328.
14. Gottesfeld, J. M., Calvin, M., Cole, R. D., Igdaloff, D. M., Moses, V., and Vaughan, W. (1972): An investigation of specific interactions of deoxyribonucleic acid and lysine-rich (F1) histone preparations. *Biochemistry,* 11:1422–1430.
15. Hurwitz, J., Yarbrough, L., and Wickner, S. (1972): Deoxynucleoside triphosphates by DNA dependent RNA polymerase in E. coli. *Biochem. Biophys. Res. Commun.,* 48:628–635.
16. Karlik, S. J., Eichhorn, G. L., Lewis, P. N., and Crapper, D. R. (1979): Aluminum interactions with DNA. Abstract presented at the XIth International Congress of Biochemistry, Toronto, Ontario, Canada.
17. Monneron, A., and Moulé, Y. (1968): Etude Ultrastructurale de Particules Ribonucleoproteiques Nucleaires Isolées a Partir du Foie de Rat. *Exp. Cell Res.,* 51:531–554.
18. Olins, D. E., and Olins, A. L. (1971): Model nucleohistones: The interactions of F1 and F2a1 histones with native T7 DNA. *J. Mol. Biol.,* 57:437–455.
19. Rao, K. G., Su, F. Y. H., Sethi, S., and Eichhorn, G. L. (1977): Effects of divalent ions on the structure and sugar fidelity of E. coli RNA polymerase. Abstract presented at the 174th Nat. Meet. Am. Chem. Soc., Chicago.
20. Rifkind, J. M., Shin, Y. A., Heim, J. M., and Eichhorn, G. L. (1976): Cooperative disordering of single-stranded polynucleotides through copper crosslinking. *Biopolymers,* 15:1879–1902.
21. Rosenberg, B. (1973): Platinum coordination complexes in cancer chemotherapy. *Naturwissenschaften,* 60:399–406.
22. Shih, T. Y., and Fasman, G. D. (1971): Circular dichroism studies of deoxyribonucleic acid complexes with arginine-rich histone IV (f2a1). *Biochemistry,* 10:1675–1683.
23. Shih, T. Y., and Fasman, G. D. (1972): Circular dichroism studies of histone-deoxyribonucleic acid complexes. A comparison of complexes with histone I (f-1), histone IV (f2a1), and their mixtures. *Biochemistry,* 11:398–404.
24. Shin, Y. A., and Eichhorn, G. L. (1977): Reversible change in ψ structure of DNA-poly(lys) complexes induced by metal binding. *Biopolymers,* 16:225–230.
25. Sincock, A. M. (1975): Life extension in the rotifer *Mytilina brevispina* var. *redunca* by the application of chelating agents. *J. Gerontol.,* 30:289–293.
26. Szer, W., and Ochoa, S. (1964): Complexing ability and coding properties of synthetic polynucleotides. *J. Mol. Biol.,* 8:823–834.
27. von Hahn, H. P., Heim, J. M., and Eichhorn, G. L. (1970): The effect of divalent ions on the isolation of proteins from rat liver nucleoprotein. *Biochim. Biophys. Acta,* 214:509–519.
28. von Hahn, H. P., Miller, J., and Eichhorn, G. L. (1969): Age-related alterations in the structure of nucleoprotein, IV. Changes in the composition of whole histone from rat liver. *Gerontologia,* 15:293–301.

Metals, DNA Polymerization, and Genetic Miscoding

Richard A. Zakour, Lawrence A. Loeb, Thomas A. Kunkel, and R. Marlene Koplitz

Department of Pathology, Gottstein Memorial Cancer Research Laboratory, University of Washington, Seattle, Washington 98195

We have accumulated evidence to support the hypothesis that mistakes in DNA synthesis made by purified DNA polymerases *in vitro* can be correlated with the initiation and progression of malignancy *in vivo*. The evidence that certain metals are mutagenic and carcinogenic has been considered in previous chapters. The causal relationship between mutagenesis and carcinogenesis forms the basis of the somatic mutation hypothesis. In this chapter, we will focus on alterations in the fidelity of DNA synthesis caused by metal ions. Since studies on fidelity by purified DNA polymerases are carried out in a homogeneous system, it has been possible to investigate the mechanism by which metals induce genetic miscoding.

RELATIONSHIP BETWEEN METAL MUTAGENESIS AND CARCINOGENESIS

In 1929, Boveri (6) postulated that somatic mutations provide the basis for carcinogenesis. Evidence to support this hypothesis is as follows: (a) the target of most carcinogens is DNA (8); (b) most chemical carcinogens, when activated to reactive electrophiles, are mutagenic (2,23,36); (c) malignant changes are frequently associated with chromosomal abnormalities (39); (d) the malignant phenotype is, in nearly all cases, permanent and heritable in cells (40); (e) certain human diseases with defects in DNA repair have an unusually high incidence of malignancy (43,57); and (f) certain inherited diseases are associated with malignancy (24). Most of this evidence can also be used to support an alternative hypothesis: that alterations in the expression of genes in the absence of mutations are the underlying basis for malignancy. Even with this alternative hypothesis, DNA may still be the critical target. Practically, the rapid technology for analyzing mutagenesis *in vitro* and for cloning individual genes has provided the experimental tools for testing the somatic mutation hypothesis.

The observations that most carcinogens can be identified as mutagens provide strong support for the concept of somatic mutations as the basis of carcinogenesis.

Some of the metals that are unequivocal carcinogens (As, Be, Cd, Co, Cr, Mn, Ni, Pb), either in experimental animals or in humans, have been assayed and are mutagenic in bacteria (As, Cr, Mn) or produce chromosomal abnormalities in eukaryotes (As, Cd, Cr, Ni, Pb) (16,18,56). Thus, metal carcinogens are no exception to the general postulate that carcinogens can be detected by their ability to cause mutations.

THE EFFECT OF METALS ON *IN VITRO* DNA SYNTHESIS

The function of divalent metal ions in DNA polymerization and the mutagenicity of certain metals suggests that exposure of cells to mutagenic metal ions could diminish the accuracy of DNA replication. *In vivo* systems are too complicated to begin to unravel the mechanism by which metal ions diminish the fidelity of DNA synthesis, and the mechanisms by which metal ions induce mutations. Our approach to this problem has been to examine DNA synthesis *in vitro*, to determine the effects of different metals ions on the fidelity of this synthesis, and then to ask whether alterations in the fidelity of DNA synthesis are related to the mutagenic and carcinogenic properties of these metals. Prior to considering these studies in detail, it is instructive to consider the mechanism of DNA synthesis *in vitro* and methods for measuring fidelity of DNA synthesis.

MECHANISM OF DNA POLYMERIZATION

Generically, DNA polymerases have similar requirements for catalysis (Fig. 1). Synthesis proceeds by a sequential addition of nucleotide monomers (deoxynucleoside monophosphates) with a concomitant release of pyrophosphate (25). DNA polymerases are part of a unique class of enzymes in that they primarily take direction from another molecule, a template (Fig. 1). In cells, the template is DNA. Only in the case of DNA polymerases from RNA tumor viruses (reverse transcriptases) has RNA been unambiguously demonstrated to serve as a template for DNA synthesis *in vivo*. Synthetic polydeoxynucleotides and polyribonucleotides can serve as templates *in vitro* for most DNA polymerases. Two DNA polymerases, reverse transcriptase (58) and *E. coli* DNA polymerase I (31), are able to copy natural RNA templates *in vitro*. In all cases, synthesis is started on the 3'-hydroxy terminus of a primer-strand hybridized onto a template strand. The primer can be an oligonucleotide, one strand of double-stranded DNA, or a hairpin loop of single-stranded DNA. Thus, DNA polymerases only elongate already existing polynucleotide chains; they cannot initiate chains *de novo* as do RNA polymerases. The substrates of all known DNA polymerases are deoxynucleoside triphosphates that are complementary to the template. Based on the similar requirements for activity and a spectrum of similar kinetic parameters, it is a reasonable expectation that there is a common mechanism for catalysis by DNA polymerases from different sources (28). The lower portion of Fig. 1

Requirements for DNA Polymerization

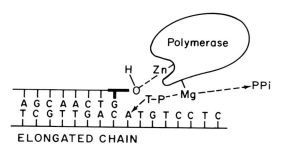

FIG. 1. DNA polymerase reaction (52).

is one of the current models for the detailed coordination of the enzyme, primer, and nucleotide substrate (52).

The added divalent ion Mg^{2+} has been shown to coordinate the enzyme with the substrate in the form of an enzyme-metal-substrate complex (53). Mn^{2+}, Co^{2+}, Ni^{2+} and in certain cases, Zn^{2+} have been shown to substitute for Mg^{2+} as activators (48). Analysis of *E. coli* DNA polymerase I-Mn^{2+}-substrate complexes indicates that in the absence of template, the enzyme alters the conformation of the deoxynucleoside triphosphate substrate to that which it would occupy in double helical DNA (53). The function of the metal activator in the DNA polymerase reaction per se has recently been extensively reviewed (34,35).

Evidence also suggests that DNA polymerases are zinc metalloenzymes. *E. coli* DNA polymerase I has been shown to contain one gm-atom of zinc per mole (55). Removal of the zinc by chelation is accompanied by a proportional loss of DNA polymerase activity. Restoration of the activity occurs upon the readdition of zinc. Similar, but less stringent, criteria indicate that DNA polymerase from avian myeloblastosis virus (AMV) is also a zinc metalloenzyme (41). The presence of zinc has been demonstrated in a variety of homogeneous DNA polymerases. Furthermore, a number of DNA polymerases have been shown to be inhibited by chelators such as o-phenanthroline but not its nonchelating analog m-phenanthroline (34,35). This latter criteria has been questioned by the studies of Sigman and co-workers (11), who found that the inhibition by o-phenanthroline was due to the formation of a complex with copper, which was a contaminant in the thiol compounds present in the assay. Presently, there is no evidence to suggest that interactions with enzyme-bound zinc alter fidelity (3).

FIDELITY OF DNA SYNTHESIS

On the basis of spontaneous mutation rates in prokaryotic and eukaryotic cells, stable misincorporation of a base during DNA replication is estimated to occur with a frequency of 10^{-8} to 10^{-11} per base-pairs synthesized (12). This accuracy appears to be achieved by a multi-step process (60). The differences between correct and incorrect Watson-Crick base-pairings involve only one or two hydrogen bonds. This difference in free energy, ΔG, has been estimated to account for an error rate of approximately 10^{-2} (22,30,34). DNA polymerases also participate in base-selection, reducing the error rate to values approaching 10^{-5} (28). Two different categories of mechanisms have been considered to account for the role of DNA polymerases in enhancing base-selection. (a) Prokaryotic DNA polymerases contain an exonuclease that is able to excise mismatched bases at or immediately after incorporation (7). Such an exonucleolytic activity provides a means by which polymerases can "proofread" for the insertion of incorrect bases during catalysis. (b) Eukaryotic DNA polymerases lack such an exonucleolytic activity (9). Thus, enhanced base-selection by these enzymes must represent increased base-selectivity at the catalytic site. At present, studies on the effects of metals on fidelity *in vitro* are limited to analysis with purified DNA polymerases.

In order to increase the accuracy achieved by DNA polymerases *in vitro* (10^{-3} to 10^{-5}) to the level that occurs during DNA replication *in vivo* (10^{-8} to 10^{-11}), additional base-selective factors must also play a role. Based on the method of catalysis by purified DNA polymerases and their known interactions with other cellular proteins, three hypothetical categories have been proposed to account for the increased accuracy:

1. In accord with the concept of error prevention, correct nucleotide selectivity may be achieved by factors which increase the rigidity of the substrate-template conformation. Candidates for such factors are possibly DNA binding proteins and/or proteins which bind to DNA polymerases.

2. By a repetitive proofreading exonuclease, any level of fidelity theoretically can be achieved. In animal cells, it is possible that exonucleases not part of the polymerase molecule serve such a function.

3. Excision of mis-matched bases by specific endonucleases could greatly enhance fidelity. In this case, the repaired segment would have to be sufficiently small so that additional mistakes would not occur during the repair synthesis of DNA. With the development of an assay for the fidelity of copying natural DNA templates, it should be possible to experimentally evaluate each of these possibilities.

ASSAYS OF FIDELITY WITH POLYNUCLEOTIDE TEMPLATES

Until very recently, all assays of the fidelity of DNA synthesis *in vitro* measured the ability of DNA polymerases to copy homopolymer or alternating copolymer templates. These templates contained only one or two nucleotides and the mismatched nucleotide was identified simply as one not complementary to the template nucleotides. Using this assay, one can observe the effects of both activating and nonactivating metals on the fidelity of DNA synthesis. The template that we have chosen for critical measurements of fidelity is poly [d(A-T)], a synthetic polynucleoside consisting of deoxythymidine and deoxyadenosine monophosphates (1,4). Poly [d(A-T)] can be synthesized to contain less than one in 2×10^6 mistakes using a *de novo* reaction with *E. coli* DNA polymerase I (1). Copied correctly, only dAMP and dTMP should be incorporated into the newly synthesized product. By using $[\alpha\text{-}^{32}P]$-dTTP, unlabeled dATP and $[^3H]$dGTP or $[^3H]$dCTP, one can simultaneously measure the incorporation of complementary and noncomplementary nucleotides (4). The incorporation of either dCTP or dGTP would represent errors. The frequency of misincorporation is obtained from the ratio of $[^3H]$ to $[^{32}P]$ in the acid-insoluble product. Control experiments are required to show that the tritium label in the reaction product is in the noncomplementary nucleotides and not in any radioactive contaminants. Also, it must be demonstrated that the noncomplementary nucleotides are covalently incorporated in phosphodiester linkage. Using nearest-neighbor analysis, one can determine the distribution of the noncomplementary nucleotides.

FIDELITY OF DNA POLYMERASES WITH POLYNUCLEOTIDES

Measurements of the frequency of misincorporation by DNA and RNA polymerases when copying polynucleotide templates carried out in this laboratory are tabulated in Table 1 (32). In these experiments, the complementary and noncomplementary nucleotides were present at equal concentrations corresponding to the Km of the complementary nucleotide. Each assay included activating concentrations of Mg^{2+} and near-saturating concentrations of template. In all cases, incorporation of correct and incorrect nucleotides was shown to be linear with time of incubation and proportional to enzyme concentration. Furthermore, nearest-neighbor analysis indicated that the noncomplementary nucleotide was invariably incorporated as a single-base substitution. The most frequent substitution was a purine for a purine or a pyrimidine for a pyrimidine. From this tabulation, one observes that different DNA polymerases can copy the same templates with differing fidelity. Perhaps the most striking feature of this analysis is that the error rates of the prokaryotic DNA polymerases, those with a $3' \rightarrow 5'$ exonuclease, are similar to those of the eukaryotic DNA polymerases, enzymes that do not have an accompanying exonuclease. Thus, the exonuclease

TABLE 1. *Fidelity of DNA polymerases in copying polynucleotide templates*

DNA polymerase	Template	Noncomplementary nucleotide	Error rate
Prokaryotic DNA polymerase			
E. coli polymerase I	poly [d(A-T)]	dGTP	1/70,000
Bacteriophage T$_4$	poly [d(A-T)]	dCTP	1/12,000
Eukaryotic DNA polymerase			
Sea Urchin nuclei-α	poly [d(A-T)]	dGTP	1/12,000
Human placenta-α	poly [d(A-T)]	dGTP	1/9,000
Human lymphocyte-α	poly [d(A-T)]	dGTP	1/14,000
Calf thymus-α	poly [d(A-T)]	dGTP	1/9,000
Human placenta-β	poly [d(A-T)]	dGTP	1/40,000
Calf thymus-β	poly [d(A-T)]	dGTP	1/30,000
RNA polymerase			
E. coli	poly [d(A-T)]	dCTP	1/2,400
E. coli	poly [d(A-T)]	dGTP	1/42,000
Reverse transcriptase			
Avian myeloblastosis virus	poly(A)·oligo d(T)	dCTP	1/300–800
Avian myeloblastosis virus	poly [d(A-T)]	dGTP	1/3,000
Rous sarcoma virus	poly(C)·oligo d(G)	dATP	1/900

in prokaryotic DNA polymerases is not necessarily the major determinant of fidelity.

In general, DNA polymerase-β from a variety of sources is more accurate than DNA polymerase-α. Interestingly, circumstantial evidence suggests that DNA polymerase-α, the major DNA polymerase activity in eukaryotic cells, participates in DNA replication and that DNA polymerase-β functions in DNA repair (5). The high error rates of DNA polymerases from RNA tumor viruses have facilitated studies of metal ion effects on the incorporation of noncomplementary nucleotides. Because of the high frequency of mistakes, it has been possible to accurately measure the distribution of noncomplementary nucleotides in the reaction product.

EFFECTS OF ACTIVATING METAL IONS ON FIDELITY

Mn^{2+}, Co^{2+}, or Ni^{2+} can substitute for Mg^{2+} as the activating metal for DNA polymerases from animal, viral, and bacterial sources (48). For example, with *E. coli* DNA polymerase I and an activated DNA template, the maximal rates of nucleotide incorporation with Mn^{2+}, Co^{2+} and Ni^{2+} were 65%, 25%, and 7% respectively, of that achieved with Mg^{2+} (51). Minimal activity has also been reported with Zn^{2+}.

The effects of Mg^{2+} and Mn^{2+} concentrations on the incorporation of complementary and noncomplementary nucleotides is illustrated in Fig. 2 for AMV DNA polymerase and a poly (rC) template (13). At activating concentrations of Mg^{2+} (6 mM), AMV DNA polymerase incorporates one molecule of dAMP for every 1,400 molecules dGMP polymerized. This error rate is invariant with

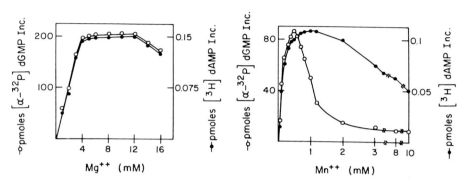

FIG. 2. Effect of Mg^{2+} and Mn^{2+} on the fidelity of DNA synthesis. With poly(rC)·oligo(dG) as a template primer for AMV DNA polymerase, fidelity assays were carried out using [α-^{32}P] dGTP as the complementary nucleotide and [^{3}H]-dAMP as the noncomplementary nucleotide (13). The reaction mixture for a typical fidelity assay using poly [d(A-T)] as a template and dGTP as the noncomplementary nucleotide is as follows: Assays are carried out in a total volume of 0.05 ml containing the following: 100 mM Tris-HCl, pH 7.5; 60 mM KCl; 1.0 mM MgCl₂; 20 μM [α-^{32}P]dTTP (13 dpm/pmol); 20 μM [^{3}H]dGTP or [^{3}H]dCTP, 0.2 mM poly [d(A-T)]; and sufficient purified DNA polymerase to incorporate 300 pmol of the complementary nucleotide. All reactions are performed in triplicate and incubated at 37°C for 15 min. Incorporation of the radioactive deoxynucleoside triphosphates into an acid-insoluble product is determined after repeatedly precipitating the polynucleotide product with 1.0 N perchloric acid, 50 mM sodium pyrophosphate, and solubilizing with 0.2 M NaOH at 20°C as previously described (4). The final precipitate is collected onto glass fiber discs and radioactivity measured by scintillation spectroscopy. After subtracting the amount of radioactivity in nonincubated controls, the error frequency is calculated from the ratio of the noncomplementary to the total complementary nucleotides incorporated.

respect to Mg^{2+} concentration. At the optimal activating concentration of Mn^{2+} (1 mM), the error rate was 1 in 800. At greater than activating concentrations of Mn^{2+}, there was a progressive decrease in the incorporation of the complementary nucleotide but not of the noncomplementary nucleotide, thus yielding a further increase in the frequency of misincorporation. At concentrations as great as 5 mM, the error rate approached 1 in 30, and nearest-neighbor analysis indicated that each misincorporation occurred as a single-base substitution. The decreased fidelity with increased Mn^{2+} concentration has been observed with all templates and noncomplementary nucleotides tested. An absolute increase in the rate of incorporation of the noncomplementary nucleotide can be demonstrated by simply using more DNA polymerase in the assay or prolonging the time of incubation. These data indicate that Mn^{2+} enhances infidelity with AMV DNA polymerase.

A tabulation of the error rates observed with Mg^{2+}, Mn^{2+}, and Co^{2+} using different DNA polymerases is shown in Table 2. These assays were carried out with alternating poly [d(A-T)] as a template and with dGTP as the noncom-

TABLE 2. Effect of metal activator on fidelity

DNA polymerase	Reference	Mg^{2+} (5 mM)	Mn^{2+}		Co^{2+}	
			0.1 mM	2 mM	0.4 mM	5 mM
AMV	47	1/1,680	1/760	1/500	1/1,100	1/200
E. coli I	51	1/20,000	1/10,000	1/1,000	1/7,500	1/7,000
Human placenta-α	46	1/6,000	1/1,900	1/300	1/1,300	1/450
Human placenta-β	46	1/20,000	1/9,000	1/2,000	1/5,000	1/1,300

plementary nucleotide. The increase in infidelity with Mn^{2+} using AMV DNA polymerse, E. coli DNA polymerase I, and DNA polymerases-α and -β is in agreement with the results of other investigations and may reflect a common aspect to the mechanism of fidelity by DNA polymerases. Substitution of Mn^{2+} for Mg^{2+} results in an increase in misincorporation by E. coli DNA polymerase I (60), T_4 DNA polymerase (20) and DNA polymerase-α (27,46). The fact that Mn^{2+} and Co^{2+} alter the fidelity of the DNA polymerases that do not have an associated exonuclease [AMV (45), DNA polymerases-α and -β (9,28)] indicates that these metal ions do not promote misincorporation by inhibiting an error correcting exonucleolytic activity. Ni^{2+} can also substitute for Mg^{2+} as a metal activator. However, the amount of synthesis achieved with Ni^{2+} as the metal activator has not been sufficient to accurately measure the changes in the fidelity of DNA synthesis with any DNA polymerase except E. coli DNA polymerase I, in which case Ni^{2+} promotes misincorporation (47).

In order to relate the measurements with alternate metal activators to a situation that would be expected to occur in cells, the effects of these activators on Mg^{2+}-activated DNA synthesis have been investigated. Co^{2+}, Mn^{2+} and Ni^{2+} have been shown to enhance misincorporation by AMV DNA polymerase (51) and E. coli DNA polymerase I (47) in the presence of activating amounts of Mg^{2+}. Thus, these metal activators could alter the fidelity of DNA polymerases in cells. In contrast to DNA polymerases, the substitution of Mn^{2+} for Mg^{2+} increases the accuracy of RNA synthesis with E. coli RNA polymerase (54). Additional studies concerning the effects of metal activators on fidelity with RNA polymerases are required before any mechanistic differences can be evaluated.

EFFECTS OF NONACTIVATING METAL CATIONS ON THE FIDELITY OF DNA SYNTHESIS

Berylium, a known carcinogen, has been shown to decrease the fidelity of catalysis with M. luteus DNA polymerase (33) and AMV DNA polymerase (49). Be^{2+} is unable to substitute for Mg^{2+} as a metal activator. However, as a nonactivating cation, Be^{2+} alters the fidelity of DNA synthesis in the presence of Mg^{2+}. Preincubation of the enzyme but not the template, primer, or substrates with high concentrations of Be^{2+} resulted in an increased error rate. This finding suggests that Be^{2+} can interact with some noncatalytic site on DNA polymerase and thereby alter the fidelity of DNA synthesis. Be^{2+} has also been shown to alter the fidelity of DNA polymerase-α from human fibroblasts (42), DNA polymerases-α and -β from human placenta and *E. coli* DNA polymerase I (46). Since the eukaryotic and viral DNA polymerases lack an exonuclease, these results mitigate against the possibility that Be^{2+} interacts with the exonucleolytic site on prokaryotic DNA polymerases.

INFIDELITY OF DNA SYNTHESIS: A BIOCHEMICAL SCREENING SYSTEM FOR METAL MUTAGENS AND CARCINOGENS

The experimental results that have been considered on the alteration in the fidelity of DNA synthesis by activating and nonactivating divalent metal cations and the dynamic nature of metal-macromolecular interactions prompted Sirover and Loeb (50) to ask if mutagenic and/or carcinogenic metals could be identified by alterations in the fidelity of DNA synthesis. So far, about 40 metal compounds have been tested in graded concentrations in this cell-free system. The method of analysis and the results are summarized in Fig. 3. Twenty-two of these metal salts were tested in a triple-blind study in which the assays, computations, and designation of each unknown compound with respect to fidelity were carried out independently (50). Compounds which increased infidelity by greater than 30% at two or more concentrations were scored as positive. Metals were designated as carcinogens or mutagens by an evaluation of the literature prior to assessment of their effects on fidelity. An enhancement in the infidelity of DNA synthesis was observed with all of the known mutagens and/or carcinogens tested (Ag, Be, Cd, Co, Cr, Mn, Ni, Pb). The evidence in the literature on the mutagenicity or carcinogenicity of three of the metal cations was considered equivocal. Of these, Cu^{2+} increased misincorporation; Fe^{2+} and Zn^{2+} did not alter fidelity. All other metal salts that were tested were considered to be neither carcinogenic nor mutagenic, and they did not increase misincorporation. Only a few of the metal salts that did not alter fidelity are listed in Fig. 3. If one considers the metals that are questionable as possible mutagens and carcinogens (Cu, Fe, Zn) to have a biological effect that is the opposite of those observed in the fidelity assay, the worst possible situation, the hypothesis that increases in fidelity and mutagenicity/carcinogenicity are independent variables can be rejected with a $p = 1.6 \times 10^{-4}$.

With only a few exceptions, these results have been confirmed by Miyaki et al. (37) and Sirover et al. (47) using *E. coli* DNA polymerase I, and by Seal et al. using DNA polymerases-α and -β from human placenta (46). Preliminary results indicate that neither arsenic (AsO_4, As_2O) nor selenium (SeO_2) diminish fidelity with *E. coli* DNA polymerase I. Furthermore, Se does not reduce the mutagenic effect of Mn in titration experiments containing these two metals ions. Pb^{2+}, Cd^{2+}, Co^{2+}, Cu^{2+}, and Mn^{2+} have been shown to stimulate chain initiation by RNA polymerases (21), whereas Zn^{2+}, Mg^{2+}, Li^+, Na^+, and K^+ were inhibitory. The similarity between the effects caused by particular

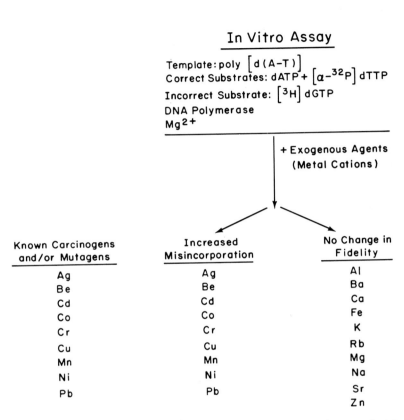

FIG. 3. Summary of results with metal ions using AMV DNA polymerase. Increased misincorporation represents a change of greater than 30% at two or more concentrations of the indicated metal cation.

metal ions on fidelity with DNA polymerases and on chain initiation with RNA polymerase could point to metal interactions with the DNA template as a common underlying mechanism *(vida infra)*.

FIDELITY OF DNA SYNTHESIS WITH NATURAL DNA TEMPLATES

All of the aforementioned studies on the fidelity of DNA synthesis depended on measuring either incorporation of noncomplementary nucleotides using synthetic polynucleotide templates of limited composition, or measuring the incorporation of nucleotide analogs. It has been assumed that the results with these model systems are similar to those that would be obtained copying natural

DNA containing all four bases. It is known, however, that slippage of the primer relative to the template can occur when primed templates of a repeating nucleotide sequence are copied (10,25). Thus, metal-mediated changes in the fidelity of DNA synthesis could result from such slippage of the primer on the template, an event that presumably does not occur during copying of natural DNA templates. Also unique to homopolymers or repeating heteropolymers is the fact that a single noncomplementary nucleotide can occupy a looped-out structure without changing the reading frame of subsequent codons. Thus, metals could enhance misincorporation by increasing the frequency of such looped-out structures. To circumvent these limitations, a system has been recently developed (62) to monitor the fidelity of *in vitro* DNA synthesis using a natural DNA template, DNA from the bacteriophage ΦX174 carrying a suppressible nonsense mutation, amber 3 (*am*3) (Fig. 4). Certain nucleotide substitu-

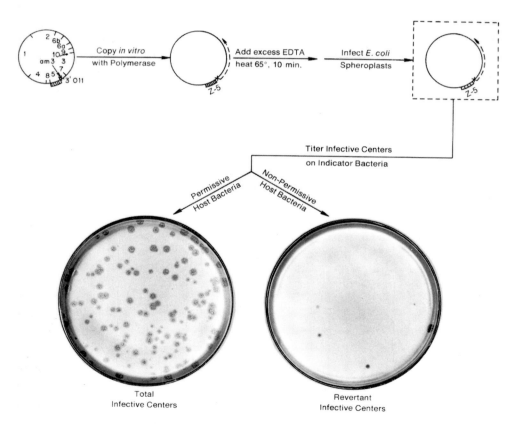

FIG. 4. Fidelity of DNA synthesis with a natural DNA template. The reversion frequency at the *am*3 locus in ΦX174 DNA is the proportion of revertant infective centers compared to the total infective centers. The error rate is calculated from the reversion frequency as previously described (62). Alterations in the fidelity of DNA synthesis are detected by changes of the error rate.

TABLE 3. Fidelity of DNA polymerases copying ΦX174 DNA[a]

Polymerase	Metal activator	Nucleotides per template	Reversion frequency	Error rate
Control	Mg^{2+}	0	2.41×10^{-5}	—
E. coli Pol I	Mg^{2+}	610	3.35×10^{-5}	1/13,800
E. coli Pol I	Mn^{2+}	962	11.7×10^{-5}	1/1,390
E. coli Pol I	Co^{2+}	1,350	5.99×10^{-5}	1/3,730
AMV	Mg^{2+}	500	20.4×10^{-5}	1/590

[a] The indicated DNA polymerases and metal activators were used to copy 6.4×10^{10} molecules of ΦX174 am3 DNA hybridized to Z-5 Hae III restriction fragment. The control DNA was subjected to all of the conditions of the reaction except the DNA polymerase was omitted. Each reaction contained 30 μM of all four deoxynucleoside triphosphates. The metal activators were present as chloride salts: 5.0 mM Mg^{2+}, 1.0 mM Mn^{2+}, and 2.5 mM Co^{2+}. The number of nucleotides per template represents a calculated average, assuming all molecules are copied. In each case, sufficient synthesis was achieved to pass the am3 mutation which is located 83 nucleotides from the Z-5 primer. The reversion frequency was determined from an infectious centers assay of E. coli spheroplasts. The error rate was calculated from the reversion frequency after subtracting the value for uncopied DNA and dividing by penetrance or frequency of expression (62) which in these experiments was 0.13. This data is from Kunkel and Loeb (26).

tions within the am3 locus that occur during in vitro replication of this DNA will cause a reversion to the wild type phenotype. Transfection with the in vitro replicated ΦX174 DNA of E. coli spheroplasts under nonsuppressive conditions permits one to assay for progeny phage revertant for the am3 mutation. Thus, measurement of the reversion frequency of the progeny phage indicates the accuracy with which the DNA in the region of this mutation was copied.

Preliminary estimates on the error rate of DNA polymerases in copying natural DNA templates in vitro are shown in Table 3 (26). With homogeneous AMV DNA polymerase, Mg^{2+}, and equal concentration of nucleotides, the in vitro mutation rate is 1 in 590 ± 300 (19). However, with less purified preparation of AMV DNA polymerase, fewer errors were observed. With E. coli DNA polymerase I, Mg^{2+}, and equal concentrations of nucleotide substrates (30 μM), the reversion rate is 1.4 times that of uncopied DNA and yields a calculated in vitro mutation rate of approximately 1 in 8,000 (45). The calculated error rates for catalysis in the presence of Mn^{2+} or Co^{2+} are higher. Thus, Mn^{2+} and Co^{2+} are mutagenic for the synthesis of biologically active DNA in vitro. By determining the sequence of the products of the reaction synthesized in the presence of Mn^{2+} and Co^{2+}, it should be possible to define the specificity of interactions of these metals with the template nucleotides.

THE MECHANISM OF GENETIC MISCODING BY METALS

The exact mechanism by which certain divalent metal ions decrease the fidelity of DNA synthesis in vitro is not known. On the basis of the available data, three alternatives can be unambiguously eliminated while three others may still be considered viable mechanisms.

The following three possibilities by which metal ions decrease the fidelity of *in vitro* DNA synthesis are no longer tenable mechanisms:

1. *Precipitation of noncomplementary nucleotides.* It can be argued that the observed increase in error frequency at high metal concentration represents the selective acid precipitation of metal ion complexes containing unincorporated noncomplementary nucleotides. However, physical and enzymatic studies of the products synthesized with AMV DNA polymerase (4), *E. coli* DNA polymerase (1), and DNA polymerases-α and -β (46), indicate that the noncomplementary nucleotides are incorporated into a polynucleotide chain, predominantly as single-base substitutions. These studies have involved isolation of the newly synthesized product by sedimentation and determination of nearest-neighbor frequencies. Since the latter procedure depends on enzymatic hydrolysis of phosphodiester bonds, it provides unequivocal evidence that the noncomplementary nucleotides are incorporated into a polynucleotide chain.

2. *Metal-substrate interactions.* Metal-induced infidelity does not appear to result from selective interactions between particular metals and particular nucleotides. For example, it could be argued that Co^{2+} selectively interacts with the noncomplementary nucleotide and reduces its effective concentration in the reaction mixture. Experiments indicate that this is unlikely. At a high concentration of Co^{2+} (5mM), the incorporation of dGTP as the complementary nucleotide with a poly (C) template is markedly inhibited. However, at the same Co^{2+} concentration, the incorporation of dGTP as a complementary nucleotide with poly [d(A-T)] as the template is undiminished (51). Similar results have been obtained with Mn^{2+} using different DNA polymerases and different template combinations.

3. *Inhibition of "proofreading" exonuclease by metal ions.* The possibility that decreases in fidelity with divalent metal ions are mediated by inhibition of $3' \rightarrow 5'$ exonucleolytic activity is also unlikely. Eukaryotic DNA polymerases and DNA polymerases from RNA tumor viruses are devoid of such an activity (28), yet mutagenic metal ions decrease the fidelity of these enzymes. Detailed studies on the effect of Mn^{2+} on fidelity, exonucleolytic activity, and monophosphate generation have been carried out with *E. coli* DNA polymerase I. Under conditions in which Mn^{2+} diminishes fidelity, there is no diminution of the $3' \rightarrow 5'$ exonucleolytic activity (47). More importantly, the effect of Mn^{2+} on nucleoside monophosphate generation is opposite to that which one would predict on the basis of a diminished exonucleolytic activity in that the production of noncomplementary nucleoside monophosphates is 40-fold greater with Mn^{2+} than with Mg^{2+} (29).

The decrease in fidelity of metal ions during *in vitro* DNA synthesis can be explained most directly by any one or more of the types of interactions which are sketched in Fig. 5.

1. *Altered substrate conformation.* The ability of Mn^{2+}, Co^{2+}, Ni^{2+}, and possibly Zn^{2+} to substitute for Mg^{2+} as a metal activator focuses (48) on the possibility that the mechanism of change in fidelity by these metals occurs by a substitution

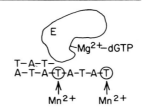

FIG. 5. Possible mechanisms for metal ion induced infidelity of DNA synthesis. (See text.)

at the substrate binding site. Using a variety of DNA polymerases, the frequency of misincorporation at activating concentrations of Mn^{2+} and Co^{2+} is two- and three-fold greater than that observed with Mg^{2+}. Magnetic resonance studies indicate that the interaction of the metal activator involves an enzyme-metal-substrate bridge complex involving the γ-phosphoryl group of the substrate (53). Studies with *E. coli* DNA polymerase I in the absence of template indicate that the bound metal changes the conformation of the substrate to that of the nucleotidyl unit in double-helical DNA. It has been pointed out that this conformation could reduce the frequency of misincorporation (53). On this basis, it can be argued that differences in conformation of the bound substrate with different metal activators might account for differences in the fidelity of DNA synthesis, particularly at activating concentrations of these cations. However, the current data are not sufficient to eliminate the possibility that differences in fidelity reflect interactions of metal cations with the template and not the enzyme even when the metal serves as an activator. Thus, the parallel incorporation of complementary and noncomplementary nucleotides at activating metal concentrations could simply indicate that polymerization is the rate-limiting event and the metal-mediated change in fidelity could be at a site other than the substrate site on the enzyme.

2. *Altered enzyme conformation.* The decrease in fidelity observed at inhibiting concentrations of metal activators suggests binding of metals at sites in addition to the catalytically active site. Ancillary binding sites for Mn^{2+} were detected on *E. coli* DNA polymerase I by nuclear magnetic resonance studies (52). The demonstrations that nonactivating metal cations alter the fidelity of other DNA polymerases is compatible with this concept. Also, evidence has been presented that Be^{2+}, a nonactivating cation, binds to AMV DNA polymerase directly and diminishes the fidelity of DNA synthesis *in vitro* (49). Multiple nucleotide binding sites on *E. coli* DNA polymerase I have been inferred from magnetic resonance (52) and kinetic studies (59). Thus, interactions of metals or metal-nucleotide complexes at distant sites could change the conformation of the polymerase so as to promote misincorporation. So far, attempts to generate an altered DNA polymerase with diminished fidelity by treatment with denaturing agents and heat have not been successful (61). Additional efforts at selective modification of DNA polymerases by tight binding metal complexing agents are required to further define enzyme sites that alter fidelity.

3. *Altered template-base specificity.* The direct interaction of metal ions with phosphates and bases on polynucleotides have been measured by a number of physical techniques (14). Studies on the interaction of Mn^{2+} with activated DNA template by para-magnetic resonance (53) indicate 5 ± two very tight sites and 52 weaker sites having an invariant association constant of 68 μM. The largest decreases in fidelity with Mn^{2+} were observed at much higher concentrations (2–5 mM). Weak Mn^{2+} binding sites on *E. coli* DNA polymerase I have been reported (53). However, it is also possible that very weak binding sites on polynucleotides are responsible for diminished fidelity, and these would not be observed in the magnetic resonance experiments. Eichhorn et al. initially observed that metal cations can cause enhanced mispairing upon renaturation of polynucleotides (14). Conceivably, the metal ions can directly interfere with complementary base-pairing or cause a shift in the keto-enol equilibria of the nucleotide. Recent studies by Murray and Flessel (38) indicate that Mn^{2+} and Cd^{2+} promote mispairing during hydridization of the synthetic templates. Moreover, the mispairing with Mn^{2+} can be demonstrated to occur at millimolar concentrations.

CONCLUSION

The results in this chapter indicate that mutagenic metal ions alter the fidelity of DNA synthesis. This has been demonstrated with purified DNA polymerases using both synthetic and natural DNA templates. We argue that in studying fidelity of DNA synthesis by DNA polymerases, one is studying mutagenesis *in vitro*. Correlations observed between alterations in fidelity *in vitro* and mutage-

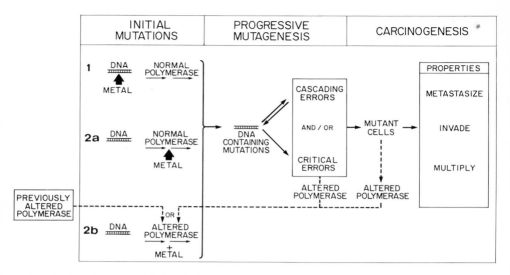

FIG. 6. Speculative models for the involvement of metal ions in mutagenesis and carcinogenesis. Metal ions may interact with (1) the DNA template or (2) the DNA polymerase. (See text.)

nicity or carcinogenicity *in vivo* are in accord with the hypothesis that infidelity during DNA synthesis may cause mutations. However, we recognize that metal ions have many other effects *in vivo*. Considerable evidence will be required to document whether or not alterations in the fidelity of DNA synthesis are causally associated with mutations and malignancy. Irrespective of a defined mechanism, the correlation between alterations in fidelity and mutagenicity and/or carcinogenicity indicates that practicality of using fidelity assays as a screen for evaluating possible mutagens and carcinogens. Since these assays are carried out *in vitro* in defined homogeneous systems, it is possible to design experiments to understand how metals alter the fidelity of DNA synthesis.

With respect to metals, diminished fidelity, and somatic mutations, the following hypothesis can be generated (Fig. 6). Metal-induced mutations may occur by the interaction of metal ions with the DNA template or with the DNA polymerase. In the latter case, a normal polymerase could be exposed to an abnormal concentration of physiologically required metals, or to exogenous metals that are usually not present during cellular metabolism. Alternatively, metal ions that are normally not used for DNA replication could serve as activators for DNA polymerases that have been previously altered. In either case, an abnormal polymerase-metal combination might decrease the fidelity of which the DNA is replicated, and thus lead to the synthesis of DNA containing mutations. This newly synthesized DNA may contain certain critical errors (e.g., genes which code for altered polymerases). Furthermore, continued replication of the DNA by an altered polymerase or in the presence of mutagenic metals could also lead to an accumulation of additional errors during subsequent rounds of replication. Such critical erros and/or cascading errors caused by an accumulation of mutations may account for the progressive change in cellular properties during tumor progression (8,30).

There is a growing realization that environmental agents can be causally associated with mutations and malignancy. One solution to the problem is the removal of these agents from our environment. The studies outlined in this chapter suggest that such an approach may be simplistic. Trace metals have been demonstrated to be required for life, and also to be mutagenic and carcinogenic. One will need a means for making judgments as to what are acceptable environmental levels of particular metals. Such judgments will require an understanding of their requirements for metabolsim as well as their toxicity, mutagenicity, and carcinogenic effects.

ACKNOWLEDGMENTS

This work was supported by Grants CA-24845 and CA-24998 from the National Institutes of Health and by Grant PC776–80439 from the National Science Foundation. R. A. Zakour and T. A. Kunkel were supported by Grants 67–0836 and CA–06168–01, respectively, from the National Institutes of Health. The data from this laboratory represent the previous contributions of Drs. D. K. Dube, G. Seal, M. A. Sirover, and L. A. Weymouth.

REFERENCES

1. Agarwal, S. S., Dube, D. K., and Loeb, L. A. (1979): On the fidelity of DNA replication. Accuracy of *Escherichia coli* DNA polymerase I. *J. Biol. Chem.*, 254:101–106.
2. Ames, B. N., Durston, W. E., Yamasaki, E., and Lee, F. D. (1973): Carcinogens are mutagens: A simple test system combining liver homogenates for activation and bacteria for detection. *Proc. Natl. Acad. Sci. USA,* 70:2281–2285.
3. Battula, N., Dube, D. K., and Loeb, L. A. (1975): Avian myeloblastosis virus DNA polymerase. Kinetic studies on the incorporation of noncomplementary nucleotides. *J. Biol. Chem.*, 250:8404–8408.
4. Battula, N., and Loeb, L. A. (1974): The infidelity of avin myeloblastosis virus deoxyribonucleic acid polymerase in polynucleotide replication. *J. Biol. Chem.*, 249:4086–4093.
5. Bertazzoni, U., Stefanini, M., Noy, G. P., Giulotto, E., Nuzzo, F., Falaschi, A., and Spadari, S. (1976): Variations of DNA polymerases -α and -β during prolonged stimulation of human lymphocytes. *Proc. Natl. Acad. Sci. USA,* 73:785–789.
6. Boveri, T. (1929): *The Origins of Malignant Tumors,* The Williams and Wilkins Company, Baltimore.
7. Brutlag, D., and Kornberg, A. (1972): Enzymatic synthesis of deoxyribonucleic acid. XXXVI. A proofreading function for the 3' $\rightarrow$ 5' exonuclease activity in deoxyribonucleic acid polymerases. *J. Biol. Chem.*, 247:241–248.
8. Cairns, J. (1975): Mutation selection and the natural history of cancer. *Nature,* 255:197–200.
9. Chang, L. M. S., and Bollum, F. J. (1973): A comparison of associated enzyme activities in various DNA polymerases. *J. Biol. Chem.*, 248:3398–3404.
10. Chang, L. M. S., Cassani, G. R., and Bollum, F. J. (1972): Deoxynucleotide-polymerizing enzymes of calf thymus gland. VII. Replication of homopolymers. *J. Biol. Chem.*, 247:7718–7723.

11. D'Aurora, V., Stern, A. M., and Sigman, D. S. (1977): Inhibition of *E. coli* DNA polymerase I by 1, 10-phenanthroline. *Biochem. Biophys. Res. Commun.*, 78:170–176.
12. Drake, J. W. (1969): Comparative rates of spontaneous mutation. *Nature,* 221:1132.
13. Dube, D. K., and Loeb, L. A. (1975): Manganese as a mutagenic agent during *in vitro* DNA synthesis. *Biochem. Biophys. Res. Commun.,* 67:1041–1046.
14. Eichhorn, G. L., and Shin, Y. A. (1968): Interaction of metal ions with polynucleotides and related compounds. XII. The relative effect of various metal ions on DNA helicity. *J. Am. Chem. Soc.,* 90:7323–7328.
15. Enterline, P. E. (1974): Respiratory cancer among chromate workers. *J. Occup. Med.,* 16:523–526.
16. Flessel, C. P. (1978): Metals as mutagens. *Adv. Exp. Med. Biol.,* 91:117–127.
17. Foulds, L. (1954): The experimental study of tumor progression: A review. *Cancer Res.,* 14:327–339.
18. Furst, A. (1978): An overview of metal carcinogenesis. *Adv. Exp. Med. Biol.,* 91:1–12.
19. Gopinathan, K. P., Weymouth, L. A., Kunkel, T. A., and Loeb, L. A. (1979): On the fidelity of DNA replication: Mutagenesis *in vitro* by DNA polymerases from an RNA tumor virus. *Nature,* 278:857–859.
20. Hall, Z. W., and Lehman, I. R. (1968): An *in vitro* transversion by a mutationally altered T_4-induced DNA polymerase. *J. Mol. Biol.,* 36:321–333.
21. Hoffman, D. J., and Niyogi, S. K. (1977): Metal mutagens and carcinogens affect RNA synthesis rates in a distinct manner. *Science,* 198:513–514.
22. Hopfield, J. J. (1974): Kinetic proofreading: A new mechanism for reducing errors in biosynthetic processes requiring high specificity. *Proc. Natl. Acad. Sci. USA,* 71:4135–4139.
23. Irving, C. C. (1973): Interaction of chemical carcinogens with DNA. In: *Medthods in Cancer Research, Volume 7,* edited by Harris Busch, pp. 189–244. Academic Press, New York.
24. Knudson, A. G. (1971): Mutation and cancer: Statistical study of retinoblastoma. *Proc. Natl. Acad. Sci. USA,* 68:820–823.
25. Kornberg, A. (1974): *DNA Synthesis.* W. H. Freeman and Co., San Francisco.
26. Kunkel, T. A., and Loeb, L. A. (1979): On the fidelity of DNA replication: Effect of divalent metal ion activators and deoxyribonucleoside triphosphate pools on *in vitro* mutagenesis. *J. Biol. Chem.,* 254:5718–5725.
27. Linn, S., Kairis, M., and Holliday, R. (1976): Decreased fidelity of DNA polymerase activity isolated from aging human fibroblasts. *Proc. Natl. Acad. Sci. USA,* 73:2818–2822.
28. Loeb, L. A. (1974): Eucaryotic DNA polymerases. In: *The Enzymes,* edited by P. D. Boyer, pp. 173–209. Academic Press, New York.
29. Loeb, L. A., Dube, D. K., Beckman, R. A., and Gopinathan, K. P. On the fidelity of DNA replication: Nucleoside monophosphate generation during polymerization. (In preparation.)
30. Loeb, L. A., Springgate, C. F., and Battula, N. (1974): Errors in DNA replication as a basis of malignant changes. *Cancer Res.,* 34:2311–2321.
31. Loeb, L. A., Tartof, K. D., and Travaglini, E. C. (1973): Copying natural RNA's with *E. coli* DNA polymerase I. *Nature,* 242:66–69.
32. Loeb, L. A., Weymouth, L. A., Kunkel, T. A., Gopinathan, K. P., Beckman, R. A., and Dube, D. K. (1979): On the fidelity of DNA replication. *Cold Spring Harbor Symp. Quant. Biol. XLIII* (in press).
33. Luke, M. Z., Hamilton, L., and Hollocher, T. C. (1975): Beryllium-induced misincorporation by a DNA polymerase. *Biochem. Biophys. Res. Commun.,* 62:497–501.
34. Mildvan, A. S. (1974): Mechanism of enzyme action. *Ann. Rev. Biochem.,* 43:357–399.
35. Mildvan, A. S., and Loeb, L. A. (1979): The role of metal ions in the mechanisms of DNA and RNA polymerases. *CRC Crit. Reve. Biochem.,* 6:219–244.
36. Miller, E. C., and Miller, J. A. (1974): Biochemical mechanisms of chemical carcinogenesis. In: *The Molecular Biology of Cancer,* edited by H. Busch, pp. 377–402. Academic Press, New York.
37. Miyaki, M., Murata, I., Osabe, M., and Ono, T. (1977): Effect of metal cations on misincorporation by *E. coli* DNA polymerases. *Biochem. Biophys. Res. Commun.,* 77:854–860.
38. Murray, M. J., and Flessel, C. P. (1976): Metal-polynucleotide interactions. A comparison of carcinogenic and non-carcinogenic metals *in vitro. Biochim. Biophys. Acta,* 425:256–261.
39. Nowell, P. (1974): Chromosome changes and the clonal evolution of cancer. In: *Chromosomes and Cancer,* edited by J. J. German, pp. 267–287. John Wiley, New York.

40. Pitot, H. C. (1974): Neoplasia: A somatic mutation or a heritable change in cytoplasmic membranes. *J. Natl. Cancer Inst.,* 53:905–911.
41. Poiesz, B. J., Seal, G., and Loeb, L. A. (1974): Reverse transcriptase: Correlation of zinc content with activity (avian myeloblastosis virus/RNA-directed DNA polymerase/RNA virus). *Proc. Natl. Acad. Sci. USA,* 71:4892–4896.
42. Radman, M., Villani, G., Boiteux, S., Defais, M., Caillet-Fauquet, P., and Spadari, S. (1977): On the mechanism and genetic control of mutagenesis due to carcinogenic mutagens. In: *Origins of Human Cancer,* edited by J. D. Watson, and H. Hiatt, pp. 903–922. Cold Spring Harbor Laboratory, Cold Spring Harbor, New York.
43. Robbins, G. H., Kraemer, K. H., Lutzner, M. A., Festoff, B. W., and Coon, H. G. (1974): Xeroderma pigmentosum: An inherited disease with sun sensitivity, multiple cutaneous neoplasms, and abnormal DNA repair. *Ann. Intern. Med.,* 80:221–248.
44. Roy-Chowdhury, A. K., Mooney, T. F., Jr., and Reeves, A. L. (1973): Trace metals in asbestos carcinogenesis. *Arch. Environ. Health,* 26:253–255.
45. Seal, G., and Loeb, L. A. (1976): On the fidelity of DNA replication. Enzyme activities associated with DNA polymerases from RNA tumor viruses. *J. Biol. Chem.,* 251:975–981.
46. Seal, G., Shearman, C. W., and Loeb, L. A. (1979): On the fidelity of DNA replication: Studies with human placenta DNA polymerases. *J. Biol. Chem.,* 254:5229–5237.
47. Sirover, M. A., Dube, D. K., and Loeb, L. A. (1979): On the fidelity of DNA synthesis. VIII. Metal activation of *E. coli* DNA polymerase I. *J. Biol. Chem.,* 254:107–111.
48. Sirover, M. A., Loeb, L. A. (1976): Metal activation of DNA synthesis. *Biochem. Biophys. Res. Commun.,* 70:812–817.
49. Sirover, M. A., and Loeb, L. A. (1976): Metal-induced infidelity during DNA synthesis. *Proc. Natl. Acad. Sci. USA,* 73:2331–2335.
50. Sirover, M. A., and Loeb, L. A. (1976): Infidelity of DNA synthesis *in vitro*: Screening for potential metal mutagens or carcinogens. *Science,* 194:1434–1436.
51. Sirover, M. A., and Loeb, L. A. (1977): On the fidelity of DNA replication. Effects of metals on synthesis with avian myeloblastosis DNA polymerase. *J. Biol. Chem.,* 252:3605–3610.
52. Slater, J. P., Tamir, I., Loeb, L. A., and Mildvan, A. S. (1972): The mechanism of *E. coli* deoxyribonucleic acid polymerase I: Magnetic resonance and kinetic studies of the role of metals. *J. Biol. Chem.,* 247:6784–6794.
53. Sloan, D. L., Loeb, L. A., Mildvan, A. S., and Feldmann, R. J. (1975): Conformation of deoxynucleoside triphosphate substrates on DNA polymerase I from *E. coli* as determined by nuclear magnetic relaxation. *J. Biol. Chem.,* 250:8913–8920.
54. Springgate, C. F., and Loeb, L. A. (1975): On the fidelity of transcription by *E. coli* ribonucleic acid polymerase. *J. Mol. Biol.,* 97:577–591.
55. Springgate, C. F., Mildvan, A. S., Abramson, R., Engle, J. L., and Loeb, L. A. (1973): *Escherichia coli* deoxyribonucleic acid polymerase I, a zinc metallo-enzyme. Nuclear quadrupolar relaxation studies of the role of bound zinc. *J. Biol. Chem.,* 248:5987–5993.
56. Sunderman, F. W., Jr. (1978): Carcinogenic effects of metals. *Fed. Proc.,* 37:40–46.
57. Swift, M., Sholman, L., Perry, M., and Chase, C. (1976): Malignant neoplasms in the families of patients with ataxia telangiectasia. *Cancer Res.,* 36:209–215.
58. Temin, H. M., and Baltimore, D. (1972): RNA-directed DNA synthesis and RNA tumor viruses. *Adv. Virus Res.,* 17:129–186.
59. Travaglini, E. C., Mildvan, A. S., and Loeb, L. A. (1975): Kinetic analysis of *Escherichia coli* deoxyribonucleic acid polymerase I. *J. Biol. Chem.,* 250:8647–8656.
60. Trautner, T. A., Swartz, M. N., and Kornberg, A. (1962): Enzymatic synthesis of deoxyribonucleic acid. X. Influence of bromuracil substititions on replication. *Proc. Natl. Acad. Sci. USA,* 48:449–455.
61. Weymouth, L. A., and Loeb, L. A. (1977): Infidelity of DNA synthesis by temperature-sensitive DNA polymerases from RNA tumor viruses. *Biochim. Biophys. Acta,* 478:305–315.
62. Weymouth, L. A., and Loeb, L. A. (1978): Mutagenesis during *in vitro* DNA synthesis. *Proc. Natl. Acad. Sci. USA,* 75:1924–1928.

Trace Metals in Health and Disease, edited by
N. Kharasch. Raven Press, New York © 1979.

Heavy Atom Labeling in Atomic Microscopy

Michael Beer, Christian Stoeckert, Rex Hjelm, Jr., James Resch, David Tunkel, Robert Hyland, and J. W. Wiggins

Department of Biophysics, Johns Hopkins University, Baltimore, Maryland 21218

Biologists have required knowledge of the organization of living systems at both the cellular and subcellular levels. The principal tool used to gain such necessary information has been the conventional electron microscope. Such electron microscopy has been applied to biological systems for many years with little change, and has provided a great deal of valuable information. However, in recent years important developments in the electron microscope have yielded new approaches to the study of subcellular systems. In this chapter, we shall discuss these new instruments and the novel structural approaches they have stimulated.

THE SCANNING TRANSMISSION ELECTRON MICROSCOPE (STEM)

In all electron microscopes, image contrast reflects some aspects of the interaction of the beam electron with the molecules of the specimen. While conventional electron microscopes yield but the crudest statement about the collisions of the electron with the sample, the STEM, developed by A. V. Crewe at Chicago, permits a far more detailed analysis (5).

Two types of scattering can occur when an electron strikes a specimen molecule (9). In the first type, called elastic scattering, the electron interacts with the much heavier atomic nuclei and undergoes a relatively large angular deviation but a negligible loss in energy. The second type, inelastic scattering, occurs when the beam electron interacts with the electron clouds of the specimen molecules, losing some of its energy and exciting specimen atoms and molecules. Inelastic scattering is associated with a small angular deviation. The STEM allows study of the details of both types of collision events.

During operation of the STEM, electrons emerge from a field emission source, are accelerated by a specially shaped anode, and are focused by lenses to a spot of 2 to 3 Å in diameter. High resolution is achieved with the production of such a narrow beam. Deflection plates are used to move the beam over the specimen, giving a scanning electron microscope with resolution near atomic dimensions. The transmitted electrons which have been elastically scattered, and thus have a large angular deviation, can be collected on a ring detector. The electrons with smaller angular deviations, resulting from inelastic collision

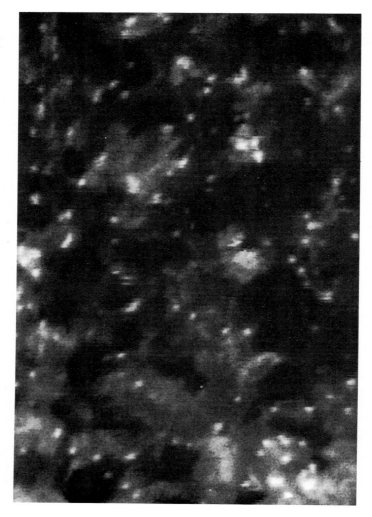

FIG. 1. Mercury atoms adsorbed to a thin carbon support film. The full width of the picture is 186 Å.

events, can pass through the ring detector into a magnetic field, where they are sorted according to their energies. The energies of these electrons can be correlated to the amount of energy lost in the specimen.

Thus, with the STEM, a rather complete statement about the nature of the electron beam-specimen collision can be made. First, the elastically scattered electrons can be separated and their relative intensities measured. Then, the energy losses of inelastically scattered electrons can be determined, immediately inferring the excited energy levels of the specimen molecules. The STEM thus functions in many respects as an incredibly fine spectrometer which can obtain a spectrum of a specimen a few angstroms in diameter and a few tens of angstroms thick.

One of the early successes of the STEM was the imaging of the larger atoms (6). Figure 1 displays this capability with a micrograph of mercury atoms scattered over a carbon film. The ability to image heavy atoms suggests an approach to biological structure determination using an old concept at much greater resolution. This concept is to establish the spatial positions of certain molecules or functional groups within assemblies of molecules by developing chemical reactions that will specifically attach heavy atoms to the functional groups that must be localized. After such reaction, the appropriately labeled structures can be examined with the STEM, and the regions where the heavy atoms are observed indicate the position of the labeled entity.

STEM STUDIES OF NUCLEIC ACIDS

Specific reactions have been developed to elucidate the structures of nucleic acids through heavy atom attachment and STEM analysis. It has been known that the bases of nucleic acids react with osmium tetroxide with some degree of specificity (1,2). In the presence of this reagent, thymidine reacts rapidly while cytidine does so somewhat more slowly. Adenosine and guanosine are not affected by OsO_4. Thus, osmium tetroxide offers a pyrimidine-specific reaction with a rate for thymine higher than that for cytosine. Osmium tetroxide reacts with carbon–carbon double bonds and forms a ring ester. The product is subject to a hydrolysis yielding a glycol which contains no osmium. This hydrolysis is particularly important in acidic medium. However, the intermediate ring ester can react with various ligands—for example, a pyridine—to give a new addition product, formed by binding two ligands per osmium atom. The ligands stabilize the osmium-containing ring ester against hydrolysis. The pyrimidines form osmium–ligand addition products at the 4–5 bond, as diagrammed in Fig. 2. Other ligands which have been used with success are bipyridine and cyanide.

If these reactions are to be utilized in labeling nucleic acids with heavy atoms, it is important to show that the osmium atoms remain bound to the macromolecule and that the integrity of the polymer is not damaged by the labeling reactions. The tight binding of osmium atoms can be inferred from the dramatic increase

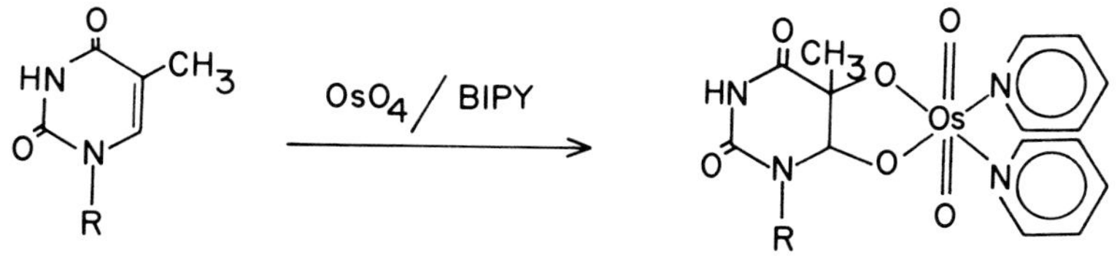

FIG. 2. Reaction of thymine with osmium tetroxide (OsO_4) and a bipyridine ligand.

in density of labeled DNA observed by its behavior on a cesium chloride density gradient. The osmium reactions are not accompanied by extensive fragmentation of the polynucleotide chain, evidenced by survival of ΦX-174 DNA circles after labeling (8). It is possible to produce modified unbroken nucleic acid molecules in which one osmium is tightly bound to each of the pyrimidines and none to the purines.

The first polynucleotide analyzed with the scanning transmission electron microscope after such a labeling procedure was polyuridylic acid. Using well-established procedures for depositing the modified poly-U on specimen supports, STEM micrographs were obtained (4). In such a STEM micrograph (see Fig. 3 *top*), some confusion exists between the organic polymer of low atomic number, which gives broad diffuse density, and the much sharper attached heavy atoms. A computer-filtering procedure can be used to emphasize the positions of the heavy atoms. The computer program eliminates slowly varying densities by plotting the difference in density between any point and the average density of the points' immediate environment. Sharp density changes, associated only with heavy atoms, are emphasized.

Figure 3 *(bottom)* is the computer-filtered image of a STEM micrograph of modified poly-U. The distribution of sharp spots presumably represents the individual osmium atoms attached to the nucleotides of the poly-U chain. The number of atoms per micron is consistent with expectation based on extension of the polynucleotide chain. This can be independently estimated by examining

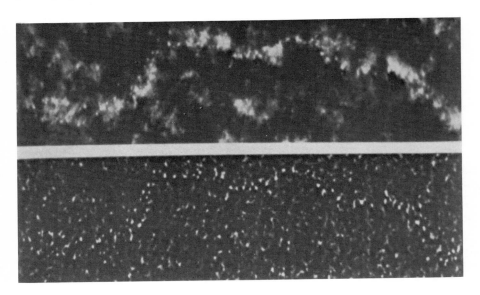

FIG. 3. STEM micrographs of polyuridylic acid reacted with osmium tetroxide and bipyridine to an extent of one atom per base. **Top:** unfiltered image. **Bottom:** filtered image. Total width of micrographs is 630 Å.

single-stranded nucleic acid molecules of known length. Measuring the contour length and knowing the number of bases gives the average spacing.

The use of STEM at such high resolution forces a consideration of the artifacts resulting from the actual act of observation. One method to evaluate this is by taking micrographs of successive scans in the STEM. The degree to which one can superimpose these successive micrographs indicates the extent of movement of the atoms during one scan. Using this method, it was found that most of the displacements of atoms were in the range of 3 Å, some of the movements were as large as 6 Å, but very few beyond 9 Å. On the basis of this and similar studies, one can conclude that such biological investigations with the STEM allow structural statements about chemical groupings down to perhaps 5 to 10 Å.

Several other methods of nucleic acid labeling with heavy atoms have been developed. These procedures involve an initial selective modification of the bases with a metal binding reagent, followed by attachment of an appropriate heavy atom.

It has been shown that O-methylhydroxylamine, CH_3ONH_2, forms specific addition compounds with cytosine in the presence of bisulfite (18). In collaboration with Dr. S. D. Rose, we have adapted this reaction by using a furan ring for the methyl group. This compound accepts two osmium atoms at the furan ring, after addition to cytosine residues. Thus, DNA or RNA can be modified so that each cytosine contains two tightly bound heavy atoms (16).

A final example of base modification prior to heavy atom attachment uses the known reaction of chloracetaldehyde with adenine and cytosine to form an etheno bridge (9,17). The product accepts osmium in a subsequent reaction (12).

A family of specific reactions indeed exists which will permit a variety of labeling patterns, each of which would reflect the underlying nucleotide sequence of the nucleic acid. These reactions were originally developed for an electron microscopic approach to sequencing. Excellent biochemical sequencing methods now exist. Nevertheless, the labeling procedures are expected to permit one to recognize sequences and so identify binding sites of macromolecules with precision. We are presently attempting to find the binding sites of ribosomal proteins with ribosomal RNA using this approach.

STEM STUDIES OF PROTEINS

Procedures are also being developed to attach heavy atom labels to proteins, for structural analysis in the STEM. We have examined the interactions of platinum compounds with macromolecules. Early studies were hampered by crosslinking of the macromolecules by the platinum reagents. This probably resulted from the binding of one platinum atom to two or more separate macromolecules. To avoid this, we deliberately attempted to block three of the four coordination sites on platinum (II) with some chelate, leaving only one site

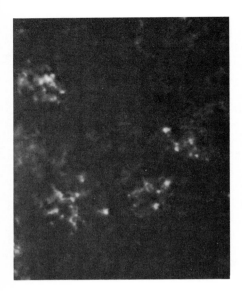 

FIG. 4. STEM micrographs of ribonuclease after labeling with methyl-4-mercaptobutyrimidate followed by Pt(Gly-L-Met)-Cl. Platinum assay indicated five heavy atoms per enzyme molecule. **Left:** unfiltered image. **Right:** computer-filtered image. Total width of micrographs is 200 Å.

for a labeling reaction. A suitable platinum compound is formed when the peptide glycyl-L-methionine is added to K_2PtCl_4 (12). This compound was particularly attractive, because it is stable and because the modified macromolecules retain the solubilities characteristic of biological macromolecules.

Our first attempt at protein modification with heavy atoms was a determination of the number of chelated platinum atoms which could be attached to the simple well-characterized protein, pancreatic ribonuclease. Kinetic studies indicated a saturation level of three bound platinum atoms. The reaction is relatively slow and takes place under mild conditions. It is noteworthy that the enzymatic activity of the modified enzyme is largely retained. This suggests that modification of the protein's tertiary structure is not extensive, a very encouraging result for a reagent required in structural studies (7).

Another approach for binding heavy metals to proteins uses reagents which show some specificity for the proteins and also bind metals. An example is the use of the well-known coupling of imidoesters to the primary amines in lysines. In our first attempt, we used methyl-4-mercaptobutyrimidate (14)—a SH-containing reagent—to label pancreatic ribonuclease. When six SH groups were bound—about half the maximal amount—the enzyme was still about 50% active. Subsequent binding of (Gly-L-Met)Pt to the modified enzyme appears slightly greater than accounted for by the SH present from the imidates, presumably due to a small amount of platinum binding to methionines as well. Most of the platinum, however, is bound to the SH.

In certain structural studies, one must determine the location of a particular protein in a complex of several macromolecules. First, one must establish that metal-labeled proteins are detectable in the STEM. Figure 4 shows ribonuclease molecules after an average of five platinum atoms have been attached.

TOWARD A STEM ANALYSIS OF NUCLEOSOME STRUCTURE

We are using the ability of the STEM to recognize labeled proteins in an analysis of the nucleosome, the packaging unit of DNA and histone proteins which occurs in the chromosomes of all eukaryotes. It is known that nucleosomes are approximately 100 Å in diameter; that each consists of eight histone proteins (two each of four different kinds) and that they contain about 200 nucleotide pairs of DNA. It is further known that the DNA is wound around the outside of the histone core. However, it is not known what the organization of the histones is within the nucleosome core.

We set out to analyze the distribution of the histones in the nucleosome using a new platinum-binding imidate. In this one, methyl-(methylthio) acetimidate, the sulfur is in a thioether form which avoids aggregation of nucleosomes due to S–S bond formation.

On the average, about 10 imidate molecules are attached to each histone when nucleosomes are reacted to saturation. From the earlier ribonuclease stud-

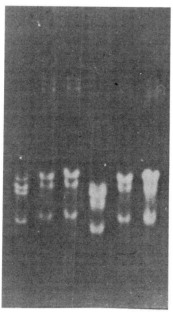

FIG. 5. Acetic acid-UREA gel electrophoresis of histones extracted from control and chromatin reacted with methyl(methy-thio)-acetimidate. (a) and (d) control histones from top to bottom, the histones are H3, H2B, H2A, H4. (b) and (e) 44% imidated histones (c) and (f) 67% imidated histones.

ies, 10 platinum atoms should be sufficient for the localization of a histone. Imidation appears not to disrupt nucleosome structure. Micrographs of reacted chromatin show the beaded structure characteristic of nucleosomes as seen in unreacted samples. Another criterion for the integrity of these nucleosomes is based on the characteristic fragments of DNA which are obtained by nuclease digestion (3). Imidation leaves that pattern of nucleotide fragments largely unaltered.

A gel electrophoretic run of the histones obtained from imidated nucleosomes indicates an even distribution of imidate to each of the four types of histones. After imidation, three of the four histones run as slightly higher in molecular weight (Fig. 5). The relatively tight band indicates a small variation in the number of imidates per histone. The positions of the histone are similar to unreacted histone positions only shifted up. Extensive imidation led to the loss of one histone band but also to the appearance of a family of new bands. It is our guess that these bands represent dimers of histone 3 with the other histones. In the future, nucleosomes will be reconstituted in which one histone is imidated and the remainder are not. In this way, a nucleosome is obtained in which one particular histone can be labeled with platinum. Then, perhaps the position of the labeled histone can be determined.

STEM STUDIES OF POLYSACCHARIDES AND GLYCOPROTEINS

Specific reactions have been developed for the attachment of heavy atom labels to glycols that occur in biological systems, using compounds of osmium (VI) to achieve the same ligand-stabilized osmium-containing ring esters as were used to modify nucleic acids. Osmium (VI) in the form of potassium osmate can be synthesized by a mild reduction of an alkaline OsO_4 solution with ethanol (11). Stable addition products of osmium (VI), some appropriate ligands, and glycols such as those found in sugars could be formed under mild conditions.

Cell membranes are known to be rich in glycoproteins, each of which carries characteristic polysaccharide chains. Certain lipids also contain sugar residues. The distribution of glycoproteins in membranes and the functions of these macromolecules are largely unknown. Procedures are needed to localize the glycoproteins within membranes and follow their behavior during various cell processes, such as cell–cell interactions or capping phenomena. We are hopeful that high resolution microscopy with the STEM, along with specific labeling of sugar glycols with osmium (VI) reagents, can provide such a tool.

The osmium (VI) reaction with sugar glycols in the presence of appropriate ligands was first demonstrated in simple sugars. Thin-layer chromatography was used to isolate and identify these sugar–osmium (VI) adducts (15). In sugars with two available glycol pairs, two products were isolated which corresponded to one osmium atom per sugar and two osmium atoms per sugar. In methyl glycosides, with only one available glycol pair for osmium addition, only one product can be isolated.

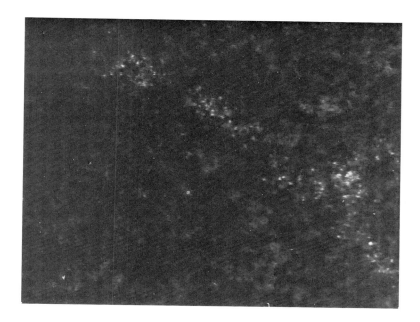

FIG. 6. High resolution scanning transmission electron micrograph of amylose reacted with dicarboxybipyridine and potassium osmate for 48 hr. Osmium assay indicates osmium attachment to 16% of the sugar residues. **Top:** unfiltered image. **Bottom:** computer filtered image. Total width of micrographs is 630 Å.

Cyclodextrins, compounds consisting of six or seven glucose residues forming a ring, react with osmium (VI) compounds as simple sugars do. However, even if large molar excesses of osmium (VI) are used, no more than three atoms of osmium are added to each ring. This suggests that adjacent sugar residues cannot both react, or nearest neighbor exclusion occurs. This hypothesis is supported by space-filling models. Amylose, a polymer of glucose, reacts to a saturation level of 39%, which approximates the value of 43% calculated for random addition and nearest neighbor exclusion.

Amylose was used as a model polysaccharide for heavy atom attachment and STEM analysis. Several ligands were used in our attempts to label this polymer. The first ligand, dicarboxybipyridine, formed a charged addition product with osmium (VI) and amylose. The negative charge acquired made amylose suitable for specimen preparation techniques designed for nucleic acids. Unreacted amylose could not be spread on a carbon grid using such methods, but reaction at even very low levels made specimen preparation with strand extension possible.

The addition of osmium (VI) using the dicarboxybipyridine ligand was very slow. Saturation of amylose with osmium adducts was achieved with a noncharged ligand, tetramethylethylenediamine (TEMED). A neutralized equimolar mixture of TEMED and potassium osmate was allowed to crystallize, and the product TEMED osmate was reacted with amylose. Chemical assay revealed 39% of the sugar residues were labeled with osmium. Since these labeled amylose molecules remained uncharged after TEMED osmate addition, methods for spreading polyanions were not effective in extending this modified amylose on a grid. We used dicarboxybipyridine to provide anionic character and TEMED osmate to increase the labeling level of the amylose. If the STEM is used to examine such extended osmium-labeled amylose at high resolution, one observes characteristic rows of dots which presumably indicate the individual osmium atoms bound to the glucose residues (Fig. 6).

Heavy atom labeling of sugar residues might be used in several ways. The simplest use, perhaps, is to visualize glycoprotein membrane components that have been isolated by some extraction technique. We have isolated, using published techniques, glycophorin from human erythrocytes, one of the best characterized membrane glycoproteins. Labeling glycophorin with osmium (VI) using the TEMED ligand results in a saturation level of 45 to 50 osmium atoms per molecule of glycophorin. We have started to analyze such labeled molecules of glycophorin in the STEM at both low and high resolution.

It is our expectation, though not yet established, that osmium (VI) treatment of membranes as we have done with glycophorin will enable us to detect glycophorin within the membrane. The distribution of these molecules and their possible associations may be better understood using such a technique. Eventually, these methods could be extended to more interesting molecules, perhaps those involved in cell interactions of membrane phenomena such as capping.

Heavy atom labels on glycoproteins, in combination with high resolution electron microscopy, may visualize these molecules while engaged in their given tasks.

REFERENCES

1. Beer, M., Stern, S., Carmalt, D., and Mohlhenrich, K. H. (1966): Determination of base sequence in nucleic acids with the electron microscope. V. The thymine-specific reactions of osmium tetroxide with deoxyribonucleic acid and its components. *Biochemistry,* 5:2283–2288.
2. Burton, K., and Riley, W. T. (1966): Selective degradation of thymidine and thymine deoxynucleotides. *Biochem. J.,* 98:70–77.
3. Camerini-Otero, R. D., Sollner-Webb, B., and Felsenfeld, G. (1976): The organization of histones and DNA in chromatin: evidence for an arginine-rich histone kernel. *Cell,* 8:333–347.
4. Cole, M. D., Wiggins, J. W., and Beer, M. (1977): Molecular microscopy of labeled polynucleotides: stability of osmium atoms. *J. Mol. Biol.,* 117:387–400.
5. Crewe, A. V., and Wall, J. (1970): A scanning microscope with 5 Å resolution. *J. Mol. Biol.,* 48:375–393.
6. Crewe, A. V., Wall, J., and Langmore, J. (1970): Visibility of single atoms. *Science,* 168:1338–1340.
7. Germinario, L. T., Reed, R., Cole, M. D., Rose, S. D., Wiggins, J. W., and Beer, M. (1978): Structural analysis of macromolecular assemblies with STEM. *Scanning Electron Microscopy,* 1:69–76.
8. DiGiamberardino, L., Koller, T., and Beer, M. (1969): Electron microscopic study of the base sequence in nucleic acids. IX. Absence of fragmentation and of cross-linking during reaction with osmium tetroxide and cyanide. *Biochim. Biophys. Acta,* 182:523–529.
9. Kochetkov, N. K., Shibaev, V. N., and Kost, A. A. (1971): New reaction of adenine and cytosine derivatives, potentially useful for nucleic acids modification. *Tetrahedron Latt.,* 22:1993–1996.
10. Lenz, F. (1954): Zur Streung mittelschneller Elektronen in Kleinste Winkel. *Z. Naturforsch,* 9a:185–204.
11. Lott, K. A. K., and Symons, M. C. R. (1960): Structure and reactivity of the oxyanions of transition metals. X. Sexivalent ruthenium and osmium. *Chem. Soc.,* pp. 973–976.
12. Marzilli, L. G., Hanson, B. E., Kapili, L., Rose, S. D., and Beer, M. (1979): Osmium-labeled polynucleotides: reaction of osmium with poly-1, N^6-ethenoadenylic acid. *Bioinorg. Chem.,* 8:531–534.
13. Mogilevkina, M. F., Revein, L. B., Rar, V. I., Cheremisina, I. M., and Logivinenko, V. A. (1976): Isomeric compounds of platinum (II) with glycyl-DL-methionine and DL-a-alanyl-DL-methionine. *Russ. J. Inorg. Chem.,* 21:1345–1348.
14. Ottensmeyer, F. D., Whiting, R. F., and Korn, A. P. (1975): Three-dimensional structure of herring sperm protamine Y-I with the aid of dark field electron microscopy. *Proc. Natl. Acad. Sci. USA,* 72:4953–4958.
15. Resch, J., Tunkel, D., Stoeckert, C., and Beer, M. (1979): Osmium-labeled polysaccharides for atomic microscopy. *(in press).*
16. Rose, S. D., and Beer, M. (1979): Modification of cytosine nucleotides: preparation and metalation of a furyl derivative. *Bioinorg. Chem.,* 9:231–243.
17. Secrist, J. A., Barrio, J. R., Leonard, N. J. (1972): A fluorescent modification of adenosine triphosphate with activity in enzyme systems: 1,N^6-ethenodenosine triphosphate. *Science,* 175:646–647.
18. Verdlov, E. D., and Budowsky, E. I. (1974): Modification of cytitine residues with a bisulfite-0-methylhydroxylamine mixture. *Biochim. Biophys. Acta,* 340:153–165.

The Role of Zinc in Prenatal and Neonatal Development

Lucille S. Hurley

Department of Nutrition, University of California, Davis, California 95616

Zinc is a required element for plants and animals, including man. One of the most striking effects of zinc deficiency is its effect on growth retardation. Rats given a zinc-deficient diet from the time of weaning show very little growth, as well as alopecia (loss of hair) and dermatitis (37). The dermatitis is especially apparent in skin lesions around the mouth and paws. In monkeys, zinc deficiency also produces the characteristic dermatitis on the face, around the eyes, on the limbs, and abdomen (38,40). In humans, zinc deficiency in an extreme form also produces a severe dermatitis. In the genetic disease called acrodermatitis enteropathica, which has recently been shown to be a disorder of zinc metabolism, there is severe dermatitis around the face, mouth, and in the diaper region, as well as diarrhea (33). Treatment with zinc brings about disappearance of all signs of the zinc deficiency (35).

If zinc deficiency is imposed upon an animal during its prenatal life rather than its growth period, however, there are very striking effects on the development of the embryo and fetus. In completely normal rats, fed a normal diet, and then subjected to a zinc deficiency regime throughout the period of pregnancy, half of the fertilized embryos died during pregnancy. Of the other half, those who survived to term, the number of young was less than normal and their body weight was smaller; they weighed about half the size of normal full-term fetuses. Furthermore, from 90–100% of them showed gross malformations. Even shorter periods of zinc deficiency during pregnancy resulted in congenital malformations. When the dietary deficiency occurred during the first 10 days of pregnancy only, as many as 22% of the full-term fetuses showed malformations. When the animals were deficient from day 6 to day 14 of gestation, half of their full-term fetuses were malformed. Thus, zinc deficiency during pregnancy had a rapid and very dramatic effect on the development of the offspring (4,25,29).

All organ systems of the body could be affected by zinc deficiency. For example, 47% of full-term fetuses showed brain malformations, 42% had small or missing eyes, 74% had small or missing lower jaw, 64% had syndactyly, and 72% had tail abnormalities. Fifty-four percent of the fetuses at term showed lung malformations. Because of this, we were interested to see whether there might be some biochemical abnormalities of the lungs as well as these morphological malformations. We therefore investigated the development of surfactant chemicals in the lungs of the rat. Surfactant is necessary for the function of the lungs during neonatal life in order to keep the alveoli of the lungs distended after the infant takes its first breath. If the surfactant is not present in normal concentration, the lungs collapse after each breath—a clinical problem in premature infants. Since the major component of lung surfactant is lecithin, we measured lecithin in the lungs at days 17 through 21 of gestation in the fetuses of female rats fed the normal control diet *ad libitum,* or the zinc deficient diet, or the normal control diet in amounts restricted to those eaten by the zinc-deficient group. Zinc deficiency in itself causes a reduction of food intake, so that one must always be careful to make sure that the effects one is observing are due to the zinc deficiency itself, and not to the restricted food intake.

In both groups of controls, there was a considerable increase in the surfactant level between 19, 20, and especially between days 20 and 21, but this was very much retarded in the zinc-deficient animals. There was thus a significantly lower than normal concentration of lecithin in the zinc-deficient fetuses at days 20 and day 21 of gestation. This was reflected also in the ratio of lecithin to sphingomyelin, which are the two major components of lung surfactant and appear in the amniotic fluid. The ratio of lecithin to sphingomyelin in the amniotic fluid of normal control rats from day 17 to day 21 of gestation was very different from that in the zinc-deficient rats at the same period. On day 19, there was a very sharp rise in the ratio in the normal animals, but not in the deficient ones (45). Thus, zinc deficiency produced abnormal development not only at the morphological level, but also at the biochemical level, so that biochemical development, at least in the lung, was not normal as a result of zinc deficiency during pregnancy.

These changes that I have been describing in morphological development so far have concerned only differences at term. But these changes are also observable at earlier stages of development. At day 14 of gestation, for example, differences are apparent between normal and zinc-deficient embryos. In the zinc-deficient embryo, there are already some abnormalities of the face, and the limbs are not at the normal stage. Even earlier, changes can be seen in zinc-deficient embryos (27). Preimplantation embryos at the blastocyst stage from the zinc-deficient females, after only 4 days of the zinc deficiency regime, are clearly abnormal. Even 1 day earlier, after 3 days of the experimental conditions, there are already abnormalities in the preimplantation egg. Cell division does not occur normally (28).

Thus, the effect on the embryos of maternal zinc deficiency in rats is very rapid. We have also now shown that in the uterine fluid of the rat at this day of gestation, after 4 days of a zinc-deficient diet, the zinc concentration of the uterine fluid is significantly lower than normal. This means, of course, that the egg, even at this early stage, in order to have normal development must derive zinc from the uterine and oviductal fluids in which it is bathed (16).

How can this happen so quickly? The reason it happens so quickly is that the plasma level of zinc of the mother falls very rapidly when the mother is subjected to the zinc deficiency regime. After only 1 day of dietary zinc deficiency, the plasma zinc level falls to almost half its original value and then continues to go down. In the controls, there is a drop during the last third of pregnancy which also occurs in pregnant women, but we have no explanation for that fall. Since the only source of zinc for the developing embryo and the fetus is the plasma zinc of the mother, this sharp fall in the plasma zinc level of rats subjected to zinc deficiency explains the very rapid effect on the embryo and the fetus (7,30,31,43).

We were also interested in establishing the mechanism by which these very profound disturbances of embryonic development can occur. The early effects of zinc and the types of malformations and their incidence led us to believe that some of the basic problems produced by zinc deficiency occurred at a very early stage in development. In addition, the variety of these effects, which altered every organ system, also suggested that the basic defect must be occurring at a very fundamental level. Because of this work, as well as the work of others (15), we thought that zinc deficiency might be affecting the synthesis of nucleic acids and this, in turn, would influence the development of the embryo. The first experiment that we did in this respect simply involved the uptake of tritiated thymidine in rat embryos on day 12 of gestation (41). The zinc-deficient embryos showed a very markedly lower uptake of tritiated thymidine than did the controls, and this could be prevented by prior injection of zinc to the mother (12). Further work also suggested that some specific enzymes related to DNA synthesis might be involved. Thymidine kinase is an enzyme which is important for the synthesis of DNA in rapidly proliferating tissues, and we therefore investigated the thymidine kinase activity in embryos of normal *ad libitum* fed and restricted intake controls, as well as of zinc-deficient females, on days 9 through 12 of gestation (6,9). On all of these days, the zinc-deficient embryos showed significantly lower activity of thymidine kinase than did either of the controls. Previous work by Duncan and Dreosti (8), using regenerating liver as a model system, showed that there was a depression in thymidine kinase activity before there was any depression in DNA synthesis, or in protein synthesis. This suggested that the depression in thymidine kinase activity might be a causative factor in producing the depression of DNA synthesis and, consequently, depression of protein synthesis.

Another enzyme that is involved in DNA synthesis is DNA polymerase. The activity of this enzyme also was significantly lower in zinc-deficient embryos than in controls, but the degree of reduction in the enzyme activity was not as great as it was with thymidine kinase (9).

Another effect of zinc deficiency is on chromosomal aberrations. We found significantly more chromosomal aberrations in zinc-deficient rats than in controls, both in maternal bone marrow and in fetal liver (1).

To summarize, we believe that a possible mechanism of teratogenesis in zinc deficiency comes about through impaired synthesis of nucleic acids. This, in turn, leads to alterations in differential rates of growth which are essential for normal embryonic morphogenesis. These alterations in differential rates of growth lead to asynchrony in histogenesis and organogenesis. Because of the very tight synchronization that is necessary in these processes in order to produce a normal end product—normal morphogenesis, asynchrony could lead to malformation. In addition, there are chromosomal aberrations, which may or may not be related to the impaired synthesis of nucleic acids. This is our present view of the effect of zinc deficiency in producing malformation. The work of Falchuk and Vallee, showing that zinc is required for every stage of the cell cycle in *Euglena gracilis* (14), is also consistent with this view.

So far I have been talking about very extreme conditions of zinc deficiency. It was necessary to use severe conditions in order to find out what zinc deficiency could do, establishing a limit. We were also interested, however, in determining the effects of milder deficiencies of zinc during pregnancy in order to have a situation more relevant to human problems. One way of doing this was to use an experimental procedure in which the period of zinc deficiency was short, only from day 6 to day 14 of gestation. With this kind of experimental regime, the plasma zinc level of the mother falls very rapidly, remains low during the time the deficient diet is fed, and then immediately goes back to normal and even above normal levels when the normal diet is fed. In other words, the embryos are subjected to zinc deficiency only during this period, day 6 to day 14 of gestation. Under these conditions, there was a very high rate of stillbirths. Thirty-five percent of the fetuses were born dead, but those that were alive were only a little smaller than normal. However, there was a very marked effect on their postnatal survival; only 18% of those born alive survived to weaning, compared to 88% of the controls (26). This experiment shows very clearly that even a relatively short period of zinc deficiency during pregnancy had profound and long-lasting effects on the development of the offspring.

Another way of investigating the effects of milder conditions of zinc deficiency was to look at the influence of various levels of zinc in the maternal diet. When we did that, we found, for example, that with 9 ppm of dietary zinc there were no malformations at all, but the survival of the offspring was extremely poor (23). We wanted then to determine whether the effect of the deficiency was on the mother, on her ability to nurse the offspring, on the development of the mammary gland, or on the neonates themselves. We investigated this question by an experiment in which we cross-fostered the offspring, using mothers who were fed during pregnancy either a stock diet or a diet containing 9 ppm zinc, our level of marginal deficiency. All of the mothers were fed a stock diet from the time of parturition; the only variable was the prenatal diet of the mother. When the prenatal diet of the natural mother was stock diet, and the prenatal diet of the foster mother was the marginally zinc-deficient diet, only 50% of the offspring survived to 6 weeks. When it was reversed—that is, when the prenatal diet of the natural mother was marginally zinc-deficient and the prenatal diet of the foster mother was the stock diet, survival was about the same, 45%. When the prenatal diet of the natural mother and the prenatal diet of the foster mother were both marginally deficient, survival was also 46%. Thus, the prenatal diet of the mother affected the development of her offspring, and it also affected the development of the mother herself, that is, her ability to nurse the offspring. Both of these factors seemed to be of about equal importance. One could not compensate for the other (22).

Another aspect of our interest in zinc is in interactions of zinc with other factors. In terms of human problems, the kind of extreme zinc deficiency that I have been describing in our experimental work is unlikely, but there may be interactions of mild, or marginal, zinc deficiencies with other factors, which promote deleterious effects on the development of the offspring.

One such interaction that we have investigated is the effect of ethylene-diamine-tetraacetic acid (EDTA) in development. The full-term fetuses of rats given EDTA in fairly large amounts in the diet are very similar in appearance to zinc-deficient fetuses. With 3% EDTA in the diet, 100% of the implantation sites were affected—that is, the fetus was either malformed or dead. The full-term fetuses were also much smaller than normal. All of the deleterious effects of EDTA on the fetuses could be completely prevented by adding zinc to the diet (39).

Another type of interaction that we have investigated relates to the genetic background of the animals. The A/J strain of mice have a rather high incidence of cleft lip or cleft palate. When A/J females were given a stock diet, 13% of their offspring showed cleft lip or cleft palate (CLP), and 21% showed malformations including CLP and other anomalies as well. The same results were seen with the purified control diet. When A/J females were given the marginally deficient zinc diet, however, the incidence of CLP remained the same, but the incidence of other malformations was higher (22). This did not occur with a hybrid strain of mice. Thus, this genetic strain of mice is more sensitive to a marginal deficiency of zinc than are the hybrid animals. We have also found that A/J mice are also more sensitive to a high level of zinc than are the hybrid mice. Both zinc deficiency and zinc excess can produce deleterious effects, and both effects are more pronounced in the A/J strain than in the hybrids (44).

Thus, there are interactions of zinc nutrition with genetic factors as well as with EDTA and other chemicals. We also know that there are interactions of zinc with other metals as well as with other nutrients, such as vitamin A (10). It is possible that all of these interactions may work in a multifactorial fashion to produce abnormalities in humans (20,21). Recently, a study in Sweden suggested that there was a correlation between the incidence of congenitally malformed infants and low plasma zinc levels in their mothers (32).

I would like now to turn to some of our more recent work with neonatal effects, which stemmed from observations of others concerning the genetic disease acrodermatitis enteropathica (AE). One of the most interesting aspects of this disease is that it was known for over 30 years that this condition could be alleviated and the symptoms could be almost completely cured by feeding these infants breast milk (2,3). Nobody knew why, however. When it became known, just a few years ago, that the problem in AE was one of zinc metabolism (33), we thought there had to be something different between cow's milk and human milk in the way zinc was bound, and its availability to the infant. We investigated this hypothesis by separating samples of human milk and cow's milk by gel filtration. In cow's milk, most of the zinc comes out in a high molecular weight fraction. In human milk, there is a low molecular weight fraction which is present only in a small amount in bovine milk (13).

Our hypothesis also went on to state that in human infants this zinc-binding ligand (ZBL) might enhance the absorption of milk during the neonatal period. We thought that possibly there were mechanisms for the intestinal absorption of zinc which did not become mature until after birth. Using the same gel filtration techniques, we found a similar type of compound in the intestine of the adult rat. However, in the newborn rat, this material was not present and did not appear definitely until day 16 of postnatal life. Thus, there appears to be an intestinal ZBL of low molecular weight in the adult rat which is not present in the newborn and which appears at about 16 days of age (24).

We then did some functional studies (11). We gave zinc-65-chloride to suckling rats at day 18 of postnatal life and found in the intestine two peaks for zinc, a high molecular weight peak and a low molecular weight peak. In the 10-day old rat, however, before the appearance of the intestinal ZBL, there was only the one high molecular weight peak; the low molecular weight peak was absent. We also did absorption studies in which we used free zinc-65-chloride and also labeled various fractions of rat milk, human milk, and bovine milk with zinc-65. These were fed to baby rats at 10 days and at 18 days of age, and after removing the gastrointestinal tract, we measured whole body zinc-65, an indication of the absorption of the zinc into the body of the baby rat. Rat milk ZBL, either *in vivo* labeled or *in vitro* labeled, enhanced the absorption of zinc very significantly. Bovine milk fractions did not change absorption at all; it was the same as with zinc chloride. Absorption of zinc from human milk ZBL fractions was intermediate.

At 18 days of age, however, there were no differences. Whichever fraction of milk we gave, the absorption of zinc was the same. We believe this indicates that the presence of the intestinal ZBL allowed the absorption of zinc, whether it was in the free chloride form or in bovine milk fractions, or the rat ZBL, absorption was just the same. Furthermore, in the intestinal fractions of 10-day old rats given zinc-65-chloride, or bovine milk fractions, there were no counts in the ZBL fraction, whereas with rat milk or human milk ZBL, there were counts in this fraction. At 18 days of age, no matter which fraction was fed, the counts in the intestinal ZBL were the same. This provides evidence that the intestinal low molecular weight zinc-binding fraction does play a role in the absorption of zinc in the intestine.

But what is the possible importance of zinc absorption in the neonate? What difference does it make? We know from rat studies that zinc deficiency during lactation, the suckling period in rats, is extremely important. The suckling rats become deficient very quickly and their survival is much diminished (34). In terms of humans, we know that zinc deficiency does occur. Hypogonadal dwarfism occurs in fairly high incidence in the Middle East (36). In this country, Hambidge and his co-workers have shown that the hair zinc level in infants is very low at 1 year of age—in fact, it *falls* significantly from a concentration similar to that of adults to one year of age, whereas in certain other countries, this does not occur. In Thailand, for example, hair zinc at one year of age is at the same level as it is at birth. Similarly, plasma zinc levels in infancy are lower in this country than they are in adults, whereas this is not the case in Sweden or in Germany (17–19). This is consistent, in my opinion, with the idea that the relatively low incidence of breast feeding in this country has resulted in this effect. Furthermore, these investigators have shown that zinc supplementation of infant formulas produced significantly higher growth of male infants than in those fed unsupplemented formula (46). Preliminary data also indicate that even in infants fed zinc-supplemented formulas, the plasma zinc levels are not as high as they are in breast-fed infants.

In conclusion, I think we have shown that zinc is an important element essential for normal development of the embryo and the fetus and of the newborn as well.

REFERENCES

1. Bell, L. T., Branstrator, M., Roux, C., and Hurley, L. S. (1975): Chromosomal abnormalities in maternal and fetal tissues of magnesium or zinc deficient rats. *Teratology,* 12:221–226.
2. Danbolt, N. (1948): Acrodermatitis enteropathica. *Acta derm.-vener., Stockh.,* 28:532–543.
3. Danbolt, N., and Closs, K. (1942): Akrodermatitis enteropathica. *Acta. derm.-vener., Stockh.,* 23:127–159.
4. Diamond, I., and Hurley, L. S. (1970): Histopathology of zinc-deficient fetal rats. *J. Nutr.,* 100:325–329.
5. Diamond, I., Swenerton, H., and Hurley, L. S. (1971): Testicular and esophageal lesions in zinc-deficient rats and their reversibility. *J. Nutr.,* 101:77–84.
6. Dreosti, I. E., and Hurley, L. S. (1975): Depressed thymidine kinase activity in zinc deficient rat embryos. *Proc. Soc. Exp. Biol. Med.,* 150:161–165.
7. Dreosti, I. E., Tao, S., and Hurley, L. S. (1968): Plasma zinc and leukocyte changes in weanling and pregnant rats during zinc deficiency. *Proc. Soc. Exp. Biol. Med.,* 127:169–174.
8. Duncan, J. R., and Dreosti, I. E. (1975): The site of action of zinc in DNA synthesis. *Proc. S. Afr. Biochem. Soc.,* 1:52.
9. Duncan, J. R., and Hurley, L. S. (1978): Thymidine kinase and DNA polymerase activity in normal and zinc deficient developing rat embryos. *Proc. Soc. Exp. Biol. Med.,* 159:39–43.
10. Duncan, J. R., and Hurley, L. S. (1978): An interaction between zinc and vitamin A in pregnant and fetal rats. *J. Nutr.,* 108:1431–1438.
11. Duncan, J. R., and Hurley, L. S. (1978): Intestinal absorption of zinc: a role for a zinc-binding ligand in milk. *Am. J. Physiol.,* 235:E556–559.
12. Eckhert, C. D., and Hurley, L. S. (1977): Reduced DNA synthesis in zinc deficiency: Regional differences in embryonic rats. *J. Nutr.,* 107:855–861.
13. Eckhert, C. D., Sloan, M. V., Duncan, J. R., and Hurley, L. S. (1977): Zinc binding: A difference between human and bovine milk. *Science,* 195:789–790.

14. Falchuk, K. H., Fawcett, D. W., and Vallee, B. L. (1975): Role of zinc in cell division of *Euglena gracilis. J. Cell. Sci.,* 17:57.
15. Fujioka, M., and Lieberman, I. (1964): A Zn^{++} requirement for synthesis of deoxyribonucleic acid by rat liver. *J. Biol. Chem.,* 239:1164.
16. Gallaher, D., and Hurley, L. S. (1979): Zinc concentration of rat uterine fluid after brief dietary zinc deficiency. *(In preparation.)*
17. Hambidge, K. M., and Walravens, P. A. (1976): Zinc deficiency in infants and preadolescent children. In: *Trace Elements in Human Health and Disease,* edited by A. S. Prasad, p. 29. Academic Press, New York, New York.
18. Hambidge, K. M., Walravens, P. A., Kumar, V., and Tuchinda, C. (1974): Chromium, zinc, manganese, copper, nickel, iron, and cadmium concentrations in the hair of residents of Chandigarh, India, and Bangkok, Thailand. In: *Trace Substances in Environmental Health,* edited by D. D. Hemphill, p. 39. University of Missouri Press, Columbia, Missouri.
19. Hambidge, K. M., Walravens, P. A., and Neldner, K. H. (1977): Zinc and copper in acrodermatitis enteropathica. *Trace Element Metabolism in Man and Animals—III.* Freising-Weihenstephan, West Germany.
20. Hurley, L. S. (1975): Trace elements and teratogenesis. In: Symposium on biochemical and nutritional aspects of trace elements. *Med. Clin. N. Am.,* 60:771–778, AAAS, New York, New York.
21. Hurley, L. S. (1976): Interaction of genes and metals in development. In: Symposium on Interaction of Nutritional and Genetic Factors. *Fed. Proc.,* 35:2271–2275.
22. Hurley, L. S. (1976): Zinc deficiency in prenatal and neonatal development. In: *Zinc Metabolism: Current Aspects in Health and Disease,* edited by G. Brewer and A. Prasad, pp. 47–58. Alan R. Liss, Inc., New York, New York.
23. Hurley, L. S., and Cosens, G. (1974): Reproduction and prenatal development in relation to dietary zinc level. In: *Second International Symposium on Trace Element Metabolism,* edited by W. G. Hoekstra, J. W. Suttie, H. E. Ganther, and W. Mertz, pp. 516–518. University Park Press, Baltimore, Maryland.
24. Hurley, L. S., Duncan, J. R., Sloan, M. V., and Eckhert, C. D. (1977): Zinc binding ligands in milk and intestine: A role in neonatal nutrition? *Proc. Natl. Acad. Sci. (USA),* 74:3457–3459.
25. Hurley, L. S., Gowan, J., and Swenerton, H. (1971): Teratogenic effects of short-term and transitory zinc deficiency in rats. *Teratology,* 4:199–204.
26. Hurley, L. S., Mutch, P. B. (1973): Prenatal and postnatal development after transitory gestational zinc deficiency in rats. *J. Nutr.,* 103:649–656.
27. Hurley, L. S., and Shrader, R. E. (1972): Congenital malformations of the nervous system in zinc-deficient rats. In: *Neurobiology of the Trace Metals Zinc and Copper,* edited by C. C. Pfeiffer, pp. 7–51. Academic Press, Inc., New York, New York.
28. Hurley, L. S., and Shrader, R. E. (1975): Abnormal development of preimplantation rat eggs after three days of maternal dietary zinc deficiency. *Nature,* 254:427–429.
29. Hurley, L. S., and Swenerton, H. (1966): Congenital malformations resulting from zinc deficiency in rats. *Proc. Soc. Exp. Biol. Med.,* 123:692–697.
30. Hurley, L. S., and Swenerton, H. (1971): Lack of mobilization of bone and liver zinc under teratogenic conditions of zinc deficiency in rats. *J. Nutr.,* 101:597–604.
31. Hurley, L. S., and Tao, S. (1972): Alleviation of teratogenic effects of zinc deficiency by simultaneous lack of calcium. *Am. J. Physiol.,* 222:322–325.
32. Jameson, S. (1976): Effects of zinc deficiency in human reproduction. *Acta Medica Scand.,* Suppl. 593.
33. Moynahan, E. J., and Barnes, P. M. (1973): Zinc deficiency and a synthetic diet for lactose intolerance. *Lancet,* 1:676–677.
34. Mutch, P. B., and Hurley, L. S. (1974): Effect of zinc deficiency during lactation on postnatal growth and development. *J. Nutr.,* 104:828–842.
35. Neldner, K. H., and Hambidge, K. M. (1975): Zinc therapy for acrodermatitis enteropathica. *New Engl. J. Med.,* 292:879–882.
36. Prasad, A. S., editor (1976): *Trace Elements in Human Health and Disease.* Academic Press, New York, New York.
37. Swenerton, H., and Hurley, L. S. (1968): Severe zinc deficiency in male and female rats. *J. Nutr.,* 95:8–18.

38. Swenerton, H., and Hurley, L. S. (1970): Zinc deficiency in pregnant rhesus monkeys. *Teratology,* 3:209.
39. Swenerton, H., and Hurley, L. S. (1971): Teratogenic effects of a chelating agent and their prevention by zinc. *Science,* 173:62–64.
40. Swenerton, H., and Hurley, L. S. (1972): Zinc deficiency in the bonnet monkey *(Macaca Radiata). Fed. Proc.,* 31:667.
41. Swenerton, H., Shrader, R., and Hurley, L. S. (1969): Zinc-deficient embryos: Reduced thymidine incorporation. *Science,* 166:1014–1015.
42. Swenerton, H., Shrader, R., and Hurley, L. S. (1972): Lactic and malic dehydrogenases in testes of zinc-deficient rats. *Proc. Soc. Exp. Biol. Med.,* 141:283–286.
43. Tao, S.-H., and Hurley, L. S. (1975): Effect of dietary calcium deficiency during pregnancy on zinc mobilization in intact and parathyroidectomized rats. *J. Nutrition,* 105:220–225.
44. Theriault, L. L., and Hurley, L. S. Congenital malformations in A/J mice fed low and high dietary zinc. *(In preparation.)*
45. Vojnik, C., and Hurley, L. S. (1977): Abnormal prenatal lung development resulting from maternal zinc deficiency in rats. *J. Nutr.,* 107:862–872.
46. Walravens, P. A., and Hambidge, K. M. (1976): Growth of infants fed a zinc supplemented formula. *Am. J. Clin. Nutr.,* 29:1114.
47. Walravens, P. A., and Hambidge, K. M. (1977): Nutritional zinc deficiency in infants and children. In: *Zinc Metabolism: Current Aspects in Health and Disease,* edited by G. J. Brewer and A. S. Prasad, pp. 61–70. Alan R. Liss, Inc., New York.

The Role of Zinc in the Biochemistry of the *Euglena Gracilis* Cell Cycle

Kenneth H. Falchuk

The Howard Hughes Medical Institute, Biophysics Research Laboratory, and Department of Medicine, Harvard Medical School, and the Division of Medical Biology, Affiliated Hospitals Center, Inc., Boston, Massachusetts 02115

Zinc is essential for the growth and development of all living organisms (17). The manifestations of zinc deficiency vary depending upon the particular species though, in all cases, proliferating tissues are usually affected (18). In spite of the advances in the past 20 years establishing the participation of zinc in the function of over 80 metalloenzymes, the roles of this element in cell division, the metabolic events associated with its deficiency which result in genetic or teratological defects (10), or, in fact, the consequences of decreases in its content in human serum, as in alcoholic cirrhosis (20) and other diseases (2), have not been examined widely.

Nearly 15 years ago, the eukaryote, *Euglena gracilis*, strain Z, was chosen as a suitable single cell organism to study the biochemical basis for the requirement of zinc for growth, and particularly, to define the metabolic consequences secondary to its deficiency (15). Its growth is sensitive to the zinc content of the medium and it can be obtained in sufficient quantities to allow measurements on the effects of zinc on its nucleic acids, proteins, carbohydrates, etc. Decrease of the zinc concentration of the culture medium from 10^{-5} M to less than 10^{-7} M arrests its growth (Fig. 1A and B). Raising the zinc content of the

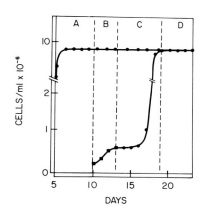

FIG. 1. Growth of zinc-sufficient (●) and zinc-deficient (■) *E. gracilis* grown in the dark.

TABLE 1. *Metal content of (+Zn) and (−Zn) E. gracilis*[a]

Metals	(+Zn)	(−Zn)
Zn	5.80	0.80
Mn	0.01	0.35
Mg	3.00	15.00
Ca	0.20	4.00
Fe	1.60	12.00
Ni	0.09	0.24
Cr	0.12	0.36

[a] $\mu g \times 10^7$/cell.

medium to 10^{-5} M completely restores normal growth within 36 to 48 hr (Fig. 1C and D). In zinc deficient (−Zn) cells, the zinc content decreases while that of other metals increases remarkably compared to that of zinc sufficient (+Zn) cells (Table 1) (4,21).

As part of our aim to determine the basis for the proliferative arrest, we undertook to detail, delineate, and define those steps of the cell cycle of *E. gracilis* affected by Zn deprivation. Towards this end, we have examined first the DNA content of (+Zn) and (−Zn) cells by means of laser excitation cytofluorometry (Fig. 2).

Aliquots of cells were collected for analysis of DNA content by flow cytofluorometry. The cells were stained with propidium diiodide solution and were analyzed in a cytofluorograph (Model 4801, Bio/Physics System, Inc., Mahopac, New York). The flow system of this instrument allows passage of one cell at a time through a 100-μm orifice (Restrictor valve) into a flow chamber where laminar flow is induced by a sheath of water. The cell traverses through exciting

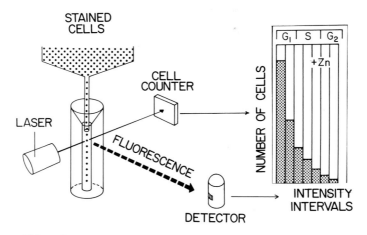

FIG. 2. Schematic diagram of flow cytofluorometer (DNA anaylsis).

monochromatic radiation from an Argon ion laser beam tuned to emit at 488 nm. The resultant fluorescence of propidium diiodide-DNA complexes of the cell nucleus is converted to an electrical signal by a photomultiplier, the output being displayed on the horizontal axis of a cathode ray tube. The signal also enters a multichannel pulse height distribution analyzer (Model 2100, Bio/physics Systems, Inc., Mahopac, New York) where the frequency distribution of the pulses as a function of the magnitude of the signal is stored in a memory unit and subsequently displayed as a histogram. One hundred channels are used, and the abscissa of the histogram reflects increasing linear values of the fluorescence signal. The numbers of cells recorded in each channel are registered simultaneously on a print-out tape system, allowing quantitation of the number of cells fluorescing at a characteristic intensity.

Incubation of mammalian cells with RNAse, prior to staining, has been shown to obviate interference by RNA in DNA analysis (1), and was shown to be true also for *E. gracilis* (3).

The histograms of DNA content of cells in each phase of the cell cycle have been identified by using synchronized cell populations (3). Log phase cultures contain organisms in all stages of the cell cycle. A histogram from a log phase culture of *E. gracilis* stained with propidium diiodide and analyzed in the cytofluorograph is shown in Fig. 2. The major fraction of cells are in G_1, with an unreplicated genome, the remainder are in S, G_2, or M phases of the cell cycle.

The DNA content of early stationary nondividing cells was examined next. Figure 3A and B compares the pattern of early stationary phase (+Zn) cells with that of (−Zn) cells, obtained when cell division ceases. The histogram of

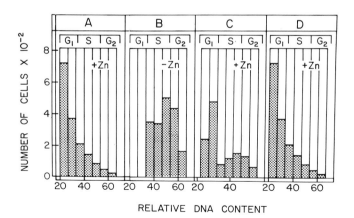

FIG. 3. Comparison of DNA histograms of zinc-sufficient with zinc-deficient *E. gracilis*. The majority of zinc-sufficient cells are in G_1 with a small fraction in S *(Panel A)*. In contrast, nondividing zinc-deficient cells are mostly in S or G_2 *(Panel B)*. On addition of zinc the number of cells blocked in S or G_2 decreases and histograms typical of dividing log phase cells result *(Panels C and D)*.

(+Zn) cells demonstrates that the majority—though not all—of the cells in the stationary phases are in G_1, with smaller numbers in S. In contrast, the pattern of nondividing (−Zn) cells is typical of that of S/G_2. The latter cells have previously been shown to cease dividing on depletion of Zn in the growth media (4,15,21)

The resulting histogram of DNA content suggests that as the nonsynchronously growing cells in the (−Zn) media are deprived of Zn, those cells which are in S do not continue to G_2, while those which reach G_2 do not proceed to mitosis. Addition of zinc to the growth media of (−Zn) results in progression of these blocked cells through the cell cycle (Fig. 3C and D). We observed a small fraction of the (−Zn) cells to be blocked in G_1. To further detail the effect of Zn deprivation of the G_1 to S transition, early stationary phase (+Zn) cells, known to be mostly in G_1, were incubated in media deficient in Zn. Following incubation in this medium, there is a 35% increase in cell number followed by cessation of growth. Addition of Zn to the medium confirms that the absence of Zn is responsible for the inhibition of cell growth. Within 24 to 36 hr of the addition of Zn, the cell numbers increase by 200%, reaching those expected for a (+Zn) culture. The cytofluorometric analysis of these cells demonstrates that, prior to the addition of Zn, when cell division has ceased, almost all cells incubated in zinc-deficient media are in G_1 phase. Hence, Zn deprivation of cells in G_1 blocks their progression into S. Addition of Zn to these cells reverses the block of their cell cycle, restoring the normal pattern of division of cells which becomes identical to that shown in Fig. 2. Clearly, the biochemical processes essential for cells to pass from G_1 into S, from S to G_2, and from G_2 to mitosis depend on the presence of zinc, and its deficiency can block all three phases of the growth cycle of *E. gracilis*.

Indirect evidence suggests that Zn is essential for the function of DNA polymerase of *E. gracilis* (12). However, the results of our cell cycle studies led to the conclusion that the limiting steps leading to the abnormalities of the cell cycle and the consequent proliferative arrest cannot be restricted solely to impaired DNA synthesis.

Studies of *E. gracilis* RNA Metabolism

Relative to Zn-sufficient cells, Zn-deficient *E. gracilis* incorporate ^{3}H-uridine into RNA at a reduced rate (4). However, they accumulate peptides and amino acids and their protein content is reduced (21). These observations focused on derangements at the level of translation in Zn-deficient cells. Such derangements could potentially be responsible both for the observed blocks of the cell cycle and the proliferative arrest since on-going RNA and protein synthesis are required for G_1, S, and G_2. Alterations in their synthesis could then block the cell cycle at each of these stages (9).

Accordingly, we next focused on the details of the role of Zn in RNA metabolism of *E. gracilis* as a possible basis for the observed chemical lesions. We first examined the RNA polymerases from (+Zn) *E. gracilis*.

The enzymes were isolated from cells harvested in the log phase of growth. A cellular homogenate was precipitated with ammonium sulfate and the pellet was dissolved in 0.15 M ammonium sulfate buffer. At this stage, the RNA polymerases are bound to DNA. The DNA was precipitated with protamine sulfate leaving the RNA polymerases in the supernatant. The preparation was then chromatographed on DEAE-Sephadex A-25. The enzymes, now free of DNA, are purified further by chromatography on phosphocellulose followed by affinity chromatography on DNA cellulose. RNA polymerases I and II have been purified to homogeneity while the isolation of polymerase III is in progress (7,8). All three polymerases are entirely dependent on an exogenous DNA template for activity. The product of their enzymatic reaction is RNA, as evidenced by an absolute substrate requirement for ribonucleotide triphosphates and by digestion of the product by ribonuclease. As with other polymerases, the *E. gracilis* enzymes are inactive in the absence of Mg^{+2} or Mn^{+2}. Both DNA dependent RNA polymerases I and II are homogeneous on polyacrylamide gels, and their estimated molecular weights, determined on SDS gels, are between 650,000 and 700,000 for both polymerases. They are composed of multiple subunits.

α-Amanitin differentiates the three RNA polymerases. The activity of the polymerase I is not inhibited by α-amanitin at concentrations up to 200 μg/ml. In contrast, increasing concentrations of α-amanitin progressively decrease and at 0.1 μg/ml nearly abolish activity of RNA polymerase II. Polymerase III is not inhibited by 50 μg/ml but at higher concentrations its activity decreases so that at 200 μg/ml of α-amanitin, only 25% of enzyme activity remains.

Inhibition by 1,10-Phenanthroline

The answer to the question of metal dependence clearly has to be approached by undertaking studies of inhibition with metal binding agents which—while conserving material—give first a valuable indication of the involvement of a metal in activity.

The chelating agent, 1,10-phenanthroline (OP) has proved exceptionally suitable to study the inhibition of zinc metalloenzymes (19). To determine its effect on *E. gracilis* RNA polymerase I and II, the effect of OP concentration on enzyme activity was studied. A stock solution of OP, 10^{-2} M, pH 7.5, was diluted variously to range from 10^{-2} to 10^{-7} M. The concentrations of template, nucleotide, and other components were standard in all assays. Throughout, Mg^{2+} was the only activating cation. The effects of the nonchelating isomers, 1,7- or 4,7-phenanthroline in the concentration range from 1 to 3×10^{-4} was determined also.

Inhibition Studies with Other Chelating Compounds

The effect of other chelating agents on activity was studied as a function of their concentrations. Stock solutions of 8-hydroxyquinoline-5-sulfonic acid,

TABLE 2. *Inhibition of RNA polymerases I and II by chelating agents*

Agent	I	II
	($V_i/V_c \times 100$)	
Chelating		
1,10-Phenanthroline	0	0
EDTA	0	0
8-Hydroxyquinoline	—	0
8-Hydroxyquinoline 5-Sulfonate	—	70
2,2'-Bipyridine	—	65
Nonchelating analogs		
1,7-Penanthroline	100	100
4,7-Phenanthroline	100	100

EDTA, α-α'-bipiridyl or 8-hydroxyquinoline, all 5×10^{-2} M, were diluted with metal-free water to prepare dilutions ranging from 10^{-6} to 10^{-2} M, adjusted to pH 8. Assays were performed with 10 μg enzyme and under standard conditions. Magnesium was the activating metal in all cases.

Both polymerase I and II are inhibited by saturating amounts of chelating agents (Table 2). 1,10-phenanthroline and EDTA inhibit both their activities completely. Other chelating agents such as 8-hydroxyquinoline 5-sulfonic acid, EDTA, and α-bipyridine, also at saturating concentrations, reduce the RNA polymerase II activity from 70% to 50%.

At saturation amounts, the nonchelating analogs of 1,10-phenanthroline, 1.7- or 4,7-phenanthroline, do not inhibit either polymerase. Hence, the inhibition by the 1,10-isomer must be due to chelation of a functional metal atom.

Collectively, these results were almost diagnostic of the presence and functional essentiality of zinc rather than any other metal, though, of themselves, such studies cannot be decisive.

Studies of Metal Content of *E. gracilis* RNA Polymerases

The presence of stoichiometric quantities of metal is, of course, essential to verify that chelating agents exert their effect by binding to a functional and/or structurally essential metal.

The presence of Zn, Cu, or Fe could account for the observed inhibition of the *E. gracilis* RNA polymerases; and these elements and Mn were determined by microwave excitation emission spectrometry (see below) after removal of metal quenching agents and of low molecular weight protein contaminants by gel exclusion chromatography of dialysis against metal-free buffers. The elements and protein were measured quantitatively with high precision in fractions containing maximal activity when absolute amounts of metal varied from 10^{-11} to 10^{-14} g-atom and utilizing ~0.6 μg of enzyme for analysis.

The *E. gracilis* RNA polymerases I and II contain 0.2 μg of Zn per mg of protein. Keeping in mind the as yet provisional nature of the molecular weights

TABLE 3. *Metal content of RNA polymerases I and II as measured by microwave excitation spectroscopy*[a]

RNA polymerase	Protein (mg/ml)	ZN (μg/mg protein)	g-atom/mole
I	0.12	0.19	2.0
II	0.10	0.21	2.2

[a] Metal content is expressed as g-atom per mole of 650,000 and 700,000 for the RNA polymerases I and II, respectively.

of these polymerases, which form the basis of the metal/protein ratio, the stoichiometry of both is essentially the same—i.e., 2 g-atom of Zn per mole. (Table 3) (7,8).

The sum of Cu, Fe, and Mn is less than 0.2 g-atom/mole. Thus, both RNA polymerase I and RNA polymerase II from Zn sufficient *E. gracilis* are zinc metalloenzymes.

Studies with Zinc Deficient (−Zn) *E. Gracilis*

Our demonstration that RNA polymerase I and II from (+Zn) *E. gracilis* were both zinc enzymes (7,8) suggested that the metabolism of these enzymes might be affected in (−Zn) cells and prompted us to study RNA polymerase function in these organisms. We have found that (−Zn) *E. gracilis* contain a single unusual RNA polymerase which is also a zinc metalloenzyme (Falchuk et al., *in preparation*). This difference in the content of RNA polymerase in (−Zn) cells confirmed that deficiency of this metal indeed altered the metabolism of RNA polymerases and, further, highlighted the need to investigate its RNA products. Consequently, we compared the amounts of RNAs synthesized by both (+Zn) and (−Zn) cells. The total RNA from (+Zn) and (−Zn) cells was isolated by a standard phenol ethanol extraction.

The various RNA classes in the total RNAs were separated by utilizing a series of affinity columns. Cytoplasmic mRNA contains a poly A segment which is absent in other RNAs. Oligo-(dT)-cellulose columns will bind only the poly A containing mRNA and, thus, afford a rapid purification method. The RNAs which did not bind to the Oligo-(dT)-cellulose column were chromatographed on a dihydroxyborylamino ethyl (DBAE) cellulose affinity column. This column binds transfer RNA and separates it from the bulk of ribosomal RNA and the minor amino acylated transfer RNA fractions. By sequential use of these columns, the three major classes of RNA were obtained. Each fraction was hydrolyzed in preparation for analyzing their base composition using high pressure liquid chromatography.

(+Zn) *E. gracilis* contain 20 μg of RNA/10^6 cells. This value is virtually unaltered by zinc deficiency (Table 4). In both cases, the total RNA content is consistent with values obtained for *E. gracilis* grown to early stationary phase.

TABLE 4. *RNA content of zinc sufficient, (+Zn), and deficient, (−Zn), E. Gracilis*[a]

Zinc	Total RNA ($\mu g \times 10^6$/cell)	Ribosomal (%)	Transfer (%)	Messenger (%)
(+Zn)	20 ± 5	79 ± 5	15 ± 3	6 ± 2
(−Zn)	19 ± 5	74 ± 4	15 ± 3	11 ± 3

[a] Values are the mean ± 1 SD of six preparations.

The total RNA is resolved into three fractions by sequential chromatography on Oligo-(dT) and DBAE celluloses. The mRNA fraction binds to Oligo-(dT)-cellulose while approximately 90% or more of the RNA does not bind. This larger RNA fraction is then applied to a DBAE cellulose column which separated the tRNA from the ribosomal fraction. The amounts of ribosomal, transfer, and messenger RNA obtained by these methods are similar to those reported by others for eukaryotic organisms, including *E. gracilis*. In (−Zn) cells, the fraction of the total RNA present as rRNA is slightly less, while that present as tRNA is essentially the same as in (+Zn) cells (Table 4). In contrast, the mRNA content of (−Zn) cells, 11% of the total RNA, is almost twice that of (+Zn) cells (Table 4) (6).

The base composition for ribosomal, transfer, and messenger RNA from (+Zn) and (−Zn) cells was determined by high pressure liquid chromatography (5). For rRNA (Table 5), the purine and pyrimidine contents are identical for (+Zn) and (−Zn) cells. For tRNA, we analyzed only for the four major bases and found that the guanine content decreased in (−Zn) cells from 34% to 24% while the cytosine content increases from 27% to 38%. The adenine and uracil contents are identical (Table 5).

The base composition of mRNA from (−Zn) cells differ strikingly. Figure 4 compares the chromatogram of a mRNA from (+Zn) cells with that of a mRNA sample from (−Zn) cells eluted from the cation exchange gel of the high pressure liquid chromatograph (HPLC). In the former, only four bases are found. They are identified as uracil (U), guanine (G), cytosine (C), and

TABLE 5. *Base composition of E. Gracilis RNA*[a]

Base	+Zn			−Zn		
	Ribosomal	Transfer	Messenger	Ribosomal	Transfer	Messenger
Guanine (G)	36	34	35	36	24	25
Cytosine (C)	26	27	26	26	38	49
Adenine (A)	21	19	21	22	20	10
Uracil (U)	17	20	18	16	18	16

[a] The data are expressed as (μg of each base)/(μg G + C + A + U) × 100. The mRNA from (−Zn) cells contain a number of additional bases (Fig. 4) which are not included in this calculation. Each value is the mean of three analyses.

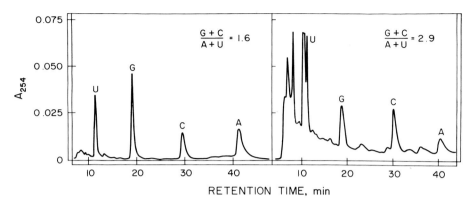

FIG. 4. The bases of *E. gracilis* mRNA from (+Zn) cells *(left)* and (−Zn) cells *(right)* were analyzed by high-pressure liquid chromatography. A 3 µl sample of an acid hydrolysate was used. The retention time and UV spectra differ for each base, allowing for base separation and identification. The mRNA from (+Zn) cells contains only the four major bases, uracil, guanine, cytosine, and adenine. In contrast, the mRNA from (−Zn) cells contains additional, unknown bases. Moreover, the ratios of (G + C)/(A + U), 1.6 and 2.9 for (+Zn) cell and (−Zn) cell mRNA, respectively, differ strikingly.

adenine (A), respectively, based on their elution volumes from the (HPLC) system and the UV spectra of each fraction. In contrast, the mRNA from (−Zn) cells contains seven major peaks and several minor ones. The uracil, guanine, cytosine, and adenine fractions in this chromatograph also have been identified by their characteristic UV spectra. The remaining fractions represent bases whose identity is presently unknown but which are not found in mRNA from (+Zn) cells. The calculated G + C/A + U ratios of the mRNA samples, 1.6 for (+Zn) and 2.9 for (−Zn) cells, differ strikingly. Thus, mRNA from (−Zn) contains additional bases and also the ratio of the known major purines and pyrimidines is nearly double compared to the mRNA from (+Zn) cells (6).

There are various mechanisms through which zinc deficiency could affect RNA and DNA metabolism (7). In particular, we have emphasized recently the essentiality of zinc for the function of both DNA and RNA polymerases. We have also noted the importance of Mn(II) [or Mg(II)] for the activity of these enzymes. However, the possible synergism and/or antagonism of these and other metals in nucleic acid polymerase action in general, and particularly on the base composition of the resultant RNA product, has not been examined critically. This is of interest, since one consequence of zinc deficiency in *E. gracilis* is a 35-fold increase in intracellular Mn and a five-fold increase in Mg content (4,21). This, together with earlier experiments utilizing micrococcal DNA polymerase (11), *E. coli* RNA polymerase (13), and viral reverse transcriptases (16), demonstrating that the base composition of the nucleic acid product synthesized *in vitro* varied according to the particular activating cation employed,

TABLE 6. *The effect of [Mn(II)] on the relative UMP/CMP incorporation by* E. gracilis *RNA polymerases*[a]

Mn(II) (mM)	RNA polymerase		
	I	II	Single enzyme from (−Zn) cells
1	1.7	2.1	3.5
5	1.4	1.7	1.5
10	1.0	0.8	0.4

[a] Analogous results are obtained on comparing the relative incorporations of UMP/AMP.

prompted us to examine the role of these metals in determining the composition of the products. Thus, we have determined the effects of varying Mn(II) concentrations on the incorporation of bases into RNA produced by RNA polymerases I and II from (+Zn) cells and the single unusual RNA polymerase from (−Zn) *E. gracilis*. These experiments were carried out in cell-free systems, and the incorporation of CMP, UMP, or AMP served as the criterion of the base composition of the resultant RNA.

The effects of various Mn concentrations on the relative incorporation of UMP and CMP into RNA by the different *E. gracilis* RNA polymerases are shown in Table 6. Increasing the Mn concentration from 1 to 10 mM in assays with RNA polymerase I or II from (+Zn) cells decreases the ratio of UMP to GMP incorporated from 1.7 to 1.0 and 2.1 to 0.8, respectively. Similarly, in assays of the single enzyme from (−Zn) cells, this ratio decreases from 3.5 to 0.4. Thus, the base composition of RNA synthesized by polymerases from either cell type varies as a function of Mn concentration (6).

These studies show that the total RNA content per cell is not altered by zinc deficiency (Table 4). Moreover, each of the RNA classes, ribosomal, transfer, and messenger are present in these cells. Thus, changes in the amount of each RNA class synthesized would not appear to be responsible for the biologic effects of zinc deficiency in *E. gracilis*. A remarkable difference does exist, however, between the composition of mRNA from (−Zn) and (+Zn) cells. The mRNA from (−Zn) cells has an unusual base composition as demonstrated by the twofold increase of its (G + C)/(A + U) base ratio and the presence of significant amounts of bases other than uracil, guanine, cytosine, and adenine (Fig. 4). Though the sequence of bases in mRNA coding for different proteins does vary, such differences in mRNA base composition are most unusual since the ratio of G + C/A + U bases has been found to be consistently uniform for nearly all mRNAs.

The changes in mRNA composition of the degree observed between (+Zn) and (−Zn) cells have not been reported previously in cells deprived of essential nutrients or as a function or growth or cell cycle stage. Therefore, the changes in composition of mRNA from (−Zn) cells reveal critical differences in their mRNA metabolism which must be the consequence of either an alteration in processes which normally regulate the incorporation of bases into mRNA or the accumulation of mRNA molecules coding for large amounts of a specific protein(s). We have obtained evidence suggesting that the increases in the content of other metals such as Mn in (−Zn) cells could play a role in the production of such mRNAs (Table 6). These results, together with the finding of a single RNA polymerase in (−Zn) organisms, represents the first major metabolic difference between these and (+Zn) organisms whose manifestations would provide a basis for the arrest of cell division in this organism. Thus, mRNA plays a central role in the translation of information from the genome into proteins, which in turn determine the phenotype. The mechanism by which genetic information is faithfully translated into proteins is dependent on the base composition and the sequence of mRNA molecules. These two variables are involved in the binding of mRNA to ribosomes and determine the amino acid composition of the proteins synthesized. Our demonstration of the unusual composition of mRNA from (−Zn) cells suggests that, in these organisms, translational processes may be altered, leading to the formation of products of translation with unusual amino acid composition and/or to changes in the rate of synthesis of specific proteins which may be either essential for or inhibitory to cellular function. Indeed, abnormalities of protein metabolism in (−Zn) E. gracilis (21) and plants (14) have been documented, albeit only in terms of total protein and changes in amino acid contents. The effects of such alterations in protein metabolism would be decisive and result in the arrest of cell division in (−Zn) organisms since on-going protein synthesis is required for this process (9).

Similarly, if extended to other systems, derangements in the metabolism of proteins involved in the formation of tissues and organs could lead to the developmental abnormalities characteristic of (−Zn) mammals (10).

These studies are part of an on-going investigation aiming at the identification of the biochemical basis for the role of zinc in cell division. They have led us to explore details of both DNA and RNA metabolism in both (+Zn) and (−Zn) cells whose importance remain to be further elucidated. In general, however, the present systematic studies are relevant to the understanding of the essential role of zinc in cell division and development and also direct attention to a novel mode of regulation of the metabolism of nucleic acids and proteins with profound implications for their mechanism.

ACKNOWLEDGMENT

This work was supported, in part, by Grant-in-Aid GM-15003 from the National Institutes of Health of the Department of Health, Education, and Welfare.

REFERENCES

1. Crissman, H. H., and Steinkamp, J. A. (1973): Rapid, simultaneous measurement of DNA, protein, and cell volume in single cells from large mammalian cell populations. *J. Cell Biol.*, 59:766.

2. Falchuk, K. H. (1977): Effects of acute disease and ACTH on serum zinc proteins. *New Engl. J. Med.,* 296:1129.
3. Falchuk, K. H., Krishan, A., and Vallee, B. L. (1975): DNA distribution in the cell cycle of *Euglena gracilis.* Cytofluorometry of zinc deficient cells. *Biochemistry,* 14:3439.
4. Falchuk, K. H., Fawcett, D., and Vallee, B. L. (1975): Role of zinc in cell division of *Euglena gracilis. J. Cell Science,* 17:57.
5. Falchuk, K. H., and Hardy, C. (1978): Determination of *E. gracilis* mRNA base composition using high pressure liquid chromatography *Anal. Bioch.,* 89:385.
6. Falchuk, K. H., Hardy, C., Ulpino, L., and Vallee, B. L. (1978): RNA metabolism manganese, and RNA polymerases of zinc sufficient and zinc deficient *Euglena gracilis. Proc. Natl. Acad. Sci.,* 75:4175.
7. Falchuk, K. H., Mazus, B., Ulpino, L., and Vallee, B. L. (1976): *Euglena gracilis* DNA dependent RNA polymerases II: a zinc metalloenzyme. *Biochemistry,* 15:4468.
8. Falchuk, K. H., Mazus, B., Ulpino, L., and Vallee, B. L. (1977): *E. gracilis* RNA polymerases I: a zinc metalloenzyme. *Biochem. Biophys. Res. Commun.,* 74:1206.
9. Gelfant, S. (1966): Method. *Cell Physiol.,* 2:359.
10. Hurley, L. S., Gowan, J., and Swenerton, H. (1971): Teratogenic effects of short term and transitory zinc deficiency in rats. *Teratology,* 4:199.
11. Litman, R. M. (1971): The differential effect of magnesium and manganese ions on the synthesis of poly (dGd·C) and *Micrococcus luteus* DNA by *Micrococcus luteus* DNA polymerases *J. Mol. Biol.,* 61:1–23.
12. McLennan, A. G., and Keir, H. M. (1975): Deoxyribonucleic acid polymerases of *Euglena gracilis* purification and properties of two distinct deoxyribonucleic acid polymerases of high molecular weight. *Biochem. J.,* 151:227.
13. Pogo, A. O., Littau, V. C., Allfrey, C. C., and Mirsky, A. E. (1967): Modification of ribonucleic acid synthesis in nuclei isolated from normal and regenerating liver. Some effects of salt and specific divalent cations. *Proc. Natl. Acad. Sci. USA,* 57:743.
14. Possingham, T. V. (1956): Effects of mineral nutrition in the content of free amino acids and amides in tomator plants. I. A comparison of the effects of deficiencies of Cu, Zn, Mn, Fe, and Mg. *Aust. J. Biol. Sci.,* 9:539–551.
15. Price, C. A., and Vallee, B. L. (1962): *Euglena gracilis,* a test organism for study of zinc. *Plant Physiol.,* 37:428–433.
16. Sirover, M. A., and Loeb, L. (1976): Metal activation of DNA synthesis. *Biochem. Biophys. Res. Commun.,* 70(3):812–817.
17. Vallee, B. L. (1959): Biochemistry, physiology, and pathology of zinc. *Physiol. Rev.,* 39:443.
18. Vallee, B. L. (1977): Recent advances in zinc biochemistry. In: *Biological Aspects of Inorganic Chemistry,* edited by D. Dolphin, pp. 37–70. Wiley-Interscience, New York, New York.
19. Vallee, B. L., and Wacker, W. E. C. (1970): Metalloproteins. In: *The Proteins,* Vol. 5, pp. 1–192.
20. Vallee, B. L., Wacker, W. E. C., Bartholomay, A. F., and Robin, E. D. (1956): Zinc metabolism in hepatic dysfunction. I. Serum zinc concentrations in Laennec's cirrhosis and their validation by sequential analysis. *New Engl. J. Med.,* 225:403–408.
21. Wacker, W. E. C. (1962): Nucleic acids and metals III. Changes in nucleic acid, protein, and metal content as consequences of zinc deficiency in *Euglena gracilis. Biochemistry,* 1:859–865.

Vitamin D Metabolism and Function

Hector F. DeLuca

Department of Biochemistry, College of Agricultural and Life Sciences, University of Wisconsin-Madison, Madison, Wisconsin 53706

The vitamin D endocrine system is the major factor controlling the overall calcium economy of the body. It is one of the major factors regulating plasma calcium concentration and phosphorus metabolism. This report will describe the vitamin D endocrine system, and will present some new developments in the metabolism and function of vitamin D (20,25).

Figure 1 shows the structure of vitamin D, but even more important it will remind you that vitamin D is in fact a steroid; it is a secosteroid, to be sure, because the B ring is replaced by a 5,7-diene bridge which is part of a *cis*-triene system. This triene chromophore dominates the chemistry of the vitamin, contributing to its instability relative to other steroids and undoubtedly also contributes to its binding and functional characteristics. Figure 1 also illustrates the unusual numbering system of the vitamin D molecule. Of particular importance are the 1 position in the A ring, and the 24, 25, 26, and 27 positions on the side chain, since they are the sites of metabolic modification or of labeling. The numbering system is derived from the parent compound, 7-dehydrocholesterol, which upon photolysis results in the production of vitamin D.

The known direct functions of vitamin D in the body are shown in Fig. 2. It has been known for more than 40 years that vitamin D stimulates intestinal calcium absorption (21). This is an active transport process in which metabolic energy is required for the thermodynamically unfavorable transfer of calcium from the lumen of intestine to the extracellular fluid compartment. In addition to this transport system, vitamin D stimulates a phosphate-oriented transport system in intestine (14,46) especially evident in the distal small intestine. This system is not related to the calcium transport system and, in fact, does not require calcium.

A direct action of vitamin D on the mobilization of calcium and phosphate from bone to maintain normal plasma levels is well known since the work of Carlsson in 1951 (46). This would appear to be quite the opposite of what might be expected from vitamin D since we know that the overall effect of vitamin D is to mineralize bone.

The fourth known function is that vitamin D or its active forms stimulate the osteoclasts to resorb bone (100,111). This is probably different from the calcium homeostatic function of vitamin D and probably represents the initial

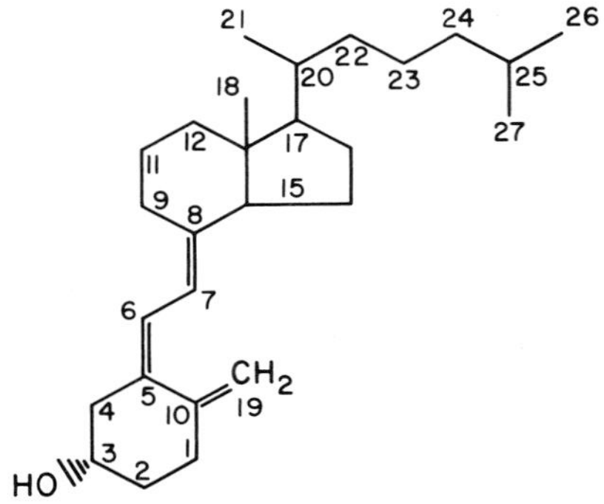

FIG. 1. Structure of vitamin D_3 (a prohormone) and its numbering system.

event in the remodeling and modeling processes of bone (35). As an organism grows, bone must be shaped, and thus certain areas are resorbed while other areas are formed: this is known as modeling. In addition, as we use our skeleton, we inflict on it microfractures. There must be a mechanism whereby these are repaired. This mechanism is known as bone remodeling. In this process, osteoclasts are activated to resorb the damaged or microfractured bone in a sort of

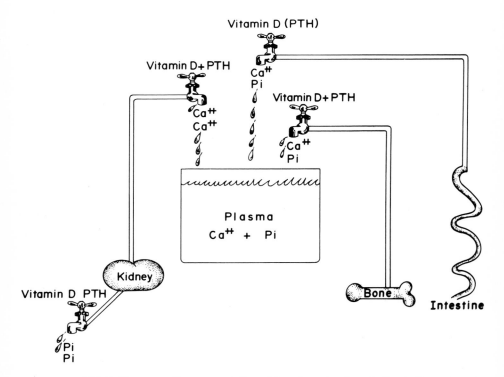

FIG. 2. Diagrammatic representation of the sites of action of vitamin D.

cutting-cone mechanism (35). This cone area is replaced by new bone. The active forms of vitamin D are necessary at least for the initial event in this osteoclastic-mediated resorption, and it can be studied in organ cultures of embryonic bone (100,111). One of the overall aspects of these systems is that by stimulating intestinal calcium transport, intestinal phosphate transport, and the mobilization of calcium from bone and possibly renal reabsorption of calcium, vitamin D functions to elevate plasma calcium and plasma phosphate concentrations. It is the concentration of calcium and phosphate in the plasma at normal levels which is necessary for normal mineralization of newly forming bone (18,71). The normal plasma calcium and phosphorus level which represents a super-saturated solution of calcium and phosphorus is important not only to mineralization of bone but also to the normal functions of nerve and muscle (20,21,25). We know that as a direct consequence of vitamin D dysfunction, muscle strength is lacking (12) and the disease tetany—which is a failure of normal operation of the neuromuscular junction—presents itself. In addition, there may be a direct action of vitamin D on the mineralization sites of bone but so far there is no direct experiment which demonstrates this (20,21,25). Furthermore, there are other suggestions that vitamin D may function directly on the parathyroid glands to suppress parathyroid hormone secretion (15,49). However, convincing evidence for this suggestion is lacking at the present time.

VITAMIN D METABOLITES

We have learned over the course of the last decade and a half that vitamin D does not act directly on the above systems but must be converted to a hormone before it can function (19,20,25). The essential metabolic conversions which result in the production of the hormonal form are shown in Fig. 3. Vitamin D should not be considered a true vitamin, since when sufficient exposure to ultraviolet light takes place, vitamin D is not required in the diet (24). In the epidermis of skin, an abundant supply of the 5,7-diene sterol, 7-dehydrocholesterol, is found (24) that undergoes a photolysis reaction under the influence of 280–300 nm ultraviolet light, producing a compound, previtamin D (not shown). This compound is slowly converted to vitamin D_3 (28,53). This is strictly a chemical photolysis reaction, and no proteins or enzymes are involved in the conversion to previtamin D_3. The conversion rate of previtamin D_3 to vitamin D_3 in skin is similar to that found in the absence of tissue (51). Thus, these reactions are apparently not regulated.

Since the antirachitic factor produced in skin on ultraviolet irradiation has now been clearly identified as vitamin D_3 (28,53), it must be regarded as the natural form of the vitamin. Another common form of vitamin D is vitamin D_2 formed from the plant sterol, ergosterol, and is used to fortify foods and as a supplement. Its metabolism is identical to that of vitamin D_3 (60) but it must be regarded primarily as a synthetic form of vitamin D. The skin production of vitamin D_3 is an efficient process, and if there is sufficient direct sunlight

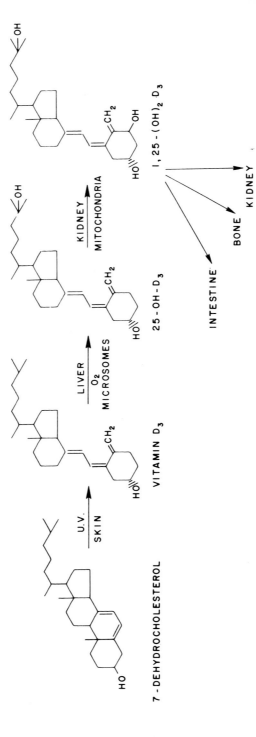

FIG. 3. The metabolism of vitamin D to its functional hormone, 1,25-(OH)$_2$D$_3$.

exposure—i.e., 30 min per day—sufficient amounts of vitamin D_3 are made and it is not required in the diet. Vitamin D, whether absorbed in the intestine along with the lipids in the lacteal system or whether injected or produced in the skin, uniquely accumulates in the liver (79,87). Only vitamin D and its metabolites will accumulate in the liver in a rather specific way. In the liver, vitamin D undergoes its first obligatory reaction in the endoplasmic reticulum where it is converted to 25-hydroxyvitamin D_3 (25-OH-D_3) (1,59). Some 25-hydroxylation occurs in tissues other than liver, but because vitamin D accumulates in the liver and because the activity in other tissues is quite small relative to liver, we believe that the major—if not exclusive—site of synthesis of 25-OH-D is the liver (79). 25-OH-D_3 is the major circulating form of vitamin D, being present at 30 ng/ml plasma, whereas vitamin D itself is found in human plasma only at a level of 1 ng/ml (59). The reaction which produces the 25-OH-D is feedback regulated by the product itself (1). This is of limited capacity and can be overcome by simply adding greater amounts of substrate. The administration of large amounts of vitamin D will result in the elevation of the circulating level of 25-OH-D (96). 25-OH-D_3 does not act directly in any target tissue as far as is currently known. Instead, it must be further metabolized before function (7,114). Further metabolism occurs in the kidney where it undergoes 1α-hydroxylation to form the final active hormone, i.e., 1,25-dihydroxyvitamin D_3 (1,25-$(OH)_2D_3$) (30,42,55). This reaction occurs exclusively in the kidney and exclusively in the mitochondrial fraction of that organ (43). The exact cellular localization of the hydroxylase remains unknown, although indirect evidence suggests that it is present in the proximal convoluted tubule epithelial cells of that organ (36). If kidneys are removed from vitamin D-deficient animals, the animals fail to respond to physiologic doses of either vitamin D_3 or 25-OH-D_3, whereas they respond fully to the administration of 1,25-$(OH)_2D_3$ (7,54,114). These responses include intestinal calcium transport, intestinal phosphate transport, and the mobilization of calcium from bone. Since 1,25-$(OH)_2D_3$ is made exclusively in the kidney, it is evident that it must be regarded as a hormone being produced in one organ and having its function in other organs, namely intestine and bone (22). These experiments demonstrated that 1,25-$(OH)_2D_3$ or a further metabolite must be the metabolically active form of the vitamin in those systems while 25-OH-D_3 and vitamin D cannot be regarded as tissue active forms. A major question remaining to be answered, however, is whether 1,25-$(OH)_2D_3$ is the final active hormone in the intestine and bone. Work carried out with 1,25-$(OH)_2$-[26,27-^{3}H]D_3 by two laboratories provided evidence that perhaps 1,25-$(OH)_2D_3$ is the final active form (33,112). In these experiments, the target organ responses of intestine and bone to radioactive 1,25-$(OH)_2D_3$ were examined. At the time the intestine and bone responded to this substance, tissues were removed, extracted with methanol and chloroform, and chromatographed to reveal which metabolites were present. In the chloroform phase of the extract, the only metabolite of significance proved to be 1,25-$(OH)_2D_3$. However, poor recoveries of the administered radioactivity located in the 26 and 27 positions were found. Furthermore, significant amounts of radioactivity from 1,25-$(OH)_2$[2-^{3}H]D_3 injected into these vitamin D-deficient animals in the water-soluble fraction were noted (34,112). Because of the poor recoveries, we became interested in whether in fact the terminal side chain label of 1,25-$(OH)_2D_3$ was not revealing the true metabolic picture. We, therefore, chemically synthesized 25-OH-[26,27-^{14}C]D_3 to determine the fate of the side chain (45,68). Seven percent of the injected dose of the 25-OH-[26,27-^{14}C]D_3 appeared in expired CO_2 of vitamin D-deficient animals. Nephrectomy prevented this CO_2 expiration, which suggested that the 25-OH-D must be converted to the 1,25-$(OH)_2D_3$ before undergoing side chain oxidative cleavage. This was confirmed by the administration of radioactive 1,25-$(OH)_2$[26,27-^{14}C]D_3. As much as 30% of the injected dose appeared as expired carbon dioxide. It is, therefore, clear that there is a very rapid metabolic conversion of 1,25-$(OH)_2D_3$ to a metabolite that has lost its terminal side chain piece as carbon dioxide and water. We have recently completed the isolation and identification of this metabolite using 1,25-$(OH)_2$[3α-^{3}H]D_3 as the label. We learned that within 2 to 3 hr after administration of this material, a large proportion of the labeled 1,25-$(OH)_2D_3$ appeared in watersoluble fractions (26). This compound was isolated in pure form and its structure is shown in Fig. 4. Thus, an important metabolite of 1,25-$(OH)_2D_3$ is its corresponding 23-carboxylic acid with loss of the remaining side chain carbons. At the present time, we do not know if this metabolite possesses biological activity. However, it is certainly formed very rapidly in vitamin D-deficient animals upon the administration of 1,25-$(OH)_2D_3$. Likely, this new metabolite will receive considerable attention in the near future. We must, therefore, conclude that we do not know at the present time if in fact 1,25-$(OH)_2D_3$ performs all of the target tissue responses without further metabolism. It is likely, however, that at least some of the responses of the target tissues are by function of 1,25-$(OH)_2D_3$ directly.

25-OH-D_3 undergoes metabolic alteration other than conversion to 1,25-$(OH)_2D_3$. In normal animals, the major dihydroxyvitamin D metabolite found is 24R,25-dihydroxyvitamin D_3 (24R,25-$(OH)_2D_3$), as shown in Fig. 5. The enzyme system which carries out this hydroxylation is found in kidney, intestine, and cartilage cells (37,61,69). The enzyme is very low or absent in vitamin D

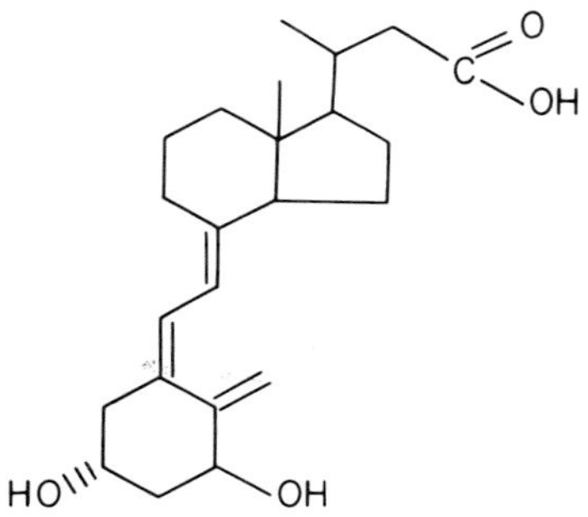

FIG. 4. Structure of a major metabolite of 1α,25-$(OH)_2D_3$ or 1α-OH-24,25,26,27-tetranor-vitamin D_3 23-carboxylic acid, or calcitroic acid.

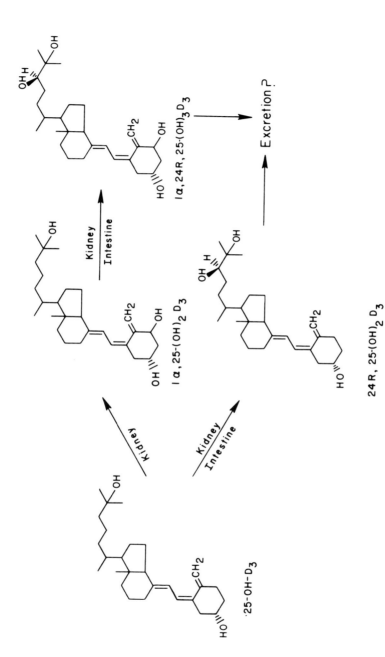

FIG. 5. The 24-hydroxylation of 25-OH-D and 1,25-(OH)$_2$D.

deficiency and must be induced by the administration of some form of vitamin D, the most active being the hormone 1,25-$(OH)_2D_3$ (109). The enzyme inserts an oxygen function from molecular oxygen into the 24R position of both 1,25-$(OH)_2D_3$ and 24,25-$(OH)_2D_3$ (107). The biological significance of this hydroxylation remains largely unknown. However, a number of suggestions have been made that 24,25-$(OH)_2D_3$ possesses unusual activity in mineralization of bone (88), suppression of parathyroid hormone secretion (15,49), and in embryonic development (48). It is important to note, however, that in the known and proven functions of vitamin D, 24,25-$(OH)_2D_3$ is less active than its immediate precursor (4,52,108). For example, 1,24R,25-trihydroxyvitamin D_3 (1,24R,25-$(OH)_3D_3$) is considerably less active than 1,25-$(OH)_2D_3$ in intestinal calcium transport, the mobilization of calcium from bone, the mineralization of bone, and the transport of phosphate across intestinal epithelium (12). Of considerable importance is our recent chemical synthesis of 24,24-difluoro-25-OH-D_3 which permits an investigation into the functional significance of 24-hydroxylation (62). We have been able to demonstrate that the 24,24-difluoro-25-OH-D_3 is equally active with 25-OH-D_3 in bone mineralization, in the prevention of rachitogenesis, in the elevation of plasma calcium at the expense of bone, and in the intestinal calcium and phosphate transport mechanisms. In the known functions of vitamin D, therefore, 24R-hydroxylation appears to play little or no significant role. It is entirely possible, however, that specialized functions may be found for this form of vitamin D, but these questions must be regarded as tenuous and requiring further verification.

In addition to the 24-hydroxylation reaction, 25-OH-D_3 also undergoes 26-hydroxylation to form 25R,26-dihydroxyvitamin D_3 (25R,26-$(OH)_2D_3$) (70, 101). This metabolite has weak biological activity only on the stimulation of intestinal calcium transport. It must be 1-hydroxylated before carrying out this function and in this respect must act as an analog of 1,25-$(OH)_2D_3$. The 26-hydroxylation reaction is known to occur in kidney, but since nephrectomy does not eliminate the 26-hydroxylation, it is likely that other sites of this enzymatic activity must exist (110).

In addition to the above described hydroxylation reactions, the vitamin undergoes other metabolic alterations which have yet to be identified. Many new metabolites are known plus several biliary metabolites which remain at this date unidentified (91–93). Much remains to be learned, therefore, in the exact metabolic pathway of vitamin D metabolism, not only for functional activation, but also degradation and excretion.

MECHANISM OF 25- AND 1-HYDROXYLATION OF THE VITAMIN D COMPOUNDS *IN VIVO*

It is indeed obvious that certainly the two initial hydroxylation reactions discussed represent activation reactions for the vitamin D molecule. Therefore, there is substantial interest in the nature of the enzymatic machinery involved and the regulation of them by physiological signals. Much work has been expended in elucidating the molecular mechanism of these hydroxylations, although much remains to be learned. In the case of the 25-hydroxylase, it is now well known that this hydroxylation occurs almost exclusively in the microsomal fraction or endoplasmic reticulum (3; Madhok and DeLuca, *unpublished results*). This reaction requires NADPH, molecular oxygen, magnesium ions, and a cytosolic fraction. The cytosolic fraction prevents destruction of the substrate and also prevents destruction of the product (3). It is unknown whether it carries out any catalytic function in this hydroxylation. The microsomal 25-hydroxylase is sensitive to metyrapone, aminoglutethemide, and (very importantly) it is inhibited by carbon monoxide when present at 90% of the atmosphere over a hydroxylation reaction occurring *in vitro* (Madhok and DeLuca, *unpublished results*). This inhibition can be partially reversed by exposure to white light, which provides strong evidence that the hydroxylase is a mixed-function monooxygenase dependent upon cytochrome P-450. Oxygen 18 experiments have been carried out which demonstrate clearly that the oxygen which is inserted on carbon 25 is derived from molecular oxygen (75). Recently, the 25-hydroxylase has been solubilized in our laboratory and shown to be composed of a least two enzymes, a flavoprotein and a cytochrome P-450 (Yoon and DeLuca, *unpublished results*). This reaction has been reconstituted, although each of the components has not yet been purified to homogeneity, thereby leaving open the possibility that other components are involved. The mechanism of this hydroxylation is shown in Fig. 6. This hydroxylase is not regulated by serum calcium, phosphorus, the parathyroid hormone, or the other signals known to regulate the 1-hydroxylase (Madhok and DeLuca, *unpublished results*). Instead, it represents an activation reaction providing the major circulating form of vitamin D.

Much work has been expended in elucidating the molecular mechanism of the 25-OH-D_3-1α-hydroxylase. This reaction is strictly mitochondrial, at least in birds where it has been studied to any degree (30,43). It requires a Krebs cycle substrate to provide reducing equivalents, molecular oxygen and magnesium ions (39,43). Oxygen 18 experiments have demonstrated that the oxygen

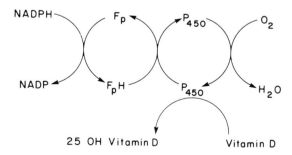

FIG. 6. Mechanism of 25-hydroxylation of vitamin D to produce 25-OH-D.

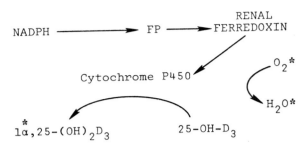

FIG. 7. Mechanism of hydroxylation of 25-OH-D to produce 1,25-(OH)₂D.

inserted in the 1α-hydroxyl function is derived from molecular oxygen (41). If the mitochondria are swollen or ruptured, the reaction can be supported by NADPH and magnesium ions (39). When the mitochondria are ruptured and NADPH is used as the reductant, oxidative phosphorylation is not required. On the other hand, intact mitochondria inhibitors of oxidative phosphorylation will inhibit the 1α-hydroxylation of 25-OH-D_3 (39,43). Inhibitors of cytochrome P-450, namely aminoglutethemide and metyrapone, strongly inhibit the 1α-hydroxylase (39). In addition, carbon monoxide–oxygen mixtures markedly suppress this hydroxylase, a suppression which is relieved by exposure to white light (39) or 450 nm wavelength light (47). This hydroxylase has been successfully solubilized, its components separated and reconstituted to form the 1α-hydroxylase (40). The components are (a) renal ferredoxin reductase; (b) renal ferredoxin, an 11,800 molecular weight iron sulfur protein containing two iron-two sulfur clusters (Yoon and DeLuca, *unpublished results*); and (c) cytochrome P-450. The reconstituted system very rapidly generates 1,25-(OH)$_2$D$_3$ *in vitro* (40,86). Therefore, the molecular mechanism of this hydroxylation is shown in Fig. 7. It is remarkably similar to the adrenal steroidogenesis reactions and the 11β hydroxylase reaction of the adrenal gland. This important parallel with the adrenal system underscores the concept that vitamin D is in fact a steroid hormone precursor.

It is of some interest that phenobarbital, a well-known inducer of cytochrome P-450 reactions, does not stimulate or enhance the vitamin D-25-hydroxylase in the liver and the renal 25-OH-D$_3$-1α-hydroxylase in the kidney (2; Madhok and DeLuca, *unpublished results*). It is likely that some other inducer might enhance these activities, but so far the classical phenobarbital induction is not seen with these two P-450 dependent enzyme systems.

The 24-hydroxylase described above is a mixed-function monooxygenase in which molecular oxygen is used for the hydroxyl placed in the 24R position (74). Little is known beyond that point regarding the 24-hydroxylase. In any case, as will be seen, physiologically the important enzyme from a regulatory point of view is the 25-OH-D$_3$-1α-hydroxylase which generates the potent calcium and phosphate mobilizing hormone, 1,25-(OH)$_2$D$_3$.

ATTEMPTS AT THE CHEMICAL SYNTHESIS OF VITAMIN D ANTAGONISTS

Because of the well known activation reactions in the liver and kidney, attempts have been made to chemically synthesize possible antivitamin D's which would block at either stage of hydroxylation. Most effort has been placed on the 25-hydroxylation stage inasmuch as 25-OH-D in large amounts can act directly on the target tissues as an analog of 1,25-$(OH)_2D_3$ (84,85). Therefore, blockage of the 1-hydroxylation reaction would not provide an absolute antagonist to vitamin D function. We have synthesized 25-aza-vitamin D_3 25-fluoro-vitamin D_3, 24-dehydro-vitamin D_3, and 25-dehydro-vitamin D_3 (81). All of these compounds can be shown to block 25-hydroxylation when given to animals and when the hydroxylase is measured either *in vivo* or *in vitro*. The most potent of these is the 25-fluorovitamin D_3 (72). Disappointingly, however, when administered to vitamin D-deficient animals, the 24-dehydro, the 25-dehydro, and the 25-fluoro vitamin D_3 all have significant biological activity in their own right, which prevents their administration as an antagonist. On the other hand, the 25-aza-vitamin D_3 showed no biological activity. This compound proved to be an effective antagonist in blocking the physiologic responses to vitamin D_3 (83). The target tissue responses of intestine and bone to vitamin D_3 are blocked by the 25-aza-vitamin D_3 compound when given in 1,500-fold excess, but that same amount of the 25-aza compound does not block the target organ responses to 25-OH-D_3. Thus 25-aza-vitamin D_3 appears to be a true antagonist, although the amounts required make it not a very useful tool.

Another compound has been described by Norman et al. (78) as being an antagonist, namely the 19S-hydroxyvitamin D_3. We have chemically synthesized both isomers of this compound and have confirmed that the 19S-hydroxyvitamin D_3 is an antagonist of vitamin D at the 25-hydroxylation state (Paaren, Schnoes, and DeLuca, *unpublished results*). It also, however, is not a very effective antagonist. Therefore, the area of chemical synthesis of vitamin D antagonists remains open and there is need for a true antagonist of potent anti-vitamin activity.

REGULATION OF VITAMIN D METABOLISM BY THE NEED FOR CALCIUM

As might be expected, plasma calcium concentration negatively regulates directly or indirectly the production of 1,25-$(OH)_2D_3$ (5,6). Under conditions of normal blood calcium, 25-OH-D_3 is converted to both 24R,25-$(OH)_2D_3$ and 1,25-$(OH)_2D_3$. When blood calcium drops below normal, the production of 1,25-$(OH)_2D_3$ markedly increases and the 24R-hydroxylation diminishes. Similarly, under conditions of hypercalcemia, 1-hydroxylase is suppressed and 24-hydroxylation is stimulated (80). Thus, the need for calcium as shown by a decrease in plasma calcium concentration stimulates production of 1,25-$(OH)_2D_3$. Parathyroidectomy eliminates the response of the hydroxylase to hypo-

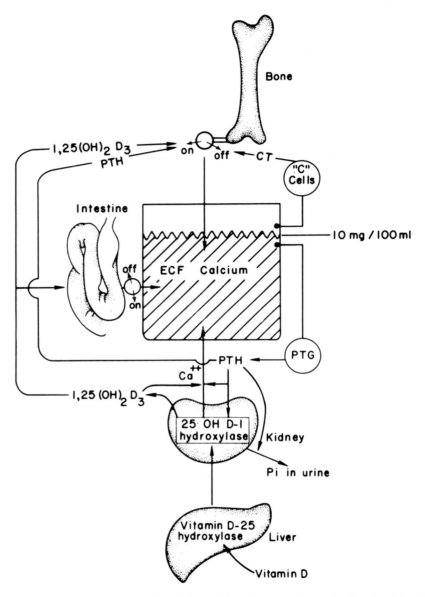

FIG. 8. Diagrammatic representation of the calcium homeostatic mechanism involving the vitamin D endocrine system and the calcium sensing organs namely, the parathyroid gland and the C-cells of the thyroid.

calcemia (36). Furthermore, the administration of parathyroid hormone to parathyroidectomized rats causes a marked stimulation of the 1α-hydroxylase (31). Thus, the sequence of events is that low plasma calcium is sensed by the parathyroid glands, which secrete parathyroid hormone in response to hypocalcemia.

This peptide hormone is bound by receptors in the kidney and in bone (116). In the kidney, it stimulates production of 1,25-$(OH)_2D_3$ in addition to its other functions. This discovery and the testing of the target organ responses to 1,25-$(OH)_2D_3$ in the presence and absence of the parathyroid hormone (38) has led to a reevaluation of the calcium homeostatic mechanism incorporating in it the vitamin D endocrine system. This system is shown in Fig. 8. When plasma calcium falls below 10 mg/100 ml, the parathyroid glands secrete parathyroid hormone. This hormone proceeds to kidney and bone. In the kidney, the parathyroid hormone stimulates phosphate diuresis in a vitamin D-independent reaction (29). It also stimulates calcium reabsorption in the distal renal tubules (103), a process which may be dependent upon the presence of 1,25-$(OH)_2D_3$ (102). Of great importance to the discussion here is that it stimulates production of 1,25-$(OH)_2D_3$. This steroid hormone then proceeds to the intestine, elsewhere in the kidney, and to bone. In the kidney, it perhaps functions together with parathyroid hormone to facilitate renal reabsorption of calcium (102). In the intestine, it functions by itself to stimulate intestinal calcium transport (38). In the bone, the mobilization of calcium from bone requires both parathyroid hormone and 1,25-$(OH)_2D_3$ (38). These three sources of calcium then elevate plasma calcium, suppressing parathyroid hormone secretion which in turn shuts down the entire calcium mobilizing system. If the animal becomes hypercalcemic, another endocrine system reacts—namely the C-cells of the thyroid—to produce the peptide hormone, calcitonin (50). This hormone functions by blocking the mobilization of calcium from the bone fluid compartment, thereby suppressing plasma calcium concentration. It does not involve vitamin D function as far as is known (73).

There are two major aspects of the calcium homeostatic mechanism that must be discussed. One involves calcium homeostasis as described above, the other is a description of regulation of overall calcium economy of the body. The vitamin D system is in itself a rather sluggish system requiring hours of parathyroid hormone stimulation before plasma 1,25-$(OH)_2D_3$ levels are elevated (36). In addition, 1,25-$(OH)_2D_3$ requires hours before its function can appear in the target organs (105). Its action in the target organs is of longer duration than the parathyroid hormone's action (20,25). Thus, the vitamin D system is not likely a candidate for minute-to-minute regulation of plasma calcium concentration. Instead, the parathyroids carry this function with endogenous levels of 1,25-$(OH)_2D_3$. Upon an immediate need for calcium, hypocalcemia would cause parathyroid hormone secretion within minutes. The parathyroid hormone would proceed to the kidney and to bone where, with endogenous levels of 1,25-$(OH)_2D_3$, calcium would be reabsorbed or mobilized. This would raise the calcium level in the plasma, shutting down the parathyroid hormone secretory system. We have, therefore, corrected the hypocalcemia but have done so at the expense of bone. Continued operation of this mechanism without the response of the intestine would certainly bring about an osteopenia resulting ultimately in an osteoporosis. However, if there is a chronic need for calcium so that parathyroid hormone secretion is continually elevated, this signal is integrated by the kidneys and in response they produce 1,25-$(OH)_2D_3$. The 1,25-$(OH)_2D_3$ is the only hormone that can stimulate the intestine. The intestine is the only organ that can bring calcium in from the environment in response to need. The 1,25-$(OH)_2D_3$ secreted in this system would also bring about an increased sensitivity of bone and kidney to the parathyroid hormone. Nevertheless, the bringing into play of the intestine by 1,25-$(OH)_2D_3$ serves an important function in protecting the skeleton against continual loss for support of plasma calcium concentration. This, therefore, can be regarded as an important aspect of calcium economy of the body, namely the ability of the intestine to elevate its efficiency of calcium absorption via the vitamin D system to meet the calcium needs of the organism. This important intestinal calcium adaptation system was discovered in the early 1930s and firmly established by Nicolaysen and his colleagues (77). To demonstrate that this system is parathyroid hormone and vitamin D dependent, animals that were given a low-calcium diet had a high efficiency of calcium transport, whereas those given a high-calcium diet had a low efficiency of calcium transport. On the other hand, when the same animals are given an exogenous source of 1,25-$(OH)_2D_3$, they lose their ability to adjust their intestinal calcium absorption according to dietary calcium (90). Similarly, if they are parathyroidectomized and the parathyroid hormone is replaced by a constant and exogenous source, again the intestine loses its ability to adjust its absorption to the dietary calcium levels (89).

There are, therefore, two different classes of diseases of the calcium homeostatic system. One is a defect in the regulation of plasma calcium concentration and the other is a disease of calcium economy of the body. The first class of diseases include hypoparathyroidism, pseudo-hypoparathyroidism, and neonatal hypocalcemia. In these diseases, there is a clear failure of the parathyroid system to signal the production of 1,25-$(OH)_2D_3$ and to work with the 1,25-$(OH)_2D_3$ to mobilize calcium to correct hypocalcemia (63,76). The best approach to clinical management of these diseases is to provide oral calcium plus 1,25-$(OH)_2D_3$ to stimulate the one calcium-mobilizing organ that does not require the parathyroid hormone—namely, the intestine. Thus 1,25-$(OH)_2D_3$ when given in small but frequent doses daily, can be used effectively in the management of these three diseases.

Work has just begun to determine whether the diseases of calcium economy are in fact defects in the calcium homeostatic system giving rise to poor calcium absorption and low plasma 1,25-$(OH)_2D_3$ levels. In particular, the disease of involutional osteoporosis and postmenopausal osteoporosis is receiving a considerable amount of attention. Figure 9 shows that as age progresses in rats, there is a fall in the ability of the intestine to concentratively transport calcium (56). Furthermore, the intestines lose their ability to adjust their efficiency of calcium absorption according to dietary levels of calcium and phosphorus. This loss in intestinal calcium transport activity with age can be directly related to a decrease in plasma 1,25-$(OH)_2D_3$ levels. A similar curve can be obtained for man although the results are not as dramatic as shown here for the rat. In addition, in the case of postmenopausal women suffering from osteoporosis, Table 1 shows that

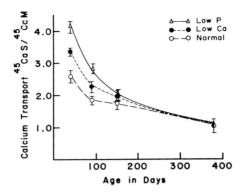

FIG. 9. The relationship between age and the concentrative transport of calcium in the duodenum of rats. Note also the influence of low dietary calcium and low dietary phosphate on the transport of calcium by the intestine in young animals versus old animals.

their plasma 1,25-$(OH)_2D_3$ levels are approximately 30% below age- and sex-matched controls not having osteoporosis (93). This decrease in plasma 1,25-$(OH)_2D_3$ level in the osteoporotic females correlates with their decreased ability to absorb calcium (94). It is, therefore, entirely possible that one aspect of the postmenopausal osteoporotic syndrome is an inadequate ability to absorb sufficient amounts of calcium because of insufficient levels of 1,25-$(OH)_2D_3$. The next question is: why would postmenopausal women have a diminished level of 1,25-$(OH)_2D_3$? This question is not entirely answered although it is known that in egg-laying birds, the sex hormones—estradiol and testosterone—when given together markedly stimulate 25-OH-D_3-1α-hydroxylase and markedly increase 1,25-$(OH)_2D_3$ levels in the plasma for the mobilization of calcium from bone, and for increased intestinal calcium absorption for egg shell formation (13,104). In fact the interrelationship between the sex hormones and 1,25-$(OH)_2D_3$ production is a very interesting and consuming subject, however, beyond the discussion here. Nevertheless, the administration of estrogen to postmenopausal osteoporotic females does bring about an increase in plasma 1,25-$(OH)_2D_3$ levels and an increased intestinal absorption of calcium (94). Of particu-

TABLE 1. *Plasma 1,25-$(OH)_2D_3$ levels in osteoporotic patients*

Subject	1,25-$(OH)_2D_3$	1,25-/25- $\times 10^{3\,a}$
Normal controls Age 50–65 ($N=20$)	35.2 ± 2.3	2.48 ± 0.25
Osteoporotic Age 50–65 ($N=25$)	25.8 ± 1.4^{b}	1.56 ± 0.17^{c}

[a] 1,25- = 1,25-dihydroxyvitamin D_3; 25- = 25-hydroxy-vitamin D_3
[b] Significantly different from control $p < 0.005$
[c] Significantly different from control $p < 0.001$

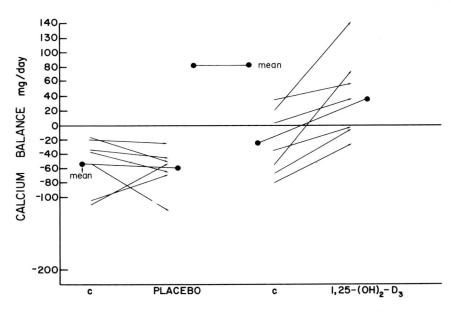

FIG. 10. The influence of 0.5 μg of 1,25-(OH)$_2$D$_3$ given orally on the calcium balance of postmenopausal osteoporotic females versus those given the placebo control.

lar importance is the fact that treatment of postmenopausal osteoporotic females with 0.5 μg of 1,25-(OH)$_2$D$_3$ orally each day results in a marked improvement in calcium absorption and a marked improvement in calcium balance. In this single blind study which was done at the Mayo Clinic in collaboration with our group, postmenopausal osteoporotic females were divided into two groups; one group received a placebo and the other group received 0.5 μg of 1,25-(OH)$_2$D$_3$ daily. Balance studies were carried out prior to the administration of the 1,25-(OH)$_2$D$_3$ and after 6 months of daily administration of the hormone. The results show that the calcium balance has gone from a very negative figure to approximately zero balance and in some cases to positive calcium balance (Fig. 10). This trend has remained for a period of one year but so far data are not available on the bone of these patients. The results, therefore, appear promising that there is some involvement of the vitamin D endocrine system in the genesis of a calcium economy disease such as postmenopausal osteoporosis. Furthermore, the results seem promising that this compound may be of use in treating patients suffering from this type of osteoporosis.

REGULATION OF 1,25-(OH)$_2$D$_3$ PRODUCTION BY 1,25-(OH)$_2$D$_3$ ITSELF AND BY PLASMA INORGANIC PHOSPHORUS CONCENTRATIONS

The administration of 1,25-(OH)$_2$D$_3$ to vitamin D-deficient chicks brings about a marked reduction in the 25-OH-D$_3$-1α-hydroxylase activity and a marked

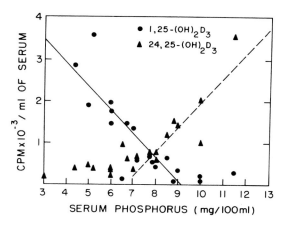

FIG. 11. The relationship between serum phosphate concentration and the accumulation of either 1,25-(OH)$_2$D$_3$ or 24,25-(OH)$_2$D$_3$ in plasma of rats. Rats were maintained on diets containing various amounts of calcium and phosphate. They were thyroparathyroidectomized, and after 48 hr their ability to convert radioactive 25-OH-D to the two indicated metabolites was examined.

elevation in the 25-HO-D$_3$-24R-hydroxylation activity of the kidney (72,109). Similar results can be obtained with rats. This change demonstrates that 1,25-(OH)$_2$D$_3$ in some manner functions to control these hydroxylases. This control appears to be blocked by the prior administration of protein and RNA synthesis inhibitors suggesting that new enzyme synthesis is involved in this regulation (72). Thus, it appears that in addition to the parathyroid system, 1,25-(OH)$_2$D$_3$ itself joins in the regulation of the 25-OH-D$_3$-1α-hydroxylase.

In parathyroidectomized animals, it can be readily demonstrated that circulating levels of 1,25-(OH)$_2$D$_3$ are markedly affected by plasma phosphate concentration (58,106) (Fig. 11). Under normal circumstances, young rats have serum inorganic phosphorus levels of 9-10 mg/100 ml. In the parathyroidectomized state, these animals make virtually no 1,25-(OH)$_2$D$_3$ and instead make 24,25-(OH)$_2$D$_3$. When they are made hypophosphatemic, plasma levels of 1,25-(OH)$_2$D$_3$ go up and the levels of 24,25-(OH)$_2$D$_3$ go down. Thus, the need for phosphate in some manner causes accumulation of 1,25-(OH)$_2$D$_3$ in the plasma. It is not clear whether this accumulation is solely due to increased production or whether there is an effect of phosphate deprivation on 1,25-(OH)$_2$D$_3$ metabolism. This remains to be determined.

Finally, it should be mentioned that other endocrine systems impinge on the vitamin D endocrine system when there are needs for calcium. For example, under conditions of rapid growth or in pregnancy and lactation, there are elevated levels of plasma 1,25-(OH)$_2$D$_3$ (57). There is evidence to suggest that growth hormone and prolactin may stimulate the 25-OH-D$_3$-1α-hydroxylase either directly or indirectly to increase circulating levels of this hormone to provide calcium for these unusual circumstances (99). This system, however, remains to be investigated more thoroughly.

OTHER DISEASES KNOWN WHICH INVOLVE DEFECTS IN THE VITAMIN D ENDOCRINE SYSTEM

Perhaps the clearest example of a defect in vitamin D metabolism in disease is renal osteodystrophy (23). In this disease, there is a loss of the organ responsible for production of the active form of vitamin D. In addition, phosphate accumu-

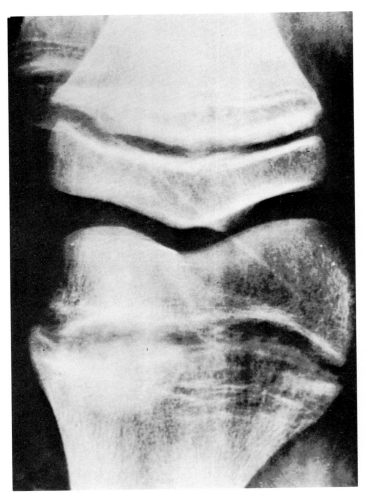

FIG. 12. The X-ray of a knee of an 8-year-old boy suffering from rickets secondary to chronic renal failure.

lates since the kidney is the major regulatory organ for inorganic phosphorus of the plasma (23,98). Both circumstances then—loss of kidney mass plus increase in phosphate levels in the extracellular fluid—depress production of 1,25-$(OH)_2D_3$. The lack of this hormone brings about a hypocalcemia and a counter-reaction by the parathyroid glands to secrete more and more parathyroid hormone. With the remaining levels of vitamin D metabolites, this increased circulating level of parathyroid hormone will mobilize bone, causing secondary hyperparathyroidism or osteitis fibrosa cystica. In addition, the lack of vitamin D metabolites brings about a failure to mineralize bone causing an osteomalacia. Administration of 1,25-$(OH)_2D_3$ to renal osteodystrophic patients brings about, in most of these subjects, marked improvement in their bone (8,97). Especially evident is what occurs in children (16). Figure 12 represents the bone of a

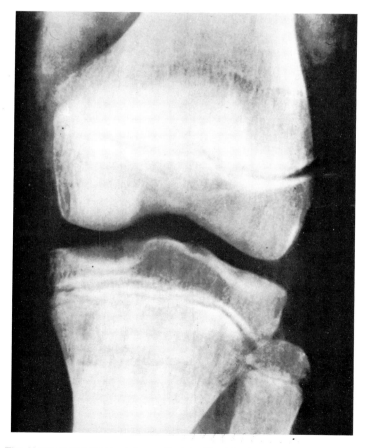

FIG. 13. The X-ray of the knee of the patient in Fig. 12 after 24 months of therapy with 1 µg of 1,25-$(OH)_2D_3$ orally per day.

child suffering from renal failure and consequent renal rickets. This child was treated for 24 months with 0.5 µg of 1,25-$(OH)_2D_3$ daily, bringing about a complete normalization of the bone as shown in Fig. 13. In addition, children suffering from renal failure receiving 1,25-$(OH)_2D_3$ show a marked growth response, correcting still another problem in the case of renal failure children.

In addition to the above, vitamin D-dependency rickets, a genetic disorder, is apparently a defect in 25-OH-D_3-1α-hydroxylation (32,95). Other vitamin D-resistant rickets have an involvement in vitamin D metabolism, although perhaps indirect (95). Nevertheless, the vitamin D metabolites are being used in the treatment of various forms of vitamin D resistant rickets.

THE FUNCTION OF 1,25-$(OH)_2D_3$ IN THE TARGET ORGANS, ESPECIALLY INTESTINE

With the chemical synthesis of high specific activity radioactive 1,25-$(OH)_2D_3$, it has been possible by frozen section autoradiography to demonstrate clearly the nuclear location of 1,25-$(OH)_2D_3$ in intestinal crypt cells and in intestinal villus cells (115). This subcellular localization occurs prior to the initiation of intestinal calcium transport and intestinal phosphate transport. Nuclear localization of 1,25-$(OH)_2D_3$ is not noted in submucosa, muscle, liver, and other nontarget organs (8). Figure 14 demonstrates a 1,000-times magnification of frozen

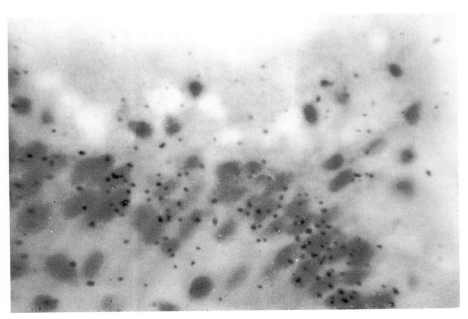

FIG. 14. The frozen section autoradiography of the small intestine of a vitamin D-deficient chicken given radioactive 1,25-$(OH)_2D_3$ (650 pmol) 2½ hours earlier.

sections of intestinal villi taken from chickens given a single physiologic dose of radioactive 1,25-$(OH)_2D_3$. Note the nuclear location of the grains resultant from the radioactivity, whereas no grains are noted in the cytoplasm or the brush border membrane. Recently, this has been considerably expanded using the rat and very high specific activity 1,25-$(OH)_2D_3$ chemically synthesized in this laboratory (Sar, Stumpf, and DeLuca, *unpublished results*). One therefore anticipates that the mechanism of action of 1,25-$(OH)_2D_3$ in the target tissues might be nuclear-mediated. In agreement with this, cytosol receptors have been detected in chick intestine (9,66), rat intestine (67), and chick and rat bone (65) using sucrose density gradient analysis and radioactive 1,25-$(OH)_2D_3$. The chick intestinal cytosol receptor is highly specific for 1,25-$(OH)_2D_3$ (27,64). Similar results can be obtained with the rat intestinal receptor and the chick bone receptor.

Figure 15 demonstrates the best possible guess regarding how 1,25-$(OH)_2D_3$ itself might function in stimulating intestinal calcium transport. It is believed that this hormone interacts with a 3.7S cytosol receptor in the case of the chick, and the receptor plus 1,25-$(OH)_2D_3$ is transferred to the nucleus where

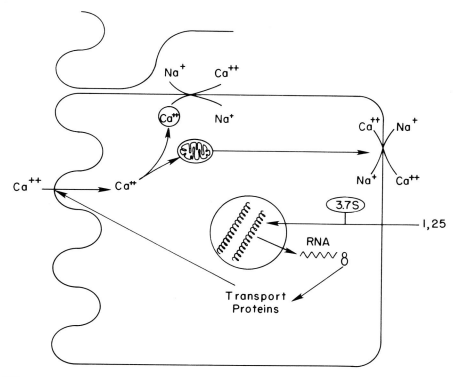

FIG. 15. Diagrammatic representation of the postulated mechanism whereby 1,25-$(OH)_2D_3$ stimulates intestinal transport of calcium and phosphorus in the intestinal epithelial cells.

it becomes bound to specific sites on chromatin (10). This leads to a transcription of messenger RNA coding for calcium and phosphorus transport proteins. The nature of these proteins is not known although a calcium binding protein of 24,000 molecular weight has been discovered by Wasserman et al. and is suggested to be one of the calcium transport components (113). This protein, as well as other proteins made in response to 1,25-$(OH)_2D_3$, appear at the brush border where they facilitate the transfer of calcium into the cell from the intestinal lumen (21). Calcium then is believed to be transferred to the basal-lateral membrane either by vesicles or by mitochondria (21). There, the calcium is released by a sodium-dependent transport system. Much remains to be learned regarding the molecular mechanism of 1,25-$(OH)_2D_3$, but at least rudiments of the system are now being seen with the tools available from the information gathered regarding the functional metabolism of vitamin D.

SUMMARY

Much has been learned regarding the metabolism and function of vitamin D, revealing that vitamin D is converted to at least one hormone which regulates calcium and phosphorus metabolism. This hormone is produced in the kidney from the major circulating form of vitamin D and is known as 1,25-$(OH)_2D_3$. The production of this hormone is regulated by the need for calcium, the need for phosphorus, by 1,25-$(OH)_2D_3$ itself, and by other hormonal systems, either directly or indirectly. 1,25-$(OH)_2D_3$ then plays an important role in the regulation of plasma calcium and in regulating calcium and phosphate economy of the body. Disturbances in this system are now being recognized and the use of the active forms of vitamin D to treat disease is well on its way. Thus, a basic investigation into understanding how vitamin D functions has led to important practical benefits to health and industry of man.

ACKNOWLEDGMENTS

This work was supported in part by a program-project grant no. AM 14881 of the NIH and the Steenbock Fund of the Wisconsin Alumni Research Foundation.

REFERENCES

1. Bhattacharyya, M. H., and DeLuca, H. F. (1973): The regulation of rat liver calciferol-25-hydroxylase. *J. Biol. Chem.*, 248:2969–2973.
2. Bhattacharyya, M. H., and DeLuca, H. F. (1973): Comparative studies on the 25-hydroxylation of vitamin D_3 and dihydrotachysterol$_3$. *J. Biol. Chem.*, 248:2974–2977.
3. Bhattacharyya, M., and DeLuca, H. F. (1974): Subcellular location of rat liver calciferol-25-hydroxylase. *Arch. Biochem. Biophys.*, 160:58–62.
4. Boris, A., Hurley, J. F., and Trmal, T. (1977): Relative activities of some metabolites and analogs of cholecalciferol in stimulation of tibia ash weight in chicks otherwise deprived of vitamin D. *J. Nutr.*, 197:194–198.

5. Boyle, I. T., Gray, R. W., and DeLuca, H. F. (1971): Regulation by calcium of *in vivo* synthesis of 1,25-dihydroxycholecalciferol and 21,25-dihydroxycholecalciferol. *Proc. Natl. Acad. Sci. USA,* 68:2131–2134.
6. Boyle, I. T., Gray, R. W., Omdahl, J. L., and DeLuca, H. F. (1972): Calcium control of the *in vivo* biosynthesis of 1,25-dihydroxyvitamin D_3: Niclaysen's endogenous factor. In: *Endocrinology 1971* (Proceedings of the Third International Symposium), edited by S. Taylor, pp. 468–476. Wm. Heinemann Medical Books, London.
7. Boyle, I.T., Miravet, L., Gray, R. W., Holick, M. F., and DeLuca, H. F. (1972): The response of intestinal calcium transport to 25-hydroxy and 1,25-dihydroxyvitamin D in nephrectomized rats. *Endocrinology,* 90:605–608.
8. Brickman, A. S., Sherrard, D. J., Jowsey, J., Singer, F. R., Baylink, D. J., Maloney, N., Massry, S. G., Norman, A. W., and Coburn, J. W. (1974): 1,25-Dihydroxycholecalciferol. Effect on skeletal lesions and plasma parathyroid hormone levels in uremic osteodystrophy. *Arch. Intern. Med.,* 134:883–888.
9. Brumbaugh, P. F., and Haussler, M. R. (1973): 1α,25-Dihydroxyvitamin D_3 receptor: Competitive binding of vitamin D analogs. *Life Sci.,* 13:1737–1746.
10. Brumbaugh, P. F., and Haussler, M. R. (1974): 1α,25-Dihydroxycholecalciferol receptors in intestine. II. Temperature-dependent transfer of the hormone to chromatin via a specific cytosol receptor. *J. Biol. Chem.,* 249:1258–1262.
11. Carlsson, A. (1952): Tracer experiments on the effect of vitamin D on the skeletal metabolism of calcium and phosphorus. *Acta Physiol. Scand.,* 26:212–220.
12. Castillo, L., Tanaka, Y., DeLuca, H. F., and Ikekawa, N. (1978): On the physiological role of 1,24,25-trihydroxyvitamin D_3. *Mineral Electrol. Metab.,* 1:198–207.
13. Castillo, L., Tanaka, Y., DeLuca, H. F., and Sunde, M. L. (1977): The stimulation of 25-hydroxyvitamin D_3-1α-hydroxylase by estrogen. *Arch. Biochem. Biophys.,* 179:211–217.
14. Chen, T. C., Castillo, L., Korycka-Dahl, M., and DeLuca, H. F. (1974): Role of vitamin D metabolites in phosphate transport of rat intestine. *J. Nutr.,* 104:1056–1060.
15. Chertow, B. S., Baylink, D. J., Wergedal, J. E., Su, M. H. H., and Norman, A. W. (1975): Decrease in serum immunoreactive parathyroid hormone in rats and in parathyroid hormone secretion *in vitro* by 1,25-dihydroxycholecalciferol. *J. Clin. Invest.,* 56:668–678.
16. Chesney, R. W., Moorthy, A. V., Eisman, J. E., Jax, D. K., Mazess, R. B., and DeLuca, H. F. (1978): Increased growth after long-term oral 1α,25-vitamin D_3 in childhood renal osteodystrophy. *New Engl. J. Med.,* 298:238–242.
17. Curry, O. B., Basten, J. F., Francis, M. J. O., and Smith, R. (1974): Calcium uptake by sarcoplasmic reticulum of muscle from vitamin D-deficient rabbits. *Nature,* 249:83–84.
18. DeLuca, H. F. (1967): Mechanism of action and metabolic fate of vitamin D. *Vitam. Horm.,* 25:315–367.
19. DeLuca, H. F. (1974): Vitamin D: The vitamin and the hormone. *Fed. Proc.,* 33:2211–2219.
20. DeLuca, H. F., editor (1978): Vitamin D. In: *The Fat-Soluble Vitamins, Vol. II of Handbook of Lipid Research,* pp. 69–132. Plenum Press, New York.
21. DeLuca, H. F. (1978): Vitamin D and calcium transport. In: *Calcium Transport and Cell Function,* Vol. 307 of Annals of the New York Academy of Sciences, edited by A. Scarpa and E. Carafoli, pp. 356–376. New York Academy of Sciences, New York.
22. DeLuca, H. F. (1978): Vitamin D metabolism and function. *Arch. Intern. Med.,* 138:836–847.
23. DeLuca, H. F., and Avioli, L. V. (1979): Renal osteodystrophy. In: *Renal Disease,* 4th edition, edited by D. Black. Blackwell Scientific Publications, Oxford (*in press*).
24. DeLuca, H. F., Blunt, J. W., and Rikkers, H. (1971): Biogenesis. In: *The Vitamins,* Chapter 7, Vol. III, edited by W. H. Sebrell, Jr. and R. S. Harris, pp. 213–230. Academic Press, New York.
25. DeLuca, H. F., and Schnoes, H. K. (1976): Metabolism and mechanism of action of vitamin D. *Ann. Rev. Biochem.,* 45:631–666.
26. DeLuca, H. F., and Schnoes, H. K. (1979): Recent advances in the metabolism and regulation of vitamin D. *Proc. Fourth Workshop on Vitamin D.* Feb. 18–22, 1979. West Berlin, Germany, (Abst.), p.115.
27. Eisman, J. A., and DeLuca, H. F. (1977): Intestinal 1,25-dihydroxyvitamin D_3 binding protein: Specificity of binding. *Steroids,* 30:245–257.
28. Esvelt, R. P., Schnoes, H. K., and DeLuca, H. F. (1978): Vitamin D_3 from rat skins irradiated *in vitro* with ultraviolet light. *Arch. Biochem. Biophys.,* 188:282–286.

29. Forte, L. R., Nickols, G. A., and Anast, C. S. (1976): Renal adenylate cyclase and the interrelationship between parathyroid hormone and vitamin D in the regulation of urinary phosphate and adenosine cyclic 3′,5′-monophosphate excretion. *J. Clin. Invest.*, 57:559–568.
30. Fraser, D. R., and Kodicek, E. (1970): Unique biosynthesis by kidney of a biologically active vitamin D metabolite. *Nature*, 228:764–766.
31. Fraser, D. R., and Kodicek, E. (1973): Regulation of 25-hydroxycholecalciferol-1-hydroxylase activity in kidney by parathyroid hormone. *Nat. New Biol.*, 241:163–166.
32. Fraser, D., Kooh, S. W., Kind, H. P., Holick, M. F., Tanaka, Y., and DeLuca, H. F. (1973): Pathogenesis of hereditary vitamin D dependent rickets: An inborn error of vitamin D metabolism involving defective conversion of 25-hydroxyvitamin D to $1\alpha,25$-dihydroxyvitamin D. *New Engl. J. Med.*, 289:817–822.
33. Frolik, C. A., and DeLuca, H. F. (1971): 1,25-Dihydroxycholecalciferol: The metabolite of vitamin D responsible for increased intestinal calcium transport. *Arch. Biochem. Biophys.*, 147:143–147.
34. Frolik, C. A., and DeLuca, H. F. (1972): Metabolism of 1,25-dihydroxycholecalciferol in the rat. *J. Clin. Invest.*, 51:2900–2906.
35. Frost, H. M. (1966): *Bone Dynamics in Osteoporosis and Osteomalacia.* Henry Ford Hospital Surgical Monograph Series, Charles C Thomas, Springfield, Illinois.
36. Garabedian, M., Holick, M. F., DeLuca, H. F., and Boyle, I. T. (1972): Control of 25-hydroxycholecalciferol metabolism by the parathyroid glands. *Proc. Natl. Acad. Sci. USA*, 69:1673–1676.
37. Garabedian, M., Lieberherr, M., Nguyen, T. M., Corvol, M. T., Dubois, M. B., and Balsan, S. (1978): *In vitro* production and activity of 24,25-dihydroxycholecalciferol in cartilage and calvarium. *Clin. Orthop. Rel. Res.*, 135:241.
38. Garabedian, M., Tanaka, Y., Holick, M. F., and DeLuca, H. F. (1974): Response of intestinal calcium transport and bone calcium mobilization to 1,25-dihydroxyvitamin D_3 in thyroparathyroidectomized rats. *Endocrinology*, 94:1022–1027.
39. Ghazarian, J. G., and DeLuca, H. F. (1974): 25-Hydroxycholecalciferol-1-hydroxylase: A specific requirement for NADPH and a hemoprotein component in chick kidney mitochondria. *Arch. Biochem. Biophys.*, 160:63–72.
40. Ghazarian, J. G., Jefcoate, C. R., Knutson, J. C., Orme-Johnson, W. H., and DeLuca, H. F. (1974): Mitochondrial cytochrome P_{450}: A component of chick kidney 25-hydroxycholecalciferol-1α-hydroxylase. *J. Biol. Chem.*, 249:3026–3033.
41. Ghazarian, J. G., Schnoes, H. K., and DeLuca, H. F. (1973): Mechanism of 25-hydroxycholecalciferol 1α-hydroxylation. Incorporation of oxygen-18 into the 1α position of 25-hydroxycholecalciferol. *Biochemistry*, 12:2555–2558.
42. Gray, R., Boyle, I., and DeLuca, H. F. (1971): Vitamin D metabolism: The role of kidney tissue. *Science*, 172:1232–1234.
43. Gray, R. W., Omdahl, J. L., Ghazarian, J. G., and DeLuca, H. F. (1972): 25-Hydroxycholecalciferol-l-hydroxylase: Subcellular location and properties. *J. Biol. Chem.*, 247:7528–7532.
44. Haddad, J. G., and Stamp, T. C. B. (1974): Circulating 25-hydroxyvitamin D in man. *Am. J. Med.*, 57:57–62.
45. Harnden, D., Kumar, R., Holick, M. F., and DeLuca, H. F. (1976): Side chain metabolism of 25-hydroxy-[26,27-^{14}C]vitamin D_3 and 1,25-dihydroxy-[26,27-^{14}C]vitamin D_3 *in vivo.* *Science*, 193:493–494.
46. Harrison, H. E., and Harrison, H. C. (1961): Intestinal transport of phosphate: Action of vitamin D, calcium, and potassium. *Am. J. Physiol.*, 201:1007–1012.
47. Henry, H. L., and Norman, A. W. (1974): Studies on calciferol metabolism. IX. Renal 25-hydroxy-vitamin D_3-l-hydroxylase. Involvement of cytochrome P-450 and other properties. *J. Biol. Chem.*, 249:7529–7535.
48. Henry, H. L., and Norman, A. W. (1978): Vitamin D: Two dihydroxylated metabolites are required for normal chicken egg hatchability. *Science*, 201:835–837.
49. Henry, H. L., Taylor, A. N., and Norman, A. W. (1977): Response of chick parathyroid glands to the vitamin D metabolites 1,25-dihydroxyvitamin D_3 and 24,25-dihydroxyvitamin D_3. *J. Nutr.*, 107:1918–1926.
50. Hirsch, P. F., Cooper, C. W., Peng, T-C., Burford, H. J., Toverud, S. U., Gray, T. K., Juan, D., Ontjes, D. A., Hennessy, J. F., and Munson, P. L. (1975): Effects, functions and endocrine interrelations of thyrocalcitonin. In: *Calcium Regulating Hormones*, edited by R. V. Talmage, M. Owen, and J. A. Parsons, pp. 89–99. Excerpta Medica, Amsterdam.

51. Holick, M. F., and Clark, M. B. (1978): The photobiogenesis and metabolism of vitamin D. *Fed. Proc.*, 37:2567–2574.
52. Holick M. F., Baxter, L. A., Schraufrogel, P. K., Tavela, T. E., and DeLuca, H. F. (1976): Metabolism and biological activity of 24,25-dihydroxyvitamin D_3 in the chick. *J. Biol. Chem.*, 251:397–402.
53. Holick, M. F., Frommer, J. E., McNeill, S. C., Richtand, N. M., Henley, J. W., and Potts, J. T. Jr. (1977): Photometabolism of 7-dehydrocholesterol to previtamin D_3 in skin. *Biochem. Biophys. Res. Commun.*, 76:107–114.
54. Holick, M. F., Garabedian, M., and DeLuca, H. F. (1972): 1,25-Dihydroxycholecalciferol: Metabolite of vitamin D_3 active on bone in anephric rats. *Science*, 176:1146–1147.
55. Holick, M. F., Schnoes, H. K., DeLuca, H. F., Suda, T., and Cousins, R. J. (1971): Isolation and identification of 1,25-dihydroxycholecalciferol. A metabolite of vitamin D active in intestine. *Biochemistry*, 10:2799–2804.
56. Horst, R. L., DeLuca, H. F., and Jorgensen, N. A. (1978): The effect of age on calcium absorption and accumulation of 1,25-dihydroxyvitamin D_3 in intestinal mucosa of rats. *Metab. Bone Dis. Rel. Res.*, 1:29–33.
57. Hughes, M. R., Baylink, D. J., Gonnerman, W. A., Toverud, S. U., Ramp, W. K., and Haussler, M. R. (1977): Influence of dietary vitamin D_3 on the circulating concentration of its active metabolites in the chick and rat. *Endocrinology*, 100:799–806.
58. Hughes, M. R., Brumbaugh, P. F., Haussler, M. R., Wergedal, J. E., and Baylink, D. J. (1975): Regulation of serum $1\alpha,25$-dihydroxyvitamin D_3 by calcium and phosphate in the rat. *Science*, 190:578–580.
59. Jones, G. (1978): Assay of vitamins D_2 and D_3, and 25-hydroxyvitamins D_2 and D_3 in human plasma by high performance liquid chromatography. *Clin. Chem.*, 24:287–298.
60. Jones, G., Schnoes, H. K., and DeLuca, H. F. (1976): An *in vitro* study of vitamin D_2 hydroxylases in the chick. *J. Biol. Chem.*, 251:24–28.
61. Knutson, J. C., and DeLuca, H. F. (1974): 25-Hydroxyvitamin D_3-24-hydroxylase: Subcellular location and properties. *Biochemistry*, 13:1543–1548.
62. Kobayashi, Y., Taguchi, T., Ikekawa, N., Morisaki, M. (1979): Synthesis of 24,24-difluoro-25-hydroxycholecalciferol. *Tetrahedron Lettr. (in press)*.
63. Kooh, S. W., Fraser, D., DeLuca, H. F., Holick, M. F., Belsey, R. E., Clark, M. B., and Murray, T. M. (1975): Treatment of hypoparathyroidism and pseudohypoparathyroidism with metabolites of vitamin D: Evidence for impaired conversion of 25-hydroxyvitamin D to 1α, 25-dihydroxyvitamin D. *New Engl. J. Med.*, 293:840–844.
64. Kream, B. E., Jose, M. J. L., and DeLuca, H. F. (1977): The chick intestinal cytosol binding protein for 1,25-dihydroxyvitamin D_3: A study of analog binding. *Arch. Biochem. Biophys.*, 179:462–468.
65. Kream, B. E., Jose, M., Yamada, S., and DeLuca, H. F. (1977): A specific high-affinity binding macromolecule for 1,25-dihydroxyvitamin D_3 in fetal bone. *Science*, 197:1086–1088.
66. Kream, B. E., Reynolds, R. D., Knutson, J. C., Eisman, J. A., and DeLuca, H. F. (1976): Intestinal cytosol binders of 1,25-dihydroxyvitamin D_3 and 25-hydroxyvitamin D_3. *Arch. Biochem. Biophys.*, 176:779–787.
67. Kream, B. E., Yamada, S., Schnoes, H. K., and DeLuca, H. F. (1977): Specific cytosol binding protein for 1,25-dihydroxyvitamin D_3 in rat intestine. *J. Biol. Chem.*, 252:4501–4505.
68. Kumar, R., Harnden, D., and DeLuca, H. F. (1976): Metabolism of 1,25-dihydroxyvitamin D_3: Evidence for side-chain oxidation. *Biochemistry*, 15:2420–2423.
69. Kumar, R., Schnoes, H. K., and DeLuca, H. F. (1978): Rat intestinal 25-hydroxyvitamin D_3- and $1\alpha, 25$-dihydroxyvitamin D_3-24-hydroxylase. *J. Biol. Chem.*, 253:3804–3809.
70. Lam, H-Y., Schnoes, H. K., and DeLuca, H. F. (1975): Synthesis and biological activity of 25ξ, 26-dihydroxycholecalciferol. *Steroids*, 25:247–256.
71. Lamm, M., and Neuman, W. F. (1958): On the role of vitamin D in calcification. *Arch. Pathol.*, 66:204–209.
72. Larkins, R. G., MacAuley, S. J., and MacIntyre, I. (1974): Feedback control of vitamin D metabolism by nuclear action of 1,25-dihydroxyvitamin D_3 on the kidney. *Nature*, 252:412–414.
73. Lorenc, R., Tanaka, Y., DeLuca, H. F., and Jones, G. (1977): Calcitonin and regulation of vitamin D metabolism. *Endocrinology*, 100:468–472.
74. Madhok, T. C., Schnoes, H. K., and DeLuca, H. F. (1977): Mechanism of 25-hydroxyvitamin

D_3 24-hydroxylation: Incorporation of oxygen-18 into the 24 position of 25-hydroxyvitamin D_3. *Biochemistry,* 16:2142–2145.
75. Madhok, T. C., Schnoes, H. K., and DeLuca, H. F. (1978): Incorporation of oxygen-18 into the 25-position of cholecalciferol by hepatic cholecalciferol 25-hydroxylase. *Biochem. J.,* 175:479–482.
76. Neer, R. M., Holick, M. F., DeLuca, H. F., and Potts, J. T. Jr. (1975): Effects of 1α-hydroxyvitamin D_3 and 1,25-dihydroxyvitamin D_3 on calcium and phosphorus metabolism in hypoparathyroidism. *Metabolism,* 24:1403–1413.
77. Nicolaysen, R., Eeg-Larsen, N., Malm, O. J. (1953): Physiology of calcium metabolism. *Physiol. Rev.,* 33:424–444.
78. Norman, A. W., Hammond, M. L., and Okamura, W. H. (1977): 19-Hydroxy-10S(19)-dihydroxyvitamin D_3-II: An analog of vitamin D_3 which possesses "Anti-vitamin" activity. *Fed. Proc.* (Abst.), 36:914.
79. Olson, E. B., Jr., Knutson, J. C., Bhattacharyya, M. H., and DeLuca, H. F. (1976): The effect of hepatectomy on the synthesis of 25-hydroxyvitamin D_3. *J. Clin. Invest.,* 57:1213–1220.
80. Omdahl, J. L., Gray, R. W., Boyle, I. T., Knutson, J., and DeLuca, H. F (1972): Regulation of metabolism of 25-hydroxycholecalciferol by kidney tissue *in vitro* by dietary calcium. *Nat. New Biol.,* 237:63–64.
81. Onisko, B. L., Schnoes, H. K., and DeLuca, H. F. (1977): Synthesis of potential vitamin D antagonists. *Tetrahedron Lettr.,* 13:1107–1108.
82. Onisko, B. L., Schnoes, H. K., and DeLuca, H. F. (1979): Inhibitors of vitamin D metabolism and action. *Fourth Workshop on Vitamin D,* Feb. 17–22, 1979, Walter de Gruyter, Berlin (in press).
83. Onisko, B. L., Schnoes, H. K., and DeLuca, H. F. (1979): 25-Azavitamin D_3, an inhibitor of vitamin D metabolism and action. *J. Biol. Chem.,* 254:3493–3496.
84. Pavlovitch, H., Garabedian, M., and Balsan, S. (1973): Calcium-mobilizing effect on large doses of 25-hydroxycholecalciferol in anephric rats. *J. Clin. Invest.,* 52:2656–2660.
85. Pechet, M. M., and Hesse, R. H. (1974): Metabolic and clinical effects on pure crystalline 1α-hydroxyvitamin D_3 and 1α, 25-dihydroxyvitamin D_3. Studies of intestinal calcium transport, renal tubular function and bone metabolism. *Am. J. Med.,* 57:13–20.
86. Pedersen, J. I., Ghazarian, J. G., Orme-Johnson, N. R., and DeLuca, H. F. (1976): Isolation of chick renal mitochondiral ferredoxin active in the 25-hydroxyvitamin D_3-1α-hydroxylase system. *J. Biol. Chem.,* 251:3933–3941.
87. Ponchon, G., and DeLuca, H. F. (1969): The role of the liver in the metabolism of vitamin D. *J. Clin. Invest.,* 48:1273–1279.
88. Rasmussen, H., and Bordier, P. (1978): Vitamin D and bone. *Metabolic Bone Dis. Related Res.,* 1:7–13.
89. Ribovich, M. L., and DeLuca, H. F. (1975): The influence of dietary calcium and phosphorus on intestinal calcium transport in rats given vitamin D metabolities. *Arch. Biochem. Biophys.,* 170:529–535.
90. Ribovich, M. L., and DeLuca, H. F. (1976): Intestinal calcium transport: Parathyroid hormone and adaption to dietary calcium. *Arch. Biochem. Biophys.,* 175:256–261.
91. Ribovich, M. L., and DeLuca, H. F. (1978): Effect of dietary calcium and phosphorus on intestinal calcium absorption and vitamin D metabolism. *Arch. Biochem. Biophys.,* 188:145–156.
92. Ribovich, M. L., and DeLuca, H. F. (1978): Adaption of intestinal calcium absorption: Parathyroid hormone and vitamin D metabolism. *Arch. Biochem. Biophys.,* 188:157–163.
93. Ribovich, M. L., and DeLuca, H. F. (1978): 1,25-Dihydroxyvitamin D_3 metabolism: The effect of dietary calcium and phosphorus. *Arch. Biochem. Biophys.,* 188:164–171.
94. Riggs, B. L., and Gallagher, J. C. (1977): Evidence for bihormonal deficiency state (estrogen and 1,25-dihydroxyvitamin D) in patients with postmenopausal osteoporosis. In: *Vitamin D: Biochemical, Chemical and Clinical Aspects Related to Calcium Metabolism,* edited by A. W. Norman, K. Schaefer, J. W. Coburn, H. F. DeLuca, D. Fraser, H. G. Grigoleit, and D. von Herrath, pp. 639–648. Walter de Gruyter, Berlin.
95. Scriver, C. R., Reade, T. M., DeLuca, H. F., and Hamstra, A. J. (1979): Serum 1,25-$(OH)_2D_3$ levels in normal subjects and in patients with hereditary rickets or bone disease. *New Engl. J. Med.,* 299:976–979.

96. Sebrell, W. H., Jr., and Harris, R. S., editors (1954): *The Vitamins, Vol. II* (first edition). Academic Press, New York.
97. Silverberg, D. S., Bettcher, K. B., Dossetor, J. B., Overton, T. R., Holick, M. F., and DeLuca, H. F. (1975): Effect of 1,25-dihydroxycholecalciferol in renal osteodystrophy. *Can. Med. Assoc. J.*, 112:190–195.
98. Slatopolsky, E., Rutherford, W. E., Hoffsten, P. E., Elkan, I. O., Butcher, H. R., and Bricker, N. S. (1972): Non-suppressible secondary hyperparathyroidism in chronic progressive renal disease. *Kidney Intern.*, 1:38–46.
99. Spanos, E., and MacIntyre, I. (1977): Vitamin D and the pituitary. *Lancet,* 1:840–841.
100. Stern, P. H., Trummel, C. L., Schnoes, H. K., and DeLuca, H. F. (1975): Bone resorbing activity of vitamin D metabolites and congeners *in vitro:* Influence of hydroxyl substituents in the A ring. *Endocrinology,* 97:1552–1558.
101. Suda, T., DeLuca, H. F., Schnoes, H. K., Tanaka, Y., and Holick, M. F. (1970): 25,26-Dihydroxycholecalciferol, a metabolite of vitamin D_3 with intestinal calcium transport activity. *Biochemistry,* 9:4776–4780.
102. Sutton, R. A. L., and Dirks, J. H. (1978): Renal handling of calcium. *Fed. Proc.,* 37:2112–2119.
103. Sutton, R. A. L., Harris, C. A., Wong, N. L. M., and Dirks, J. (1977): Effects of vitamin D on renal tubular calcium transport. In: *Vitamin D: Biochemical, Chemical, and Clinical Aspects Related to Calcium Metabolism,* edited by A. W. Norman, K. Schaefer, J. W. Coburn, H. F. DeLuca, D. Fraser, H. G. Grigoleit, and D. von Herrath, pp. 451–453. Walter de Gruyter, Berlin.
104. Tanaka, Y., Castillo, L., and DeLuca, H. F. (1976): Control of the renal vitamin D hydroxylases in birds by the sex hormones. *Proc. Natl. Acad. Sci. USA,* 73:2701–2705.
105. Tanaka, Y., and DeLuca, H. F. (1971): Bone mineral mobilization activity of 1,25-dihydroxycholecalciferol, a metabolite of vitamin D. *Arch. Biochem. Biophys.,* 146:574–578.
106. Tanaka, Y., and DeLuca, H. F. (1973): The control of 25-hydroxyvitamin D metabolism by inorganic phosphorus. *Arch. Biochem. Biophys.,* 154:566–574.
107. Tanaka, Y., DeLuca, H. F., and Ikekawa, N. (1979): An investigation into the importance of 24-hydroxylation to the function of vitamin D. *Proc. Fourth Workshop on Vitamin D.* Feb. 18–22, 1979. West Berlin, Germany, (Abstr.), p. 137.
108. Tanaka, Y., DeLuca, H. F., Ikekawa, N., Morisaki, M., and Koizumi, N. (1975): Determination of stereochemical configuration of the 24-hydroxyl group of 24,25-dihydroxyvitamin D_3 and its biological importance. *Arch. Biochem. Biophys.,* 170:620–626.
109. Tanaka, Y., Lorenc, R. S., and DeLuca, H. F. (1975): The role of 1,25-dihydroxyvitamin D_3 and parathyroid hormone in the regulation of chick renal 25-hydroxyvitamin D_3-24-hydroxylase. *Arch. Biochem. Biophys.,* 171:521–526.
110. Tanaka, Y., Shepard, R. M., DeLuca, H. F., and Schnoes, H. K. (1978): The 26-hydroxylation of 25-hydroxyvitamin D_3 *in vitro* by chick renal homogenates. *Biochem. Biophys. Res. Commun.,* 83:7–13.
111. Trummel, C. L., Raisz, L. G., Blunt, J. W., and DeLuca, H. F. (1969): 25-Hydroxycholecalciferol: Stimulation of bone resorption in tissue culture. *Science,* 163:1450–1451.
112. Tsai, H. C., Wong, R. G., and Norman, A. W. (1972): Studies on calciferol metabolism. IV. Subcellular localization of 1,25-dihydroxy-vitamin D_3 in intestinal mucosa and correlation with increased calcium transport. *J. Biol. Chem.,* 247:5511–5519.
113. Wasserman, R. H., and Feher, J. J. (1977): Vitamin D-dependent calcium-binding proteins. In: *Calcium Binding Proteins and Calcium Function,* edited by R. H. Wasserman, R. A. Corradino, E. Carafoli, R. H. Kretsinger, D. H. MacLennan, and S. L. Siegel, pp. 292–302. Elsevier North-Holland, Amsterdam.
114. Wong, R. G., Norman, A. W., Reddy, C. R., and Coburn, J. W. (1972): Biological effects of 1,25-dihydroxycholecalciferol (a highly active vitamin D metabolite) in acutely uremic rats. *J. Clin. Invest.,* 51:1287–1291.
115. Zile, M., Bunge, E. C., Barsness, L., Yamada, S., Schnoes, H. K., and DeLuca, H. F. (1978): Localization of 1,25-dihydroxyvitamin D_3 in intestinal nuclei *in vivo. Arch. Biochem. Biophys.,* 186:15–24.
116. Zull, J. E., and Repke, D. W. (1972): The tissue localization of tritiated parathyroid hormone in thyroparathyroidectomized rats. *J. Biol. Chem.,* 247:2195–2199.

A Review of the Basic and Applied Pharmacology of a New Group of Calcium Antagonists: The 2-Substituted 3-Dimethylamino-5, 6-Methylenedioxyindenes

Ralf G. Rahwan and Donald T. Witiak

Divisions of Pharmacology and Medicinal Chemistry, College of Pharmacy, The Ohio State University, Columbus, Ohio 43210

Calcium antagonists are valuable pharmacological tools for the study of cellular mechanisms of excitation-contraction coupling in smooth, cardiac, and skeletal muscles, and of stimulus-secretion coupling in exocrine and endocrine glands, neurons, and other secretory cells. Other potential investigative uses of calcium antagonists are predicated upon the universality of involvement of the calcium ion in blood clotting and coagulation mechanisms, cellular adhesion and integrity and membrane stability, bone and teeth formation, enzyme activity, control of certain aspects of cyclic nucleotide metabolism, cell division, and probably many more physiological functions. More importantly, a number of drugs which were either designed as calcium antagonists or whose pharmacological mechanism of action was later found to involve calcium antagonism are currently in clinical use for a variety of disease conditions. This subject was recently reviewed by Rahwan et al. (42).

The source of calcium for the physiological functions cited above is either the extracellular compartments (the basement membrane, the ground substance, or the extracellular fluids) or the intracellular calcium storage pools (mitochondria, endoplasmic reticulum, nucleus, inner aspect of the plasma membrane, and possibly secretory vesicles) (37–39,50). Drugs which activate calcium-dependent processes do so by increasing the level of cytoplasmic free calcium by either enhancing the influx of extracellular calcium or by mobilizing (or preventing sequestration or efflux of) intracellular calcium (37,42,50). Consequently, calcium antagonists may be classified according to whether they block the influx of extracellular calcium or block the action or mobilization (or enhance the sequestration or efflux) of intracellular calcium (42). The calcium antagonists believed to act by interfering with the influx of extracellular calcium (i.e. block the slow inward calcium current) into cells exhibiting calcium-dependent functions include local anesthetics, lanthanum, diphenylhydantoin, barbiturates, prenylamine, fendiline, verapamil, compound D600 (methoxy verapamil), nifedipine, diltiazeme, perhexiline, compound Org 6001 (3α-amino-2β-hydroxy-5α-

androstan-17-one), methadone, *l*-acetylmethadol, *l*-pentazocine, dantrolene, nitroglycerine and other nitrites and organic nitrates, indomethacin, adrenergic β-receptor agonists (on smooth muscle β_2 receptors), morphine, alcohol, aminoglycoside antibiotics (streptomycin and neomycin), SKF 525A, R33711, flunarizine, cinnarazine, and hydralazine (2,3,10,11,14,16,21,23,24,26,28,34, 43,44,46,49,50–52,54,55). Pharmacological agents which interfere with the action or mobilization of intracellular calcium in cells whose functions are calcium-dependent are magnesium (17,19,20,27,29,30,39,41,53), sodium nitroprusside (15), diazoxide (15), the ω-(*N,N*-diethylamino)alkyl-3,4,5-trimethoxybenzoates (6,7,22), and the 2-substituted 3-dimethylamino-5,6-methylenedioxyindenes (MDI compounds) developed and studied in our laboratories (31,32,40,42,56–59). The calcium antagonistic properties of these drugs and their contribution to the clinical efficacy of the individual agents used in therapeutics have recently been reviewed (42).

Drugs, which in most cases have only recently been recognized as calcium antagonists, have enjoyed a wide variety of therapeutic applications as local anesthetics, anticonvulsants, antiarrhythmics, coronary dilators, antihypertensives, and skeletal muscle relaxants among others (42). The remarkable tissue specificity of the calcium antagonists currently in clinical use may be a function of the administered dose, the microdistribution of the drug to the various tissues, or both (42).

The remainder of this chapter will be devoted to a review and update of the pharmacology and toxicology of the 2-substituted methylenedioxyindenes (MDI compounds).

CHEMISTRY AND STRUCTURE-ACTIVITY RELATIONSHIPS OF THE 2-SUBSTITUTED METHYLENEDIOXYINDENES

The structures of the 2-*n*-propyl and 2-*n*-butyl MDIs are shown below:

2-n-PROPYL   2-n-BUTYL

The 2-substituted MDIs were synthesized in our laboratories as intermediates in the synthesis of their corresponding indanones (59) which were designed as potential prostaglandin receptor antagonists. However, the initial pharmacological studies on the intermediate MDIs demonstrated a calcium antagonistic property of these compounds (40; see below). In preliminary structure-activity studies

(40), we had demonstrated that the antispasmodic action of the 2-substituted MDIs on smooth muscle preparations increased with increasing length of the 2-substituted side-chain. Thus, the 2-methyl and 2-ethyl MDIs were weaker than the 2-n-propyl and 2-n-butyl analogs in their ability to antagonize the spasmogenic action of oxytocin, prostaglandin $F_{2\alpha}$ ($PGF_{2\alpha}$), and prostaglandin E_2 (PGE_2) on the isolated rat uterus strip (40). In further studies (56–58), we found that conversion of the 5,6-methylenedioxy bridge of the MDIs into 5,6-dimethoxy groups (with resultant formation of the corresponding 2-substituted dimethoxyindenes) led to a loss of spasmolytic activity in the acetylcholine-depolarized or $PGF_{2\alpha}$-contracted isolated rat ileum preparation. Furthermore, the 2-substituted dimethoxyindenes exhibited inherent spasmogenic activity on the nonstimulated isolated rat ileum. Pharmacological studies with the 2-phenyl and 2-heptyl MDIs are currently in progress in our laboratories.

BASIC PHARMACOLOGY OF THE 2-SUBSTITUTED METHYLENEDIOXYINDENES

Since the 2-substituted MDIs were obtained as intermediates in the synthesis of potential prostaglandin antagonistic indanone end products (59), our original pharmacological experiments were performed on the isolated rat uterus (40) with the intention of screening the intermediate MDIs for potential prostaglandin receptor antagonistic activity. However, our early observation that the 2-substituted MDIs were not selective prostaglandin blockers led us to suspect that they may be acting by a more general mechanism, such as calcium antagonism. Thus the 2-n-propyl and 2-n-butyl MDIs (5×10^{-5} to 10^{-4} M) blocked the spasmogenic action on the estrogenized rat uterus of PGE_2 (10^{-7} M), $PGF_{2\alpha}$ (10^{-7} M), oxytocin (10^{-3} U/ml), barium (2.2×10^{-4} M), acetylcholine (10^{-6} M), and ergonovine (7.5×10^{-4} M), in a concentration-dependent manner (40). Furthermore, the MDIs blocked the contractile effects of histamine (10^{-6}M) on the isolated guinea pig ileum and of acetylcholine (10^{-6} M) on the isolated rat ileum. In further experiments on the isolated rat uterus (40), using the agonists acetylcholine (10^{-6} M)—which presumably contracts smooth muscle by enhancing the influx of extracellular calcium (9)—and barium (2.2×10^{-4} M)—which presumably contracts smooth muscle by mobilizing intracellular calcium (1,5,8,18,45)—a progressive increase in extracellular calcium concentration (from 9×10^{-4} to 7.2×10^{-3} M) was paralleled by progressive reversal of the blockade produced by the 2-n-propyl or 2-n-butyl MDIs (10^{-4} M). Since no tissue damage could be demonstrated in any of these experiments, as evidenced by reversal of the spasmolytic effects of the MDIs by increasing the agonist concentration (40), it was suspected that the MDIs were interfering with the basic role of calcium in excitation-contraction coupling in smooth muscles. Furthermore, because of the inhibition of the spasmogenic effect of barium by the MDIs, it was suspected that these antagonists interfered with calcium at an intracellular site.

To further characterize the calcium antagonistic mechanism of action of the 2-n-propyl and 2-n-butyl MDIs, we utilized the perfused isolated bovine adrenal medulla as a model for stimulus-secretion coupling (31), since this preparation has been shown to share many common molecular features with excitation-contraction coupling in muscle (33,37,44). Adrenal catecholamine secretion induced by carbachol is known to be mediated by extracellular calcium (47) and not intracellular calcium (4). On the other hand, acetaldehyde-evoked adrenomedullary catecholamine secretion is independent of both extracellular calcium (29,41,48) and intracellular calcium (29). The 2-n-propyl and 2-n-butyl MDIs (10^{-8} to 10^{-4} M) reversibly blocked adrenomedullary catecholamine secretion evoked by carbachol (0.03 to 3.3 mM) in a concentration-related manner. However, the MDIs (10^{-4} M) failed to block catecholamine secretion evoked by acetaldehyde (1 to 100 mM). These findings again confirm the calcium antagonistic actions of the MDI (31).

In order to determine whether the 2-substituted MDIs were interfering with calcium influx into cells, the adrenal medulla model was again employed (31). In this model, it has been previously shown that the bovine adrenomedullary cells take up ^{45}Ca when the radiolabeled cation is perfused into the gland after a period of perfusion with calcium-free medium (39). On the other hand, under similar conditions ^{14}C-sorbitol is confined to the extracellular space. If the adrenal is therefore perfused with calcium-free medium and then exposed to a brief pulse of either ^{45}Ca or ^{14}C-sorbitol, the washout of the radioactivity follows a predictable pattern with the ^{14}C-sorbitol being washed out of the gland at a significantly faster rate than ^{45}Ca (39). When the glands are exposed to a pulse of ^{45}Ca in the presence of 2-n-propyl or 2-n-butyl MDI (10^{-4} M), ^{45}Ca washes out at the same rate as is the case with glands not exposed to the MDIs (31). If ^{45}Ca uptake into the chromaffin cells was inhibited by the MDIs, the confinement of this cation to the extracellular space should have resulted in a significantly more rapid washout of radioactivity from the glands exposed to the MDIs as compared to the washout from control glands not exposed to MDIs. These findings, therefore, support our contention that the MDIs do not inhibit the influx of extracellular calcium, but rather interfere with calcium at an intracellular site (31,40).

APPLIED PHARMACOLOGY OF THE 2-SUBSTITUTED METHYLENEDIOXYINDENES

Calcium antagonists are known to decrease contractility of vascular smooth muscle and the myocardium (12). The value of this class of pharmacological agents in coronary therapeutics has been attributed to the ability of these compounds to reduce myocardial oxygen consumption, decrease arterial blood pressure, and improve myocardial oxygen supply through dilation of extramural coronary arteries, collaterals, and anastamoses (14). The 2-n-propyl and 2-n-butyl MDIs were studied for their potential coronary and cardiac effects and found to be both coronary dilators and antiarrhythmic agents (32,42,57,58) as summarized below.

The 2-n-propyl and 2-n-butyl MDIs (5×10^{-6} to 10^{-4} M) produced a concentration-dependent relaxation of potassium-depolarized strips of bovine extramural coronary vessels, which was reversible upon elevation of the calcium concentration of the medium. In the nonstimulated isolated perfused rabbit heart preparation, the 2-n-propyl MDI (3×10^{-5} M) and the 2-n-butyl MDI (3×10^{-5} and 10^{-4} M) increased coronary flow without affecting cardiac chronotropic activity. However, the MDIs produced a negative inotropic effect in this preparation (32) but not in the *in vivo* anesthetized dog (42,57). The pharmacodynamic characteristics of the MDIs (except for their subcellular site of action) resemble more closely those of the coronary dilating inhibitors of slow channel calcium influx (e.g., prenylamine) than those of the nitrites, in that relaxation of potassium-contracted coronary strips by the MDIs is prolonged and not spontaneously reversible in the continued presence of the drug (but is reversed by excess calcium).

It has been previously emphasized that the rational approach to the treatment of myocardial ischemia should involve both a reduction in myocardial oxygen consumption in addition to an increase in myocardial oxygen supply (35,36). The decrease in myocardial oxygen consumption can be achieved primarily through inhibition of adrenergic influences on the myocardium (35,36) or through direct depression of myocardial contractility brought about by calcium antagonists (13,14). A decrease in venous return would also decrease cardiac work and diminish myocardial oxygen consumption. The increase in myocardial oxygen supply can be achieved directly through coronary dilation, and it has been proposed that the beneficial effects of coronary dilation in ischemic heart disease would reside predominantly in dilation of the large conducting coronary arteries which can direct more blood to the ischemic regions (25), rather than in any action on the small resistance vessels which would respond negligibly to coronary dilators due to their already maximal dilation through autoregulatory mechanisms in the ischemic heart (35,36).

The antiarrhythmic properties of the 2-n-propyl and 2-n-butyl MDIs were also reported from our laboratories (32,42,57) using the pathophysiological model of the ouabain-toxic anesthetized dog. Cardiac arrhythmias in dogs were induced by intravenous administration of a loading dose of 55–60 μg/kg of ouabain, followed by a continuous infusion of a maintenance dose of 0.072 μg/kg/min of the cardiac glycoside. Intravenous doses of 5 to 30 mg/kg of either of the MDIs administered to arrhythmic dogs resulted in a dose-dependent conversion to sinus rhythm which was preceded by a dose-dependent drop in diastolic blood pressure. Systolic blood pressure was unaltered, and the effects of the MDIs on heart rate were variable but generally unimpressive and not dose-related. Pretreatment of dogs with single (30 mg/kg) or multiple (six injections of 10 mg/kg each) doses of the MDIs prior to ouabain administration afforded significant protection of the animals against ouabain-induced arrhythmias, with the time-to-onset of arrhythmias being 4.5 to 8 times longer in the presence of the MDI pretreatments as compared to nonprotected dogs. These findings indicate that the MDIs possess significant antiarrhythmic properties against ouabain-induced arrhythmias, and may lower blood pressure by direct arteriolar dilation without significant effects on cardiac chronotropy or inotropy.

TOXICOLOGY OF THE 2-SUBSTITUTED METHYLENEDIOXYINDENES

Toxicological investigations are still in progress in our laboratories, and therefore only a brief summary of available data will be presented here. Acute toxicity studies in mice resulted in an intravenous LD50 of 32 mg/kg and 40 mg/kg for the 2-n-butyl and 2-n-propyl MDIs, respectively. The intraperitoneal LD50 in mice for both compounds is 185 mg/kg. In rats, the intraperitoneal LD50 for the 2-n-butyl MDI was determined to be 210 mg/kg. Dogs appear to tolerate substantially higher MDI doses than rodents, although accurate figures are not available.

Different behavioral tests were performed on mice injected intravenously with single doses of 8 and 16 mg/kg of either of the MDIs. Gross observation revealed some sedation and a noticeable decrease in defecation. Spontaneous motor activity was decreased at both doses of the 2-n-butyl MDI and by the higher dose of the 2-n-propyl MDI. Barbiturate sleeping time was increased by both compounds at the 16 mg/kg dose level. The drugs had no effect on mice in the rotorod test, the inclined-screen test, or on conditioned avoidance behavior.

To date, subacute toxicity studies have been performed only with the 2-n-butyl MDI in rats and mice of both sexes. The MDI was administered intraperitoneally daily for 4 weeks to several groups of animals, the daily doses administered being 70 and 52.5 mg/kg to rats and 46.25 and 23.125 mg/kg to mice. The following parameters were monitored by blood analyses periodically and at the termination of the 4-week period of drug administration: cholesterol, calcium, glucose, bilirubin, chloride, uric acid, alkaline phosphatase, glutamic-oxalacetic transaminase, glutamic-pyruvic transaminase, lactic dehydrogenase, creatine phosphokinase, isocitric dehydrogenase, and prothrombin time. Interestingly, no changes in any of these parameters could be detected at any time in mice or rats of either sex, even in the groups of rats receiving the highest dose of the 2-n-butyl MDI. Gross pathological examination at the time of sacrifice showed that practically all animals (except for those in the low-dose groups in the early stages of drug administration) showed some degree of intestinal adhesions, slight intestinal bleeding, and pale discoloration of the liver (despite normal liver function tests). Ulceration of the small intestine was observed in a few animals, and, in a few of the severely affected animals, a yellowish-brown fluid filled the peritoneal cavity. Light microscopic and electron microscopic studies are still in progress in our laboratories.

ACKNOWLEDGMENTS

The work from the authors' laboratories reported in this chapter is supported by U.S. Public Health Service Grant 1-R01-HL-21670 from the National Heart, Lung, and Blood Institute. The authors are indebted to Dr. Mary F. Piascik, Dr. Michael T. Piascik, Mr. Timothy P. Johnson, Mr. John R. Baldwin, and Mrs. Carin Åkesson for their participation in this research program.

REFERENCES

1. Antonio, A., Rocha E Silva, M., and Yashuda, Y. (1973): The tachyphylactic effect of barium on intestinal smooth muscle. *Arch. Int. Pharmacodyn. Thér.,* 204:260–267.
2. Blaustein, M. P. (1976): Barbiturates block calcium uptake by stimulated and potassium-depolarized rat sympathetic ganglia. *J. Pharmacol. Exp. Ther.,* 196:80–86.
3. Blum, K., Hamilton, M. G., and Wallace, J. E. (1977): Alcohol and opiates: A review of common neurochemical and behavioral mechanisms. In: *Alcohol and Opiates: Neurochemical and Behavioral Mechanisms,* edited by K. Blum, pp. 203–236. Academic Press, New York.
4. Borowitz, J. L. (1969): Effects of acetylcholine on the subcellular distribution of ^{45}Ca in bovine adrenal medulla. *Biochem. Pharmacol.,* 18:715–723.
5. Caldwell, P. C., and Walster, G. (1963): Studies on the microinjection of various substances into crab muscle fibers. *J. Physiol. (London),* 169:353–372.
6. Chiou, C. Y., and Malagodi, M. H. (1975): Studies on the mechanism of action of a new Ca^{++} antagonist, 8-(N,N-diethylamino)-octyl-3,4,5-trimethoxybenzoate hydrochloride, in smooth and skeletal muscles. *Br. J. Pharmacol.,* 53:279–285.
7. Chiou, C. Y., Malagodi, M. H., Sastry, B. V. R., and Posner, P. (1976): Effects of calcium antagonist, 6-(N,N-diethylamino)-hexyl-3,4,5-trimethoxybenzoate, on digitalis-induced arrhythmias and cardiac contractions. *J. Pharmacol. Exp. Ther.,* 198:444–449.
8. Daniel, E. E. (1964): Effects of drugs on contractions of vertebrate smooth muscle. *Annu. Rev. Pharmacol.,* 4:189–222.
9. Daniel, E. E., and Janis, R. A. (1975): Calcium regulation in the uterus. *Pharmacol. Ther.,* B1:695–729.
10. Fairhurst, A. S., and Macri, J. (1975): Aminoglycoside-Ca^{++} interactions in skeletal muscle preparations. *Life Sci.,* 16:1321–1330.
11. Fleckenstein, A. (1971): Specific inhibitors and promoters of calcium action in the excitation-contraction coupling of heart muscle and their role in the prevention or production of myocardial lesions. In: *Calcium and the Heart,* edited by P. Harris and L. H. Opie, pp. 135–188. Academic Press, New York.
12. Fleckenstein, A. (1975): On the basic pharmacological mechanism of nifedipine and its relationship to therapeutic efficacy. In: *New Therapy of Ischemic Heart Disease,* edited by W. Lochner, pp. 1–13. Springer-Verlag, New York.
13. Fleckenstein, A. (1977): Specific pharmacology of calcium in myocardium, cardiac pacemakers, and vascular smooth muscle. *Annu. Rev. Pharmacol. Toxicol.,* 17:149–166.
14. Fleckenstein, A., Nakayama, K., Fleckenstein-Grün, G., and Byon, Y. K. (1976): Mechanism and sites of action of calcium antagonistic coronary therapeutics. In: *Coronary Angiography and Angina Pectoris,* edited by P. R. Lichten, pp. 297–315. Publishing Science Group, Massachusetts.
15. Gross, F., and Kreye, V. A. W. (1977): Drugs acting on arteriolar smooth muscle (vasodilator drugs). In: *Handbook of Experimental Pharmacology,* edited by F. Gross, pp. 397–476. Springer-Verlag, New York.
16. Kalsner, S., Nickerson, M., and Boyd, G. N. (1970): Selective blockade of potassium-induced contraction of aortic strips by beta-diethylamino-diphenylpropylacetate (SKF 525A). *J. Pharmacol. Exp. Ther.,* 174:500–508.
17. Kanno, T., Cochrane, D. E., and Douglas, W. W. (1973): Exocytosis (secretory granule extrusion) induced by injection of calcium into mast cells. *Can. J. Physiol. Pharmacol.,* 57:1001–1004.
18. Karaki, H., Ikeda, M., and Urakawa, N. (1967): Effects of external calcium and some metabolic

inhibitors on barium-induced tension changes in guinea pig taenia coli. *Jap. J. Pharmacol.,* 17:603–612.
19. Krnjević, K., Puil, E., and Werman, R. (1976): Intracellular Mg^{++} increases neuronal excitability. *Can. J. Physiol. Pharmacol.,* 54:73–77.
20. Lastowecka, A., and Trifaró, J. M. (1974): The effect of Na and Ca ions on the release of catecholamines from the adrenal medulla: Sodium deprivation induces release by exocytosis in the absence of extracellular calcium. *J. Physiol. (London),* 236:681–705.
21. Lee, C.-H., and Berkowitz, B. A. (1977): Calcium antagonist activity of methadone, *l*-acetylmethadol, and *l*-pentazocine in the rat aortic strip. *J. Pharmacol. Exp. Ther.,* 202:646–653.
22. Malagodi, M. H., and Chiou, C. Y. (1974): Pharmacological evaluation of a new calcium antagonist, 8-(N,N-diethylamino)-octyl-3,4,5-trimethoxybenzoate hydrochloride (TMB-8): Studies in smooth muscle. *Eur. J. Pharmacol.,* 27:25–33.
23. Malaisse, W. J., Sener, A., Devis, G., and Somers, G. (1976): Calcium antagonists and islet function. V. Effect of R33711. *Horm. Metab. Res.,* 8:434–438.
24. Mayer, C. J., Van Breemen, C., and Casteels, R. (1972): The action of lanthanum and D600 on the calcium exchange in the smooth muscle cells of guinea pig taenia coli. *Pflügers Arch. Eur. J. Physiol.,* 337:333–350.
25. McGregor, M., and Fam W. M. (1966): Regulation of coronary blood flow. *Bull. N.Y. Acad. Med.,* 42:940–950.
26. McLean, A. J., Du Souich, P., Barron, K. W., and Briggs, A. H. (1978): Interaction of hydralazine with tension development and mechanisms of calcium accumulation in K^+-stimulated rabbit aortic strips. *J. Pharmacol. Exp. Ther.,* 207:40–48.
27. Miledi, R. (1973): Transmitter release induced by injection of calcium ions into nerve terminals. *Proc. R. Soc. London, B. Biol. Sci.,* 183:421–425.
28. Northover, B. J. (1977): Indomethacin: A calcium antagonist. *Gen. Pharmacol.,* 8:293–296.
29. O'Neill, P. J., and Rahwan, R. G. (1975): Experimental evidence for calcium-independent catecholamine secretion from the bovine adrenal medulla. *J. Pharmacol. Exp. Ther.,* 193:513–522.
30. Peach, M. J. (1975): Cations. In: *The Pharmacological Basis of Therapeutics,* 5th ed., edited by L. S. Goodman and A. Gilman, pp. 787–791. Macmillan, New York.
31. Piascik, M. F., Rahwan, R. G., and Witiak, D. T. (1978): Pharmacological evaluation of new calcium antagonists: 2-substituted 3-dimethylamino-5,6-methylenedioxyindenes. Effects on adrenomedullary catecholamine secretion. *J. Pharmacol. Exp. Ther.,* 205:155–163.
32. Piascik, M. F., Rahwan, R. G., and Witiak, D. T. (1978): Pharmacological evaluation of new calcium antagonists: 2-substituted 3-dimethylamino-5,6-methylenedioxyindenes. Coronary and cardiac effects. *Pharmacologist,* 20:151.
33. Poisner, A. M. (1970): Release of transmitters from storage: A contractile model. *Adv. Biochem. Psychopharmacol.,* 2:99–108.
34. Putney, J. W., and Bianchi, C. P. (1974): Site of action of dantrolene in frog sartorius muscle. *J. Pharmacol. Exp. Ther.,* 189:202–212.
35. Raab, W. (1962): The sympathogenic biochemical trigger mechanism of angina pectoris. Its therapeutic suppression and long range prevention. *Am. J. Cardiol.,* 9:576–590.
36. Raab, W. (1963): The nonvascular metabolic myocardial vulnerability factor in coronary heart disease. Fundamental pathogenesis, treatment, and prevention. *Am. Heart J.,* 66:685–706.
37. Rahwan, R. G., and Borowitz, J. L. (1973): Mechanisms of stimulus-secretion coupling in adrenal medulla. *J. Pharm. Sci.,* 62:1911–1923.
38. Rahwan, R. G., Borowitz, J. L., and Hinsman, E. J. (1973): Evidence for secretory tubules in the adrenal medulla. *Arch. Int. Pharmacodyn. Thér.,* 206:345–351.
39. Rahwan, R. G., Borowitz, J. L., and Miya, T. S. (1973): The role of intracellular calcium in catecholamine secretion from the bovine adrenal medulla. *J. Pharmacol. Exp. Ther.,* 184:106–118.
40. Rahwan, R. G., Faust, M. M., and Witiak, D. T. (1977): Pharmacological evaluation of new calcium antagonists: 2-substituted 3-dimethylamino-5,6-methylenedioxyindenes. *J. Pharmacol. Exp. Ther.,* 201:126–137.
41. Rahwan, R. G., O'Neill, P. J., and Miller, D. D. (1974): Differential secretion of catecholamines and tetrahydroisoquinolines from the bovine adrenal medulla. *Life Sci.,* 14:1927–1938.
42. Rahwan, R. G., Piascik, M. F., and Witiak, D. T. (1979): The role of calcium antagonism in the therapeutic action of drugs. *Can. J. Physiol. Pharmacol., (in press).*

43. Refsum, H. (1975): Calcium antagonistic and antiarrhythmic effects of nifedipine on the isolated rat atrium. *Acta Pharmacol Toxicol.*, 37:377–386.
44. Rubin, R. P. (1970): The role of calcium in the release of neurotransmitter substances and hormones. *Pharmacol. Rev.*, 22:389–428.
45. Saito, Y., Sakai, Y., and Urakawa, N. (1972): Effect of cholinergic drugs and barium on oxygen consumption in guinea pig taenia coli. *Jap. J. Pharmacol.*, 22:653–661.
46. Salako, L. A., Vaughan Williams, E. M., and Wittig, J. H. (1976): Investigations to characterize a new antiarrhythmic drug, Org 6001, including a simple test for calcium antagonism. *Br. J. Pharmacol.*, 57:251–262.
47. Schneider, F. H. (1969): Drug-induced release of catecholamines, soluble protein and chromogranin A from the isolated bovine adrenal gland. *Biochem. Pharmacol.*, 18:101–107.
48. Schneider, F. H. (1971): Acetaldehyde-induced catecholamine secretion from the cow adrenal medulla. *J. Pharmacol. Exp. Ther.*, 177:109–118.
49. Sohn, B. S., and Ferrendelli, J. A. (1973): Inhibition of Ca^{++} transport into brain synaptosomes by diphenylhydantoin (DPH). *J. Pharmacol. Exp. Ther.*, 185:272–275.
50. Somlyo, A. P., and Somlyo, A. V. (1969): Vascular smooth muscle. I. Normal structure, pathology, biochemistry, and biophysics. *Pharmacol. Rev.*, 20:197–272.
51. Somlyo, A. P., and Somlyo, A. V. (1970): Vascular smooth muscle. II. Pharmacology of normal and hypertensive vessels. *Pharmacol. Rev.*, 22:249–353.
52. Swain, A. W., Kiplinger, G. F., and Brody, T. M. (1956): Actions of certain antibiotics on the isolated dog heart. *J. Pharmacol. Exp. Ther.*, 117:151–159.
53. Turlapaty, P. D. M. V., and Carrier, O. (1973): Influence of magnesium on calcium-induced responses of atrial and vascular muscle. *J. Pharmacol. Exp. Ther.*, 187:86–98.
54. Van Breemen, C., Farinas, B. R., Gerba, P., and McNaughton, E. D. (1972): Excitation-contraction coupling in rabbit aorta studied by the lanthanum method for measuring cellular calcium influx. *Circ. Res.*, 30:44–54.
55. Van Neuten, J. M., Tobia, A. J., and McGuire, J. L. (1978): Antagonism of vascular smooth muscle contraction by flunarizine. *Fed. Proc.*, 37:793.
56. Witiak, D. T., Kakodkar, S. V., Brunst, G. E., Baldwin, J. R., and Rahwan, R. G. (1978): Pharmacology on the rat ileum of certain 2-substituted 3-dimethylamino-5,6-dimethoxyindenes related to 5,6-methylenedioxyindene calcium antagonists. *J. Med. Chem., (in press)*.
57. Witiak, D. T., and Rahwan, R. G. (1978): An update on the pharmacology of 2-substituted methylenedioxyindene calcium antagonists. *Abst. Intra-Science Research Foundation Symp., (in press)*.
58. Witiak, D. T., Rahwan, R. G., and Piascik, M. F. (1978): Recent advances in the pharmacology of aminoindene calcium antagonists. *Proc. 16th Natl. Med. Chem. Symp. (Amer. Chem. Soc.)*.
59. Witiak, D. T., Williams, D. R., Kakodkar, S. V., Hite, C., and Shen, M.-S. (1974): Vilsmeir-Haack cyclizations. Synthesis of 2-substituted 3-dimethylamino-5,6-methylenedioxyindenes and their corresponding indanones. *J. Org. Chem.*, 39:1242–1247.

… *Trace Metals in Health and Disease*, edited by
N. Kharasch. Raven Press, New York © 1979.

Calcium Antagonists: Mechanisms of Action with Special Reference to Nifedipine

Philip D. Henry

Cardiovascular Division, Barnes Hospital, Washington University School of Medicine, St. Louis, Missouri 63110

It is well established that calcium ions play an important role in the regulation of cell metabolism. Calcium activates contractile processes in muscle and nonmuscle cells and is involved in the control of glycolysis and oxidative phosphorylation (18). In many systems, calcium mediates or modulates cyclic nucleotide-dependent control mechanisms (27). In most tissues, calcium is highly compartmentalized with ionic activities across cellular and subcellular membranes often differing by a factor of 1,000 or more (18,27). The active mechanisms regulating the uptake and distribution of calcium are obviously very important, but are still incompletely understood.

Recently, Fleckenstein and collaborators (11–13) have delineated a group of compounds thought to act by blocking the transmembrane inward movement of calcium. These agents, which are often referred to as calcium antagonists, have attracted considerable attention as they provide a potential tool for the elucidation of calcium-dependent regulatory mechanisms. In addition, current clinical evidence suggests that these compounds may have important therapeutic applications. However, as originally delineated by Fleckenstein, the concept of calcium antagonism has not been universally accepted. Drugs included in this group vary markedly in chemical structure, and voltage clamp experiments in mammalian myocardium have failed to reveal a unique electrophysiological mechanism of action (2,20,22). In particular, there is no evidence that calcium antagonists constitute a new group of anti-arrhythmic agents acting by blocking the slow calcium current into myocardial cells (20,29).

The chemical structure of some of the compounds tentatively included in the group of calcium antagonists are shown in Fig. 1. With the exception of hexoestrol, all compounds possess a tertiary amine, but differ otherwise markedly in structure. Several agents possess an asymmetrical carbon atom and occur as optical isomers. As will be described later, optical isomers may exert fundamentally different electrophysiological effects. Nifedipine, a dihydropyridine derivative, has a nitrate group which, unlike that of organic nitrates, is not linked through an oxygen atom and is not essential for the vasodilator property of the drug. Verapamil has some features in common with the classical vasodilator papaverine. Both diltiazem and diazoxide have a sulfur atom but are otherwise

Nifedipine
(Adalat, Bay a 1040)

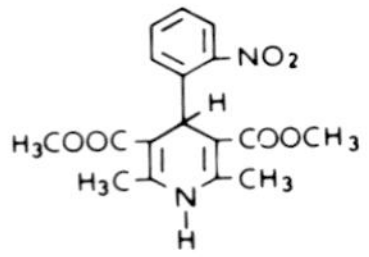

Diltiazem
(Herbesser)

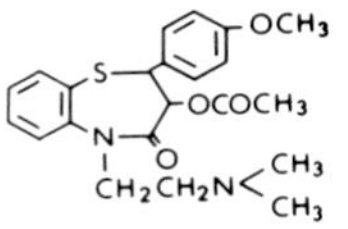

Verapamil
(Isoptin, Iproveratril)

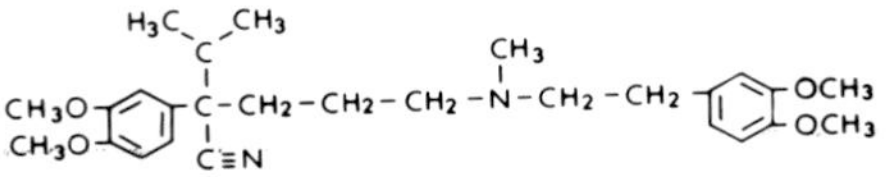

D600
(Methoxy-verapamil)

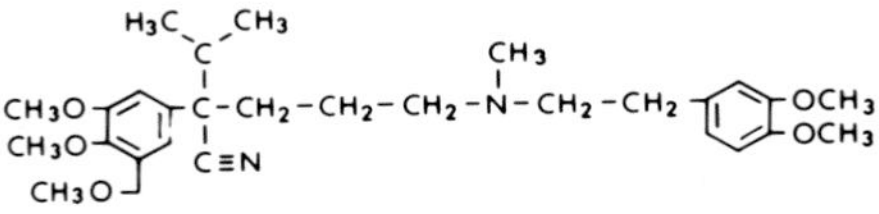

Perhexilinine Maleate
(Pexid)

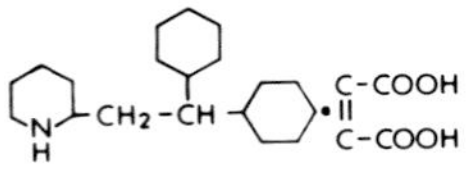

Prenylamine
(Segontin)

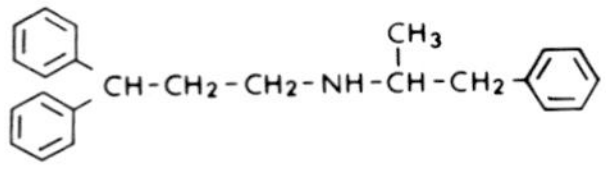

Fendiline
(Sensit)

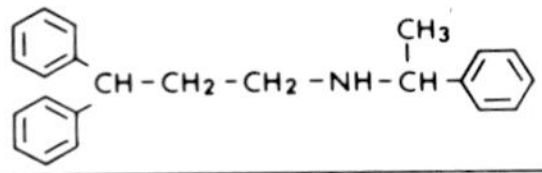

Lidoflazine
(Corflazine)

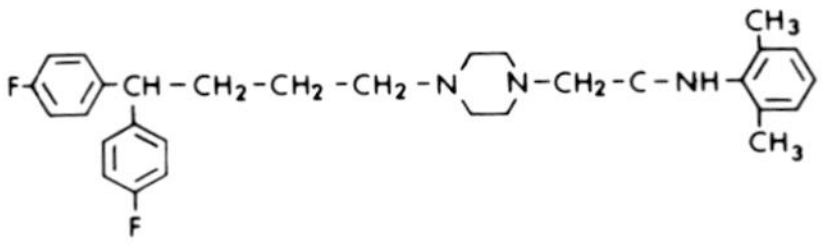

Cinepazet maleate
(Vascoril)

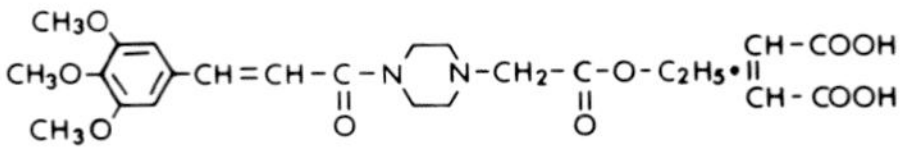

Diazoxide
(Hyperstat)

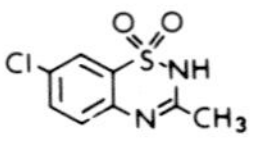

Hexoestrol
(Synthovo)

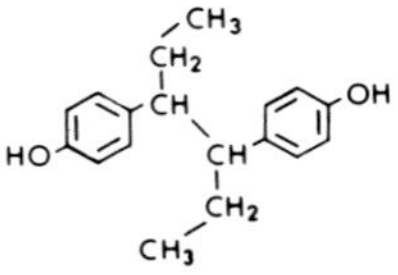

FIG. 1. Ca^{2+} antagonists.

structurally unrelated. It can be seen that calcium antagonists have in general no structural similarity with sympathomimetic agents, β-adrenergic blockers, or calcium ionophores.

Initially, calcium antagonists were defined as agents that reversibly depress the contractile activity of mammalian myocardium without appreciably affecting the monophasic action potential (13). The phenomenon was interpreted as representing a specific inhibition of the excitation-contraction coupling. However, subsequent studies have clearly demonstrated that alterations in mechanical performance induced by calcium antagonists are associated with changes in the monophasic action potential. According to current concepts, depolarization of the myocardial cell membrane during the action potential is mediated by two inward currents, an initial fast current carried by sodium ions and a subsequent slow current partly carried by calcium ions (7). The two currents are thought to occur through different channels in the cell membrane, the fast (Na^+)-channel and the slow (Ca^{2+})-channel. Calcium penetrating the cells through the slow channel appears to play an important role in triggering muscular contraction. Voltage-clamp experiments indicate that the fast and slow channels are selectively inhibited by tetrodotoxin and Mn^{2+}, respectively. It was suggested that drugs such as verapamil or prenylamine were, like manganese, selective inhibitors of the slow channel (11). Thus, calcium antagonists were thought to differ fundamentally in their electrophysiological effect from local anesthetics, drugs known to decrease the slope of the rapid upstroke of the action potential by inhibiting the sodium current through the fast channels. However, subsequent studies have clearly demonstrated this simplifying scheme to be incorrect. In fact, it has been conclusively demonstrated that a number of calcium antagonists possess local anesthetic, sodium channel inhibitory effects. Of particular interest is the observation that the (+)-isomer of racemic verapamil is predominantly a fast-channel inhibitor, whereas the (−)-isomer is predominantly a slow-channel inhibitor (1). A similar dichotomy with respect to fast and slow channel inhibition is observed with the optical isomers of D-600 and prenylamine (1,20). Similarly, diltiazem has been shown to block the fast sodium channel and to reduce the rate of rise of the monophasic action potential (23). On the other hand, nifedipine appears to block exclusively the slow channel (2,20). It is of interest that only calcium antagonists with sodium channel inhibitory effects have proved to have anti-arrhythmic properties. Thus, nifedipine and (+)-verapamil have no or limited anti-arrhythmic effects (2,14,20,21,25,26). There is considerable evidence that the slow inward calcium current plays an important role in the pacemaker activity of the sinus node and of the A-V node (7). In contrast, the automaticity of the normally polarized His-Purkinje system does not appear to depend upon the slow inward current (7). The finding that the nonanesthetic calcium antagonist nifedipine inhibits the activity of the sinus and A-V node without affecting ventricular automaticity is consistent with this concept (8,24). However, it should be emphasized that the A-V blocking effects of nifedipine occur at concentrations which are much higher than those required to produce profound arterial smooth muscle relaxation (24). In doses that produce similar changes in blood pressure *in vivo*, nifedipine does not prolong or may shorten the PR interval in contrast to racemic verapamil which tends to prolong the PR interval and produce A-V block (25).

Recent voltage clamp experiments by Kaufmann et al. (2,20) have demonstrated that there are appreciable differences between the slow channel inhibitory effects of nifedipine and those of the (−)-isomer of verapamil. Nifedipine reduces the slow inward current in a dose-dependent manner without affecting the kinetic parameters of the current. In other words, the time courses of the activation, inactivation, and recovery from the inactivation of the slow channel remains unchanged with this drug. One way to explain these findings is to assume that nifedipine reduces the slow inward current simply by diminishing the number of channels available (20). This is in sharp contrast to the (−)-isomers of verapamil and D-600 which reduce both the rate of activation and recovery of the slow inward channel (1,20,22). In addition, these substances affect the repolarizing outward potassium current. Accordingly, even if one considers only the (−)-isomer of verapamil, the electrophysiological effects of this agent appear to be complex. These findings clearly indicate that calcium antagonism in the mammalian myocardium cannot be explained on the basis of a unique electrophysiological mechanism.

Negative chronotropic and negative inotropic responses to calcium antagonists may be related to their ability to block the slow calcium inward current. It is of interest that the relative negative chronotropic and inotropic potency of calcium antagonists may vary considerably. In the isolated, isometrically contracting guinea pig atrium, nifedipine has a very strong negative inotropic effect and only a modest negative chronotropic effect (3,16). In contrast, diltiazem markedly depresses cardiac frequency, whereas negative inotropy is modest when frequency is held constant by electrical pacing. On the other hand, verapamil appears to be potent both on rate and contractility with or without pacing (3,16). The electrophysiological mechanisms underlying these apparent inotropic and chronotropic selectivities have not yet been elucidated.

In view of the varying effects of calcium antagonists, one may wonder whether or not these agents may be distinguished on the basis of specific actions on smooth muscle or cardiac muscle. Classical vasodilators such as nitroglycerin, adenosine, and nitroprusside have potent relaxing effects on vascular smooth muscle but usually depress myocardium only at very high concentrations (3,16). One distinguishing feature of calcium antagonists such as nifedipine and verapamil is that they are capable of depressing both cardiac and smooth muscle in relatively low concentrations. Thus, whereas classical vasodilators are basically selective smooth muscle depressants, calcium antagonists act as cardio-vasodepressants. In this context, it appears somewhat unfortunate to include drugs such as papaverine (28) and diazoxide (5) in the group of calcium antagonists as these drugs depress myocardium only at very high concentrations (10^{-4} M).

It should be remembered that many drugs not generally thought to be cardiovascular agents will depress myocardium at near millimolar concentrations. Recently, calcium antagonists have been shown to have qualitatively different effects on isolated coronary arteries compared to those exerted by classical vasodilators. In one study, effects of calcium antagonistic and classical vasodilators on isolated canine coronary arteries exposed to selected external potassium concentrations were studied (4,17). Classical vasodilators such as nitroglycerin, adenosine, or nitroprusside relaxed the artery at all potassium concentrations. In sharp contrast, calcium antagonists acted as potent relaxing agents only at very low (less than 1 millimolar) and at normal or supernormal potassium concentrations (more than 3 mM). Within a narrow range of low potassium concentrations (1 to 2 mM), calcium antagonists unexpectedly elicited contractions, whereas calcium evoked dose-dependent relaxations (4,17). In another study, it was found that bovine coronary arteries during sodium fluoride-induced contracture were relaxed only by nitroprusside and nitroglycerin, but not by the calcium antagonists verapamil and prenylamine (10). These findings appear to support the concept that coronary relaxation by classical vasodilators and calcium antagonists are mediated by a different mechanism.

In vivo, the pharmacological effects of calcium antagonists are dominated by the vasodilator response. In doses that produce profound vasodilation, nifedipine does not appear to depress cardiac performance. The reason why the negative inotropic effect of nifedipine is not manifest *in vivo* is not entirely clear but may simply reflect the fact that the vessels are slightly more sensitive to the drug than the heart. *In vitro*, however, we have repeatedly observed that the threshold concentration for the depression of cardiac and smooth muscle is close to the 10^{-9} M for both tissues. Another factor that may mask the negative inotropic effect is the unloading of the left heart secondary to decreased arterial impedance and reflex sympathetic discharge.

Calcium antagonists have been found to be potent coronary vasodilators and effective anti-anginal agents (12). Of considerable interest is that nifedipine may relieve angiospastic angina in patients who are not responsive to nitrates and β-blockers (9). Preliminary data indicate that nifedipine and other calcium antagonists may be useful for the vasodilator therapy of heart failure and for the treatment of arterial hypertension. Calcium antagonists have been found to protect myocardium against catecholamine-induced cardiac necrosis (11). It is well known that β-adrenergic agonists enhance the slow inward current in the mammalian myocardium. The protective effects of calcium antagonists against catecholamine-induced cardiac necrosis may thus reflect diminished entry of calcium into the cells. Indeed, treatment with calcium antagonists was associated with decreased accumulation of tissue calcium (11). More recently, nifedipine has been found to protect the severely ischemic myocardium *in vitro* and *in vivo* (15,18,19). Treatment with nifedipine prevented in the isolated rabbit heart perfused at low flow (ischemia) deterioration of mechanical performance and accumulation of calcium in the mitochondrial fraction (15,18). In additional experiments, it was demonstrated that nifedipine may exert marked protective effects on the globally ischemic heart during cardiopulmonary bypass (6).

In conclusion, calcium antagonists are very heterogeneous group of compounds with varying chemical structure and pharmacological effects. There is clearly not a single mechanism of action that mediates the multiple effects of these drugs, and current information does not rule out the possibility that Ca^{2+} antagonists act at sites other than the cell membrane. Physiological experiments and clinical trials strongly suggest that some of these agents may have important therapeutic applications.

REFERENCES

1. Bayer, R., Kalusche, D., Kaufmann, R., and Mannhold, R. (1975): Inotropic and electrophysiological actions of Verapamil and D600 in mammalian myocardium. *Naunyn-Schmiedeberg's Arch. Pharmacol.,* 290:81–97.
2. Bayer, R., Rodenkirchen, R., Kaufmann, R., Lee, J. H., and Hennekes, R. (1977): The effects of Nifedipine on contraction and monophasic action potential of isolated cat myocardium. *Naunyn-Schmiedeberg's Arch. Pharmacol.,* 301:29–37.
3. Borda, L., and Henry, P. D. (1978): Direct cardiac effects of vasodilators. *Clin. Res.,* 26:280A.
4. Borda, L., Schuchleib, R., and Henry, P. D. (1977): Effects of potassium on isolated coronary arteries: Modulation of adrenergic responsiveness and release of norepinephrine. *Circ. Res.,* 41:778–786.
5. Bristow, M. R., and Green, R. D. (1977): Effect of diazoxide, verapamil and compound D600 on isoproterenol and calcium-mediated dose-response relationships in isolated rabbit atrium. *Eur. J. Pharmacol.,* 45:267–279.
6. Clark, R. E., Ferguson, T. B., West, P. W., Shuchleib, R., and Henry, P. D. (1977): Pharmacological preservation of the ischemic heart. *Am. Thor. Surg.,* 24:307–314.
7. Coraboeuf, E. (1978): Ionic basis of electrical activity in cardiac tissues. *Am. J. Physiol.,* 234:H101–H116.
8. Endoh, M., Yanagisawa, T., and Taira, N. (1978): Effects of calcium-antagonistic coronary vasodilators, Nifedipine and Verapamil, on ventricular automaticity of the dog. *Nauyn-Schmiedeberg's Arch. Pharmacol.,* 302:235–238.
9. Endoh, M., Kanda, I., Hosoda, S., Hayashi, H., Hirosawa, K., and Konno, S. (1975): Prinzmetal's variant form of angina pectoris. *Circulation,* 52:33–37.
10. Fermum, R., Meisel, P., and Klinner, U. (1977): Versuche zum Wirkungsmechanismus von Gefässpasmolytika. *Acta Biol. Med. Germ.,* 36:245–255.
11. Fleckenstein, A. (1974): Drug-induced changes in cardiac energy. *Adv. Cardiol.,* 12:183–197.
12. Fleckenstein, A. (1975): Fundamentale Herz-und Gefässwirkungen Ca^{++}-antagonistischer Koronartherapeutika. *Med. Klinik,* 70:1665–1674.
13. Fleckenstein, A., Kammermeier, H., Döring, H., and Freund, H. J. (1967): Zum Wirkungsmechanismus neuartiger Koronardilatatoren mit gleichzeitig sauerstoffeinsparenden Myokard-Effekten, Prenylamin, und Iproveratril. *Z. Kreisl.-Forsch.,* 56:839–858, 716–744.
14. Gutovitz, A. L., Cole, B., Henry, P. D., Sobel, B. E., and Roberts, R. (1977): Resistance of ventricular dysrhythmia to Nifedipine, a calcium antagonist. *Circulation,* 56(Suppl. III):179.
15. Henry, P. D. (1976): Protection of ischemic myocardium by Nifedipine. In *3rd International Adalat Symposium. New Therapy of Ischemic Heart Disease,* edited by A. D. Jatene and P. R. Lichtlen. Excerpta Medica, Amsterdam.
16. Henry, P. D. (1978): Chronotropic and inotropic effects of coronary vasodilators. In: *4th International Adalat Symposium* (Tokyo), edited by P. R. Lichtlen. Excerpta Medica, Amsterdam.
17. Henry, P. D., Borda, L., and Schuchleib, R. (1977): Paradoxical coronary constrictor effects of calcium antagonists. *Circulation,* 56(Suppl. III):127.
18. Henry, P. D., Schuchleib, R., Davis, J., Weiss, E. S., and Sobel, B. E. (1977): Myocardial

contracture and accumulation of mitochondrial calcium in ischemic rabbit heart. *Am. J. Physiol.,* 233:H677–H684.
19. Henry, P. D., Schuchleib, R., Borda, L., Roberts, R., Williamson, J. R., and Sobel, B. E. (1978): Effects of Nifedipine on myocardial perfusion and ischemic injury in dogs. *Circ. Res.,* 43:372–380.
20. Kaufmann, R. (1977): Differenzierung verschiedener Kalzium-Antagonisten. *Münchener Medizinische Wochenschrift,* 119:6–11.
21. Kaumann, A. J., and Serur, J., (1975): Prevention of ventricular fibrillation by canine coronary artery ligation with optical isomers of Verapamil. *Proc. 6th Int. Congr. Pharmacol.,* Helsinki, July 20–25.
22. Kohlhardt, M., and Fleckenstein, A. (1977): Inhibition of the slow inward current by Nifedipine in mammalian ventricular myocardium. *Naunyn-Schmiedeberg's Arch. Pharmacol.,* 298:267–272.
23. Nabata, H. (1977): Effects of calcium-antagonistic coronary vasodilators on myocardial contractility and membrane potentials. *Jap. J. Pharmacol.,* 27:239–249.
24. Narimatsu, A., and Taira, N. (1976): Effects on atrio-ventricular conduction of calcium-antagonistic coronary vasodilators, local anaesthetics and quinidine injected into the posterior and the anterior spetal artery of the atrio-ventricular node preparation of the dog. *Naunyn Schmiedeberg's Arch. Pharmacol.,* 294:169–177.
25. Raschack, M. (1976): Differences in the cardiac actions of the calcium antagonists Verapamil and Nifedipine. *Arzneim.-Forsch.*[Drug. Res.], 26:1330–1333.
26. Raschack, M. (1976): Relationship of antiarrhythmic to inotropic activity and antiarrhythmic qualities of the optical isomers of Verapamil. *Naunyn-Schmiedeberg's Arch. Pharmacol.,* 294:285–291.
27. Rasmussen, H., and Nagata, N. (1970): Hormones, cell calcium and cyclic AMP. In: *A Symposium on Calcium and Cellular Function,* edited by A. W. Cuthbert. Macmillan, New York.
28. Rheinhardt, D., Freynik, P. and Schümann, H. J. (1977): Effects of D600 and papaverine on the inotropic response to isoprenaline on left guinea pig atria under the graded influence of changes of frequency and temperature. *Arch. Int. Pharmacodyn.,* 229:67–82.
29. Singh, B. N., and Vaughan Williams, E. M. (1972): A fourth class of antiarrhythmic action? Effect of verapamil on ouabain toxicity, on atrial and ventricular intracellular potentials, and on other features of cardiac function. *Cardiovasc. Res.,* 6:109–119.

Trace Metals in Health and Disease, edited by
N. Kharasch. Raven Press, New York © 1979.

Verapamil: Mechanisms of Pharmacologic Actions and Therapeutic Applications*

Bramah N. Singh

Department of Cardiology, Wadsworth Veterans Administration Hospital, and the Department of Medicine, UCLA School of Medicine, Los Angeles, California 90073.

When verapamil (iproveratril)—a synthetic papaverine derivative (Fig. 1)—was first introduced over 15 years ago (1), its varied pharmacological properties were initially little known, barring the appreciation that the compound was a potent peripheral as well as coronary vasodilator (1,2). The rapid advances in knowledge of the ionic basis of the cardiac membrane currents in the last decade (3) raised the possibility that verapamil may act essentially by inhibiting the second slow inward current in the mammalian myocardium (4,5) and by interfering with excitation-contraction coupling in vascular smooth muscle (4–7). These fundamental properties of the drug have been considered to constitute the basis of verapamil's therapeutic action in ameliorating myocardial ischemia and in controlling a variety of cardiac arrhythmias as well as in producing varying hemodynamic effects in experimental animals as well as in man (8–10). As indicated elsewhere in this volume (11), this somewhat simplistic scheme attributing the fundamental action of the drug to calcium antagonism at the membranes of excitable tissues is no longer tenable in light of the newer experimental findings. However, the delineation of the group of compounds which selectively inhibit calcium-dependent transmembrane potentials (12) is still a relatively useful pharmacological concept which has therapeutic relevance (13). The possibility remains that a number of chemically heterogeneous compounds may share the common property of calcium antagonism in excitable tissues while having signifi-

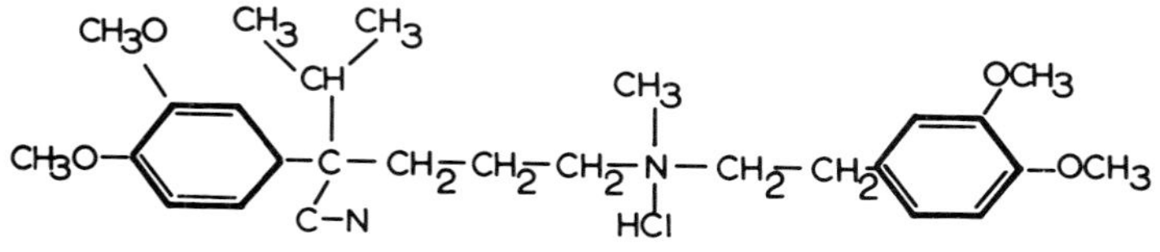

FIG. 1. Structural formula of verapamil.

* Based on a presentation at the 1978 Intra-Science Symposium, November 29, 1978, Santa Monica, California.

cantly different qualitative and quantitative net pharmacological and therapeutic effects. This may result not only from differing sensitivities of individual tissues to the fundamental actions of various calcium antagonists but also from differences in their associated pharmacological properties. For example, a compound such as nifedipine may have a greater specificity for vascular smooth muscle whereas verapamil may exhibit a somewhat more balanced action on blood vessels and the myocardium. In this chapter, the overall properties of verapamil are discussed in light of these considerations and in relation to the established role of the drug in controlling cardiac arrhythmias and in exerting a salutary effect in the ischemic consequences of coronary artery stenosis.

ELECTROPHYSIOLOGICAL EFFECTS

Although recent studies have shown that the optical isomers of verapamil may have differing effects on the myocardial membrane with respect to the main (Na-mediated) inward depolarizing current and on experimental cardiac arrhythmias (11), the bulk of the experimental and clinical data do not support the belief that the antiarrhythmic effects of the compound can reasonably be

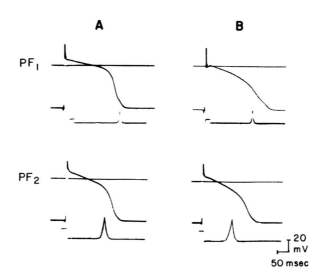

FIG. 2. Effects of verapamil on normal canine cardiac Purkinje fibers. For these and subsequent records, the upper trace is the action potential and the lower shows a 200 V/sec calibration followed by the electronic differentiation of the maximum rate of rise of phase 0 ($\dot{V}_{max}$). Panel **A** is a control record of two Purkinje fiber action potentials: PF 1 and PF 2. Panel **B** shows the same action potentials 30 min after onset of superfusion with verapamil, 1 mg/liter. Note that this concentration of verapamil has no effect on action potential amplitude or $\dot{V}_{max}$ or on resting membrane potential. However, the voltage at which the plateau originates is decreased and the slope of phase 2 repolarization is increased by verapamil. These changes are consistent with the block of a slow inward current such as that carried by calcium ion. Cycle length 800 msec; $(K^+)_o$ = 4 mM; $(Ca^{2+})_o$ = 2.7 mM; temp. = 37°C. (Reprinted from ref. 17, with permission.)

accounted for by its "quinidine-like" action in heart muscle (14). The electrophysiological actions of the agent must account for the fact that the drug predictably and promptly aborts supraventricular tachycardias reciprocating through the atrioventricular node as well as ectopic atrial tachycardias and the slowing of the ventricular response in atrial flutter and fibrillation (10,13), observations which indicate a major inhibitory effect of verapamil on atrioventricular nodal transmission and on spontaneous diastolic depolarization of potential pacemaker tissues.

Although verapamil was found to have local anesthetic potency 1.6 times that of procaine (15), in clinically relevant concentrations, it has significant actions neither on the rate of depolarization nor repolarization phases of the action potential in atrial (15), ventricular (15), or Purkinje fibers (16); the only measurable effect was a modest increase in the rate of early repolarization (Fig. 2) of the action potential (15,16). Of particular interest are the observations of Cranefield et al. (16) who demonstrated that Purkinje fibers which were spontaneously active in a media containing no sodium and 4.0 mM Ca were markedly depressed in rate and amplitude on exposure to relatively low concentrations of verapamil (Fig. 3); similarly, fibers exposed to low sodium concentrations and developing repetitive activity following bursts of sustained depolarizing

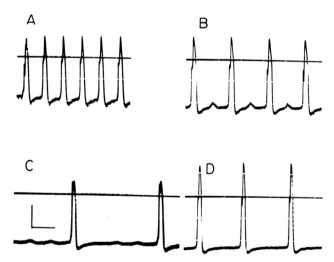

FIG. 3. Effects of verapamil on calcium-dependent "slow response" in a Purkinje fiber. Records shown are from a fiber which was spontaneously active in a solution containing zero sodium and 4.0 mM calcium; the horizontal line represents zero transmembrane potential. **A:** Control record obtained in the absence of verapamil. **B** and **C:** Records showing marked reduction in rate and loss of amplitude (most marked in **C**) 3 and 18 min, respectively, after exposure to 0.25 mg/liter of verapamil. **D:** With 0.25 mg/liter of verapamil still present, the level of calcium raised to 16.2 mM. Records taken 4.5 min later show a partial return to the control frequency of spontaneous activity and a marked increase in amplitude and overshoot. Calibrations: vertical 20 mV and horizontal 5 sec. (Reprinted from ref. 16, with permission.)

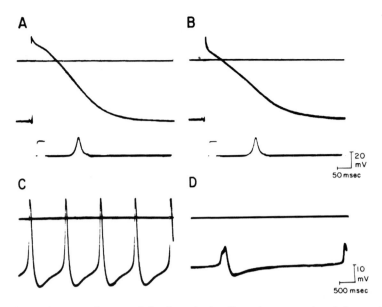

FIG. 4. Effects of verapamil on specialized conducting fibers from normal and diseased human atria. Panels **A** and **B** are from a segment of normal human atrium stimulated at a cycle length of 800 msec. **A** is a control. In **B** after 30 min of superfusion with verapamil, 1 mg/liter, resting membrane potential, action potential amplitude, and $\dot{V}_{max}$ are unchanged. However, the voltage at which the plateau originates is decreased and the slope of phase 2 is increased. Panels **C** and **D** are recorded from an isolated sample of diseased human atrium. **C** is a control record of the spontaneous rhythm occurring in this tissue. In **D**, following 7 min of superfusion with verapamil, 1 mg/liter, the maximum diastolic potential, the slope of phase 4 and the spontaneous rate have decreased, and action potential amplitude is markedly diminished. Within 1 min, the preparation became quiescent. For both preparations, $(K^+)_o = 4$ mM; $(Ca^{2+})_o = 2.7$ mM. Temp. $= 37°C$. Both samples were obtained from human right atria as part of the routine cannulation procedure for cardiac bypass. (Reprinted from ref. 17, with permission.)

impulses could be inhibited by verapamil (16). Rosen et al. (17) have also shown that spontaneous activity in diseased human atria, undoubtedly mediated by slow-channel activity, was suppressed by verapamil (Fig. 4). Perhaps of significance are the observations that catecholamine-induced delayed after-depolarization noted in simian mitral valve fibers and "triggered" (18) sustained rhythmic activity may both be abolished by relatively low concentrations of verapamil (18), as are also the after-depolarizations induced by cardiac glycosides in isolated diseased human atria (17). These findings thus raise the possibility that verapamil may have a role in the control of arrhythmias which may arise on the basis of certain types of atrial disease and in particular mitral valve prolapse, or of those occurring in association with cardiac glycoside intoxication.

However, from the standpoint of the drug's overall antiarrhythmic actions, its effect on nodal cardiac tissues is especially relevant. In *in vitro* preparations, verapamil has a significant effect on phase 4 depolarization in the sino-atrial node (19) characterized by the reduction in the rate of rise, and the slope of diastolic slow depolarization, the maximum diastolic potential, and the membrane potential at the peak of depolarization of the sino-atrial node (20). These observations and others (3) are thus consistent with the belief that slow-channel activity is involved in the generation of pacemaker potential in the sino-atrial node. Verapamil also exerts a depressant effect on AV nodal function and in low concentrations prolongs the effective refractory period, an action clearly of significance in relation to the demonstration (21) that supraventricular tachycardias may result from the continuous re-entry of impulses utilizing the AV node as a part of the re-entrant pathway (21). Unlike the mechanism of the depression of AV transmission by β-adrenoceptor blocking drugs and vagomimetic interventions—both of which alter autonomic impulse traffic through the AV node—verapamil appears to prolong AV nodal refractoriness by a direct action on the slow-channel fibers in the node (19). Figure 5 (from 22) shows that verapamil depresses the amplitude of the action potentials in the upper and midportions of the AV node without altering the resting membrane voltage. The overall electrophysiologic properties of verapamil thus results from its ability to depress the slow response.

The results from various clinical studies are in susbstantial agreement with the experimental data discussed above. For example, in patients in sinus rhythm, intravenously administered verapamil was found to have no effect on the R-R, QRS, and Q-T$_c$ intervals of the electrocardiogram (10). The P-R interval, reflecting atrioventricular conduction, however, increased after verapamil in all patients in one study. These findings are thus consistent with the fact that the drug has no significant effects on the depolarization and repolarization phases

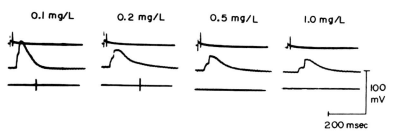

FIG. 5. Effect of verapamil on an action potential recorded from a fiber in the upper region of the AV node. The top trace in each section is an atrial electrogram and 0 potential. The middle trace is an action potential of an upper nodal cell, and the bottom trace is a His bundle electrogram. Action potentials were recorded from the nodal fiber after superfusion of the preparation for 20 min with 0.1 mg/liter of verapamil. Impalement was then maintained as the drug concentration was increased to 0.2 mg/liter, 0.5 mg/liter, and 1.0 mg/liter. Note the decrease in amplitude of depolarization with increasing drug concentration without any loss of maximum diastolic potential. At 0.2 to 1.0 mg/liter of verapamil, the peak of the action potential fell well short of reversal. After 0.5 mg/liter of verapamil, there was AV block between the recording site and the His bundle. After 1 mg/liter, there was some conduction delay between the atrial site and the AV node. (Reprinted from ref. 58, with permission.)

of the intracellularly recorded action potential in isolated cardiac tissues and that its major action in conscious man is on atrioventricular conduction through the depression of the slow response fibers in this structure. The minimal effect on heart rate is undoubtedly due to the fact that the drug's depressant action on sinoatrial nodal frequency is largely nullified by the reflex tachycardia that results from hypotension due to peripheral vasodilatation (23).

These findings are further supported by studies using His bundle electrocardiography, which has demonstrated that verapamil impedes AV conduction proximal to the His bundle without having an effect on intra-atrial or intraventricular conduction (24–26). The site of maximal delay appears to be in the atrioventricular node and is largely independent of autonomic influences (27). The blocking effect of verapamil on AV conduction is of clinical significance in that it represents the mechanism through which the ventricular response in atrial flutter and fibrillation is controlled and nodal re-entrant supraventricular tachycardia abolished (10). However, programmed stimulation techniques have revealed that the drug has minimal effect on anterograde and retrograde conduction or refractoriness in the anomalous pathway in the Wolff-Parkinson-White syndrome (28). It is, therefore, unlikely to be of value in the treatment of this syndrome complicated by atrial flutter or fibrillation (10). Clinical studies (10) are in line with these observations; the main therapeutic indication for the use of the drug is thus in the blocking of conduction across the atrio-ventricular node. This leads to prompt reversion of most cases of atrio-ventricular nodal reentrant tachycardias (Fig. 6) and in the slowing of the ventricular response in atrial fibrillation and flutter (Fig. 7) unrelated to the Wolff-Parkinson-White syndrome.

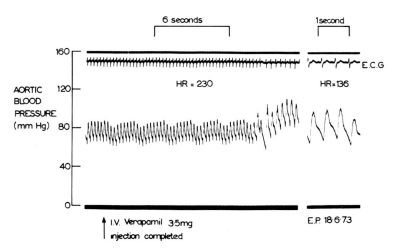

FIG. 6. Effect of verapamil on arterial pressure during conversion by the drug of rapid SVT (230/min) to sinus rhythm (136/min) during cardiac catheterization. Note the prompt response of the arrhythmia to verapamil, and the accompanying improvement in blood pressure. (Reprinted from ref. 10, with permission.)

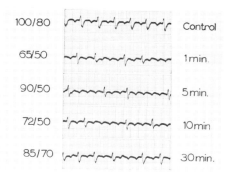

FIG. 7. Effects of intravenous verapamil in a patient with atrial flutter with 2:1 AV conduction, ventricular rate 140/min (control tracing). After verapamil, there is increased AV block, reduction in ventricular rate, and a fall in blood pressure. The maximal AV block occurred between 5 and 10 min after injection of 10 mg of verapamil; the ventricular rate at 10 min was 83/min. By 30 min, the ventricular rate had increased to about 120/min. (Reproduced from ref. 10, with permission.)

HEMODYNAMIC EFFECTS

By inhibiting excitation-contraction coupling in vascular smooth muscle (6,7), verapamil has been shown to cause marked vasodilatation in most peripheral vascular beds including the coronary (29), mesenteric, and canine hind limb preparations (8). It is, however, an extremely potent coronary vasodilator (30). For example, in animal studies it was found to increase coronary blood flow to a much greater extent than does papaverine, while it reduced cardiac oxygen uptake (1). Nayler and Szeto (31) have demonstrated that like propranolol, verapamil decreases myocardial demand for oxygen, but whereas propranolol consistently increases coronary vascular resistance (32), verapamil has the opposite effect even during the hypotensive phase of the drug's action (9). During selective coronary angiography, verapamil has been reported to produce minimal coronary vasodilatation in patients with normal coronary arteries and in those with atherosclerotic disease (33), so that the mechanism whereby the drug produces a beneficial effect in angina relates more to reduced cardiac work rather than to enhance coronary perfusion.

As might be expected, verapamil has a marked negative inotropic effect on isolated cardiac muscle (12,15)—this is in part reversed by calcium, catecholamines, or glucagon (15). In intact animals and in man, the drug's depressant action on heart muscle is not only dose-dependent but is greatly modified by extramyocardial factors. In anesthetized dogs, Ross and Jorgensen (8) reported a dose-related reduction in cardiac output and stroke volume, but the hypotensive effect of the drug could be accounted for almost entirely by peripheral vasodilatation in intravenous doses below 0.25 mg/kg body weight. More recent studies by Angus et al. (27) are in general agreement with these findings; in anesthetized dogs, they found that verapamil produced a dose-dependent peripheral vasodilatation with a reflex increase in myocardial contractility and heart rate. Propranolol pretreatment blocked the reflex cardiac stimulation but the vasodilatation was unaffected, consistent with the fact that the vascular effects of the drug were mediated through a mechanism independent of peripheral β-adrenoceptor stimulation. When doses of verapamil above the range used clinically were administered, a direct myocardial depressant action became apparent.

There is still a relative paucity of data on the hemodynamic effects of intravenously administered verapamil in man, but those available suggest that, as in the experimental animal, the drug produces a complex interplay of simultaneous alterations in preload, afterload, myocardial contractility, and, in all probability, coronary blood flow. Using right heart catheterization in 12 patients with cardiac disease, Ryden and Saetre (34) studied the hemodynamic effects of 0.1 mg/kg body weight of verapamil as an intravenous injection followed immediately by continuous infusion of 0.005 mg/kg/min for 30 min. In patients with sinus rhythm, the mean arterial pressure was reduced with a slight increase in cardiac output, but no significant reduction in stroke volume was observed except in patients who were in atrial fibrillation. Very similar results for patients in sinus rhythm have been reported by Saebra-Gomes et al. (35) who studied the effects of 0.1 mg/kg of verapamil in 10 patients (seven with coronary artery disease); in healthy adults, a slight negative inotropic action of the drug was found but it could easily be abolished by exercise (36). In another study (37) in six patients (three with valve disease) with normal ejection fractions, 10 mg bolus injections of verapamil were found to decrease left ventricular filling pressure. The effects were similar in digitalized and nondigitalized patients.

We recently evaluated the hemodynamic effects of intravenously administered verapamil (10 mg) during diagnostic cardiac catheterization in 20 patients (13 males and 7 females) with coronary artery disease and rheumatic valve lesions (23). The peak alterations in various hemodynamic variables induced by the drug occurred between 3 and 5 min after the completion of the injection with a relatively rapid dissipation of the effects, so that by 10 min the mean changes as compared to the control values were not statistically significant. The overall hemodynamic effects of the drug are summarized in Fig. 8, and records of the ventricular pressure trace from a typical drug response in a patient are shown in Fig. 9. Again, as in the study of Lewis et al. (3), no differences were noticed in the responses to verapamil in digitalized or nondigitalized patients. The mean arterial pressure fell from 97.8 ± 3.4 to 85.9 ± 2.7 mm Hg (-12%, $p < 0.01$) accompanied by a significant decrease (-21%, $p < 0.001$) in systemic vascular resistance (from $1,435 \pm 80$ to $1,131 \pm 82$ dynes/sec/cm^{-5}) with an increase in left ventricular end-diastolic pressure from 11.0 ± 0.9 to 15.0 ± 1.0 mm Hg ($+36\%$, $p < 0.01$) and a reduction in LV dp/dt max from $1,343 \pm 152$ to $1,007 \pm 102$ mm Hg (-25%, $p < 0.005$). The changes in heart rate (from 75.7 ± 3.0 to 80.2 ± 2.8 beats/min), cardiac index (from 3.17 ± 0.15 to 3.61 ± 0.17 liters/min/m^2), left ventricular minute work, and mean pulmonary artery pressures were not statistically significant. These overall data (see Fig. 8) therefore, indicate that the negative inotropic action of verapamil is substantially minimized by its effect on afterload so that cardiac index is not reduced by the drug in patients with cardiac disease. The data demonstrate that the intravenous dose (10 mg) required for the antiarrhythmic action of

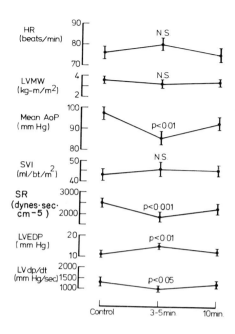

FIG. 8. Summary of hemodynamic data with verapamil. Each data point represents the mean ± standard error of the mean from 20 patients. HR = heart rate; LVMW = left ventricular minute work; AoP = aortic pressure; SVI = stroke volume index; LVEDP = left ventricular end diastolic pressure; and LV = left ventricular. NS = not significant.

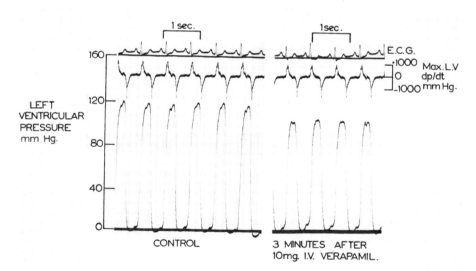

FIG. 9. Effects of verapamil on left ventricular contractility in a patient with coronary artery disease. The left ventricular pressure was increased by high-fidelity catheter tip manometry during cardiac catheterization and the first derivative of the pressure (LV dp/dt max) was obtained by electronic differentiation. Verapamil diminished the peak systolic left ventricular pressure and raised the left ventricular end-diastolic pressure; it produced only a small decrease in LV dp/dt. There was no change in heart rate. (Reprinted from ref. 23, with permission.)

verapamil in adult patients does not usually produce severe depression on hemodynamic variables—the effects on systemic arterial pressure, systemic vascular resistance, left ventricular filling pressure, and contractility being mild to moderate in severity and relatively short-lived. However, caution should nevertheless be exercised in the use of verapamil in patients with severe myocardial decompensation and in patients with acute myocardial infarction with unstable clinical state, as very few data are available in these conditions and the drug may well prove deleterious in patients with markedly compromised hemodynamics.

VERAPAMIL IN MYOCARDIAL ISCHEMIA

The potential role of verapamil in minimizing the severity of myocardial ischemic injury following experimental coronary occlusion was initially suggested by studies of Sodi-Pallares et al. (38). They reported that the drug either prevented or reversed the electrocardiographic manifestations of myocardial ischemia in the dog. Perhaps of greater clinical significance was the report by Kaumann and Aramendia (39) who found that dogs pretreated with intravenous verapamil before coronary occlusion failed to develop ventricular fibrillation and survived for many months, whereas animals similarly treated with the β-blocking drug sotalol invariably died from ventricular fibrillation within 24 hr of coronary occlusion. These results thus raised the possibility that verapamil might act by containing the extent of the initial ischemic injury (i.e., limiting "infarct size"), contributing to fewer arrhythmias and improved survival rate following coronary occlusion.

Various studies have since been undertaken to examine the effects of verapamil on experimental infarct size, but the results so far have not been entirely conclusive. For example, the drug has been found to improve hemodynamics in experimental infarction in anesthetized animals (40,41), notably preventing increases in systemic vascular resistance that occurs following coronary artery ligation (Fig. 10); the drug also reduced epicardial ST segment elevation following coronary occlusion (40,41) while having little effect on metabolism, regional blood flow, or tissue levels of enzymes in the ischemic myocardium. These somewhat inconsistent experimental data may be related to the use of anesthetized rather than conscious animals, but it is of interest that the histological study of Reimer et al. (42) has revealed that the drug does exert a beneficial effect on the extent of tissue necrosis when it was administered before coronary occlusion. For example, pretreatment of dogs with verapamil before the ligation of the circumflex coronary artery resulted in significantly less necrosis (14% treated vs. 35% untreated) with minimal hemodynamic consequences. In another study (43) in the isolated perfused heart, verapamil was found to reduce the extent of ultrastructural damage caused by hypoxic perfusion.

These results hold promise, but further work is clearly needed to decide whether the drug does indeed produce a salutary effect on infarct size in experimental animals and whether these findings are of clinical relevance. In contrast,

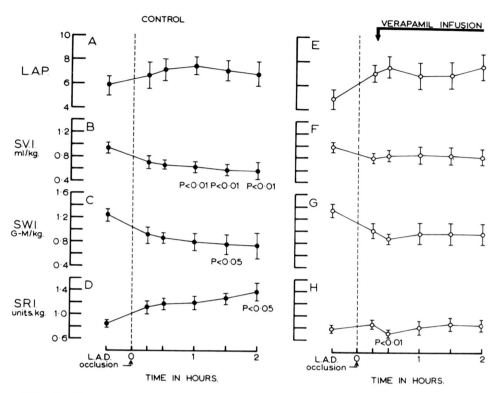

FIG. 10. Effects of verapamil on hemodynamic variables following coronary occlusion in anesthetized dogs. Each data point represents the mean ± standard error of the mean of 10 experiments. LAP = mean left atrial pressure; SVI = stroke volume index; SWI = stroke work index; and SRI = systemic resistance index. Note that verapamil prevents the increases in SRI following coronary artery occlusion.

much more encouraging data are now available on verapamil as an antianginal agent, although published experience of a controlled nature remains somewhat limited.

The prophylactic use of verapamil in angina pectoris stems from the original observations that the drug had potent coronary vasodilator properties. Increasing evidence is now accumulating to confirm the efficacy of verapamil in reducing the frequency and severity of angina (44–46). However, the precise mode of action of the drug in relieving angina is still poorly understood (46). Although the cardiocirculatory effects of intravenous verapamil in patients with coronary artery disease is established (23), little is known about the hemodynamic changes which might occur following long-term oral therapy with the drug. Thus, the results of the various clinical studies using oral regimens in angina cannot, at present, be correlated with the measured changes in myocardial oxygen supply and demand ratios. Such clinical observations are nevertheless of interest and do indicate the need for stringent double-blind clinical trials to evaluate the role of verapamil in the management of angina pectoris.

Neumann and Luisada (47) in a double-blind trial, studied the effects of oral verapamil (40 mg three times daily) on the weekly consumption of glyceryl trinitrate (nitroglycerin) in 30 geriatric patients with stable angina. They found that over a 6-week period of drug administration, there was a significant decrease in the number of glyceral trinitrate tablets required by anginal patients while on verapamil compared with the number of tablets consumed by the patients on placebo. Subsequently, in a controlled study in 16 anginal patients, Sandler et al. (48) found that the effects of 120 mg three times daily of verapamil were comparable with those of 100 mg three times daily of propranolol in terms of a significant decrease in the frequency of angina and improvement of exercise tolerance. In these doses, both drugs had a beneficial effect on the amount and duration of ischemic ST segment depression during exercise. When only 40 mg three times daily of verapamil was used, there was no objective benefit in the electrocardiographic appearances, despite appreciable subjective improvement. Unlike propranolol, verapamil had no effect on resting heart rate or exercise tachycardia in this study, but it produced a mild reduction in diastolic blood pressure.

The study of Livesley et al. (49) had also suggested no statistically significant difference between the effects of verapamil 120 mg three times daily and propranolol 100 mg three times daily in the treatment of angina of effort: both regimens were more effective than a placebo in reducing daily attacks of angina and in prolonging exercise tests. Both verapamil and propranolol reduced diastolic blood pressure as well as the resting heart rate and exercise tachycardia. The conflicting reports of Sandler et al. (48) and that of Livesley et al. (49) on the effects of verapamil on resting heart rate and exercise tachycardia are not readily explained.

Two further clinical studies have recently been reported. In the multicenter trial in Denmark (45), the effect of verapamil 80 mg three times daily on angina pectoris as compared with a placebo was studied in a double-blind crossover trial of two 4-week periods in 47 patients. The frequency of anginal attacks and glyceral trinitrate consumption was 25% lower during verapamil therapy with a significant improvement in exercise tolerance as judged by bicycle ergometry. However, the exercise-induced tachycardia was not reduced by verapamil, although the average resting heart rate on verapamil was lower than that on the placebo. The drug had no effect on the P-R interval of the electrocardiograms. The study of Balasubramanian et al. (46) from India utilized a protocol in which the efficacy of verapamil in a dose of 120 mg three times daily was evaluated in 25 patients with ischemic heart disease. The patients were subjected to serial measured multistage treadmill exercise without any drugs, and after propranolol 40 mg three times daily for 3 days; the tests were then repeated after 2, 4, and 8 weeks of verapamil therapy. The evaluation was based on computerized ST segment alterations, anginal threshold, and electrocardiographic changes. Anginal threshold was increased by verapamil in all patients, and 8 of the 13 symptomatic patients became pain-free with a significant improvement in the ischemic ECG pattern in all cases. The resting and exercise heart rates were reduced by verapamil but not to the same extent as that by propranolol. No patients developed atrioventricular block, significant hypotension, or heart failure during the trial, and side effects from the drug were limited to transient constipation in four patients.

Thus, the overall data suggest that verapamil may be an effective antianginal drug which is extremely well tolerated by the oral route. Since the drug does not alter airways resistance (50), it may have advantages over conventional β-blocking drugs in anginal patients who also have bronchial asthma.

In the context of clinical myocardial ischemia, perhaps the most exciting recent development is the recent appreciation that many cases of angina, especially that occurring at rest, may be due to primary decreases in myocardial perfusion resulting from coronary vasospasm both in atherosclerotic vessels as well as in those free of disease (51–53). Evidence is also accruing to suggest that vasospasm may also be a significant factor in the pathogenesis of acute myocardial infarction (54). The question has therefore arisen in this setting that calcium antagonists which are potent coronary vasodilators may well become therapeutic agents of significance in preventing the occurrence of reversible episodes of ischemia and infarction in certain subsets of patients with ischemic heart disease. Preliminary experience justifies considerable optimism. For example, both verapamil (55) and nifedipine (56) have been used successfully in the control of vasospastic angina, and a double blind crossover study with verapamil has been shown to be effective in the control of symptoms of unstable angina (57). If these tentative data are substantiated in larger clinical trials, the role of verapamil and presumably other calcium antagonists such as nifedipine in the management of myocardial ischemia will be considerably broadened.

SUMMARY AND CONCLUSIONS

Verapamil is a novel antiarrhythmic and antianginal agent which, although introduced in 1962, has only recently gained prominence not only as a significant agent in cardiovascular therapeutics but also as a powerful tool to examine the nature of some of the biophysical phenomena at the membrane of cardiac and other excitable tissues. Verapamil is the prototype of those agents which selectively inhibit membrane transport of calcium, an action which accounts for the drug's peripheral and coronary vasodilator properties, its effect on excitation-contraction coupling and hence its negative inotropic propensity, as well as its depressant effects on the sinus node and atrioventricular conduction. Its pharmacological effects are largely independent of the autonomic nervous system. The main therapeutic uses of the drug are in the management of atrial tachyarrhythmias and angina, especially that arising on the basis of coronary vasospasm. The overall experimental and clinical data suggest that verapamil will become an important and safe addition to existing drug regimens, especially as an agent of choice for the short-term treatment of most cases of paroxysmal supraventricular tachycardias. The initial experience in other arrhythmias and angina is also sufficiently encouraging to justify further detailed clinical trials to define its potential role in cardiovascular therapeutics.

ACKNOWLEDGMENTS

This paper was supported by the Medical Research Service of the Veterans Administration, American Heart Association, Greater Los Angeles Affiliate, and the National Institutes of Health, Bethesda, Maryland. I am much indebted to Mrs. Betty Garrigues for help in the preparation of this manuscript.

REFERENCES

1. Haas, H., and Hartfelder, G. (1962): α-isopropyl-α-(N-methyl-N-homoveratryl-α-amino propyl)-3-4 dimethoxyphenylacetonitril, eine Aulstanz mit coronasgefassen Eigenschaften. *Arzneimittel-Forschung,* 12:549–558.
2. Melville, K. I., and Benfey, B. C. (1965): Coronary vasodilatory and cardiac adrenergic blocking effects of iproveratril. *Can. J. Physiol. Pharmacol.,* 43:339–342.
3. Hauswirth, O., and Singh, B. N. (1979): Ionic mechanisms in heart muscle in relation to the genesis and the pharmacological control of cardiac arrhythmias. *Pharmacol. Rev., (in press).*
4. Singh, B. N., Ellrodt, G., and Peter, C. T. (1979): Verapamil: A review of its pharmacological properties and therapeutic use. *Drugs,* 15:169–197.
5. Carnefield, P. F. (1975): *The Conduction of the Cardiac Impulse.* Futura Publishing Company, New York.
6. Golenhofen, K., and Lammel, E. (1972): Selective suppression of some components of spontaneous activity in various types of smooth muscle by iproveratril (Verapamil). *Pflugers Archiv für die gesamte Physiologie des Menschen und die Tiere,* 331:233–243.
7. Haeusler, G. (1972): Differential effect of verapamil on excitation-contraction coupling in smooth muscle and on excitation-secretion coupling in adrenergic terminals. *J. Pharmacol. Exp. Ther.,* 180:672–679.
8. Ross, G., and Jorgensen, C. R. (1967): Cardiovascular action of iproveratril. *J. Pharmacol. Exp. Ther.,* 158:504–509.
9. Nayler, W. G., McInnes, L., Swann, J. B., Price, J. M., Carson, V., Race, D., and Lowe, T. E. (1968): Some effects of Iproveratril (isoptin) on the cardiovascular system. *J. Pharmacol. Exp. Ther.,* 161:247–261.
10. Heng, M. K., Singh, B. N., Roche, A. H. G., Norris, R. M., and Mercer, C. J. (1975): Effects of intravenous verapamil on cardiac arrhythmias and on the electrocardiogram. *Am. Heart J.,* 90:487–498.
11. Henry, P. D. (1979): Calcium antagonists: Mechanisms of action with special reference to nifedipine. *(This volume.)*
12. Fleckenstein, A. (1970): Specific inhibitors and promotors of calcium action in the excitation-contraction coupling of heart muscle and their role in the prevention or production of myocardial lesions. In: *Calcium and the Heart.* Academic Press, London.
13. Ellrodt, G., and Singh, B. N. (1979): Therapeutic implications of slow-channel blockade in cardiocirculatory disorders. *Circulation (in press).*
14. Garvey, H. L. (1969): The mechanism of action of verapamil on the sinus and AV nodes. *Eur. J. Pharmacol.,* 8:159–165.
15. Singh, B. N., and Vaughan Williams, E. M. (1972): A fourth class of antidysrhythmic action? Effect of verapamil on ouabain toxicity, on atrial and ventricular intracellular potentials and on other features of cardiac function. *Cardiovasc. Res.,* 6:109–114.
16. Cranefield, P. F., Aronson, R. S., and Wit, A. L. (1974): Effect of verapamil on the normal action potential and on a calcium-dependent slow response of canine cardiac Purkinje fibers. *Circ. Res.,* 34:204–213.

17. Rosen, M. R., Wit, A. L., and Hoffman, B. F. (1975): Appraisal and reappraisal of cardiac therapy. Electrophysiology and pharmacology of cardiac arrhythmias. VI. Cardiac effects of verapamil. *Am. Heart. J.,* 89:665–673.
18. Wit, A. L., and Cranefield, P. F. (1976): Triggered activity in cardiac muscle fibers of the simian mitral valve. *Circ. Res.,* 38:85–94.
19. Zipes, D. P., and Fischer, J. C. (1974): Effects of agents which inhibit the slow channel on sinus node automaticity and atrioventricular conduction in the dog. *Circ. Res.,* 34:184–192.
20. Okada, T. (1976): Effect of verapamil on electrical activities of SA node, ventricular muscle, and Purkinje fibers in isolated rabbit hearts. *Jap. Circ. J.,* 40:329–341.
21. Goldreyer, B. N., and Bigger, J. T., Jr. (1971): Site of re-entry in paroxysmal supraventricular tachycardia in man. *Circulation,* 43:15–26.
22. Wit, A. L., and Cranefield, P. (1974): Effect of verapamil on the sinoatrial and atrioventricular nodes of the rabbit and the mechanism by which it arrests re-entrant atrioventricular nodal tachycardia. *Circ. Res.,* 35:413–425.
23. Singh, B. N., and Roche, A. H. G. (1977): Effects of intravenous verapamil on hemodynamics in patients with heart disease. *Am. Heart J.,* 94:593–599.
24. Hussaini, M. H., Kurasnicka, J., Ryden, L., and Holmberg, S. (1973): Action of verapamil on sinus node, atrioventricular and intraventricular conduction. *Br. Heart. J.,* 35:734–737.
25. Roy, P. R., Spurrell, R. A. J., and Sowton, G. E. (1974): The effect of verapamil on the conduction system in man. *Postgrad. Med. J.,* 50:270–275.
26. Neuss, H., Nowak, F. G., Schlepper, M., and Wusten, B. (1974): The effects of antianginal drugs on AV conduction in normal subjects. *Arzneimittel-Forschung,* 24:213–216.
27. Angus, J. A., Dhumma-Upahorn, P., Cobbin, L. B., and Goodman, A. H. (1976): Cardiovascular action of verapamil in the dog with particular reference to myocardial contractility and atrioventricular conduction. *Cardiovasc. Res.,* 10:623–632.
28. Spurrell, R. A. J., Krikler, D. M., and Sowton, G. E. (1974): The effect of verapamil on the electrophysiological properties of the anomalous atrioventricular connexions in Wolff-Parkinson-White syndrome. *Br. Heart J.,* 36:256–257.
29. Melville, K. I., Shister, H. E., and Huq, S. (1964): Iproveratril: Experimental data on coronary-dilatation and antiarrhythmic action. *Can. Med. Assoc. J.,* 90:761–770.
30. Winbury, M. M., Howe, B. B., and Hefner, M. A. (1969): Effect of nitrates and other coronary dilators on large and small coronary vessels: An hypothesis for the mechanism of action of nitrates. *J. Pharmacol. Exp. Ther.,* 168:70–95.
31. Nayler, W. G., and Szeto, J. (1972): Effect of Verapamil on contractility, oxygen utilisation and calcium exchangeability in mammalian heart muscle. *Cardiovasc. Res.,* 6:120–128.
32. Nayler, W. G., McInnes, I., Swann, J. B., Carson, V., and Lowe, T. E. (1967): Effect of propranolol, a beta-adrenergic antagonist on blood flow in the coronary and other vascular fields. *Am. Heart. J.,* 73:207–216.
33. Mignault, S. H. (1966): Coronary cineangiographic study of intravenously administered isoptin. *Can. Med. Assoc. J.,* 95:1252–1253.
34. Ryden, L., and Saetre, H. (1971): The hemodynamic effect of verapamil. *Eur. J. Clin. Pharmacol.,* 3:153–157.
35. Seabra-Gomes, R., Richards, A., and Sutton, R. (1976): Hemodynamic effects of verapamil and practolol in man. *Eur. J. Cardiol.,* 4:79–85.
36. Attehog, J. H., and Ekelund, L. G. (1975): Haemodynamic effects of intravenous verapamil at rest and during exercise in subjectively healthy middle aged males. *Eur. J. Clin. Pharmacol.,* 8:317–322.
37. Lewis, B., Mitha, A., and Gotsman, M. (1976): Immediate haemodynamic effects of Verapamil in man. *Cardiology,* 60:366–376.
38. Sodi-Pallares, D., Bisteni, A., Medrano, G. A., deLeon, J. P., de Michelli, A., and Ariza, D. (1968): El efecto del Verapamil (iproveratril) sobre los cambios electrocardiograficos y de contraccion observados en el corazon del perro despues de la ligadura de la coronaria descendente anterior. *Revista Peruana de Cardiologia,* 14:3–12.
39. Kaumann, A. J., and Aramendia, P. (1968): Prevention of ventricular fibrillation induced by coronary ligation. *J. Pharmacol. Exp. Ther.,* 164:326–340.
40. Smith, H. J., Singh, B. N., Nisbet, H. D., and Norris, R. M. (1975): Effects of verapamil on infarct size following experimental coronary occlusion. *Cardiovasc. Res.,* 9:569–578.
41. Smith, H. G., Singh, B. N., Norris, R. M., Nisbet, H. D., John, M. B., and Hurley, P. J.

(1977): The effect of verapamil on experimental myocardial ischemia with a particular reference to regional myocardial blood flow and metabolism. *Aus. NZ. J. Med.,* 7:114–121.
42. Reimer, K. A., Lowe, J. E., and Jennings, R. B. (1977): Effect of the calcium antagonist verapamil on necrosis following temporary coronary artery occlusion in dogs. *Circulation,* 55:581–587.
43. Nayler, W., Grau, A., and Slade, A. (1976): A protective effect of verapamil on hypoxic heart muscle. *Cardiovasc. Res.,* 10:650–662.
44. Krikler, D. M. (1974): Verapamil in cardiology. *Eur. J. Cardiol.,* 2:3–10.
45. Andreasen, F., Boye, E., Christoffersen, D., et al. (1975): Assessment of verapamil in the treatment of angina pectoris. *Eur. J. Cardiol.,* 2:443–452.
46. Balasubramanian, V., Khanna, P. K., Naryanan, G. R., and Hoon, R. S. (1976): Verapamil in ischemic heart disease—quantitative assessment by serial multistage treadmill exercise. *Postgrad. Med. J.,* 52:143–147.
47. Neumann, M., and Luisada, A. A. (1966): Studies of iproveratril on oral administration in double blind trials in patients with angina pectoris. *Am. J. Med. Sci.,* 251:552–556.
48. Sandler, G., Clayton, G. A., and Thornicrofts, S. (1968): Clinical evaluation of Verapamil in angina pectoris. *Br. Med. J.,* 3:224–227.
49. Livesley, B., Catley, P. F., Campbell, R. C., and Oram, S. (1973): Double blind evaluation of verapamil, propranolol and isosorbide dinitrate against a placebo in the treatment of angina pectoris. *Br. Med. J.,* 1:375–378.
50. Hills, E. A. (1970): Iproveratril and bronchial asthma. *Br. J. Clin. Prac.,* 24:116–117.
51. Miller, J., Pichard, A. and Dack, S. (1976): Coronary arterial spasm in Prinzmetal's angina: A proved hypothesis. *Am. J. Cardiol.,* 37:938–940.
52. Maseri, A., Severi, S., deNes, M., et al. (1978): "Variant" angina: One aspect of a continuous spectrum of vasospastic myocardial ischemia. *Am. J. Cardiol.,* 42:1019–1035.
53. Figueras, J., Singh, B. N., Ganz, W., Charuzi, Y., and Swan, H. J. C. (1979): Mechanism of rest and nocturnal angina: Observations during continuous hemodynamic and electrocardiographic monitoring. *Circulation (in press).*
54. Johnson, A. D., and Detwiler, J. A. (1977): Coronary spasm, variant angina and recurrent myocardial infarction. *Circulation,* 55:947–950.
55. Solberg, L. E., Nissen, R. S., and Vliestra, R. E. (1978): Prinzmetal's variant angina—response to verapamil. *Mayo Clinic. Proc.,* 53:256–259.
56. Miller, J. E., and Gunther, S. J. (1978): Nifedipine therapy for Prinzmetal's angina. *Circulation,* 57:137–139.
57. Parodi, O., Simonetti, I., and Maseri, A. (1977): Management of "crescendo" angina by verapamil: A double blind cross-over study in CCU. *Circulation,* 55 and 56 (Suppl 1):867(abstract).
58. Wit and Cranfield. (1974): Effect of Verapamil on the sinoatrial and atrioventricular nodes of the rabbit and the mechanism by which it arrests re-entrant atrioventricular nodal tachycardia. *Circ. Res.,* 35:413–425.

Trace Metals in Health and Disease, edited by
N. Kharasch. Raven Press, New York © 1979.

Klaus Schwarz
1914–1978
Commemoration of a Leader in Trace Element Research

G. N. Schrauzer

Department of Chemistry, University of California at San Diego, Revelle College, La Jolla, California 92093

Klaus Schwarz was born January 9, 1914, in Barnewitz, Prussia the son of a physician and received his early education at Brandenburg, Berlin, and Davos (Switzerland). He subsequently studied chemistry and medicine from 1933 to 1939 at the Universities of Freiburg, Rostock, Berlin, and later at Heidelberg, where he received an M.D. degree in 1939. After working in Richard Kuhn's Laboratory at the Kaiser Wilhelm Institute (now Max Planck Institute) of Medical Research in Heidelberg, he obtained a "Dr.med.habil." degree in 1943, which, according to German University regulations, was a prerequisite for beginning an academic career. With interruptions due to the war, he stayed in Heidelberg until 1946. He then moved to the University of Mainz, where he became Head of the Institute of Vitamin and Hormone Research. However, in 1949 he decided to leave Germany, accepting an invitation to join the National Institutes of Health as a Special Research Fellow. He became Chief, Section on Experimental Liver Diseases, National Institute of Arthritis and Metabolic Diseases, National Institutes of Health, and an Associate of George Washington University Medical School, in 1952.

In 1963, he moved to California as the Chief of the newly established Laboratory of Experimental Metabolic Diseases, Veterans Administration Hospital, Long Beach, and became an Associate Clinical Professor of Biological Chemistry at the University of California, Los Angeles, in 1964. Klaus Schwarz died on January 30, 1978 after a brief illness. He is survived by his wife and faithful companion during his California years, Joyce Ann Schwarz, and his children.

In this chapter, the major research accomplishments of Klaus Schwarz will be described with occasional brief personal notes.

While at Heidelberg, Schwarz discovered *dietary liver necrosis* as a defined deficiency disease. It remained one of his main research interests in the following years.

It is not easy for us now to fully appreciate the enormous experimental and

FIG. 1. Klaus Schwarz, M.D., 1914–1978.

conceptual difficulties that had to be overcome before dietary liver necrosis could be rigorously differentiated from other degenerative diseases of the liver. These can be experimentally induced by vitamin deficient diets alone. Schwarz recognized earlier than any of his contemporaries that it was necessary, in the experimental work with laboratory animals, to use diets that were supplemented with all the vitamins and other cofactors then known, if any real progress was to be made. During his stay at Heidelberg, there was great activity in vitamin research. Riboflavin, vitamin B_6, and pantothenic acid had only just been discovered. It is not too well known that Schwarz discovered one additional vitamin-like growth factor for certain bacteria, p-aminobenzoic acid (PABA), in 1941 (7) quite independent from D. D. Woods, who detected the same substance as a naturally occurring metabolite antagonizing the sulfanilamides at about the same time (33).

During World War II, Schwarz saw many cases of malnutrition and liver disease while serving as a Medical Officer in the German Armed Forces. I understand that he was spared combat duty but for some time was suspected of treason and shadowed by counterespionage agents. Perhaps his war experience deepened his interest in nutritionally induced acute and chronic diseases of the liver. In 1953, he organized a symposium on "Nutritional Factors and

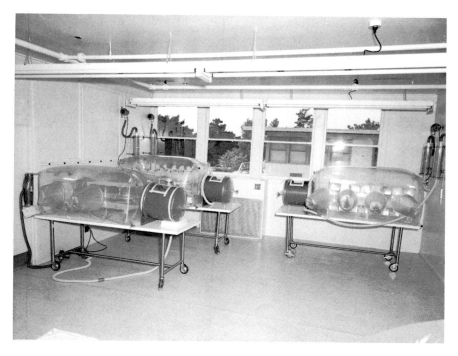

FIG. 2. Animal isolators at the Laboratory for Experimental Metabolic Diseases, Veterans Administration Hospital, Long Beach, California.

Liver Diseases" for the New York Academy of Sciences. At this meeting, progress was made in distinguishing liver necrosis from cirrhosis. Liver necrosis was at that time chiefly linked to protein malnutrition, particularly, a deficiency in sulfur amino acids such as cystine and methionine. But experimental work with animals produced contradictory results. Be feeding diets low in protein, Schwarz generated liver necrosis in rats and first used sulfur amino acids ("Factor 1") and later vitamin E ("Factor 2") to prevent its development. Both factors exerted protective action, Factor 2 more than Factor 1. But a "Factor 3", first detected in alcohol-extracted casein in 1951 (15), seemed to exhibit far greater potency.

By 1952, Schwarz (16) had achieved a 30-fold enrichment of Factor 3, but subsequent work was becoming exceedingly difficult. At last, 1.3 mg of a highly active material were obtained from 1 ton of pork kidneys. He still succeeded in separating Factor 3 into two further fractions, "α" and "β," but the extreme lability of the factor made its isolation in pure form impossible. During various stages of experimentation, Schwarz observed a peculiar garlic odor. This led him on May 17, 1957, to analyze the factor for selenium (22), an element that hitherto had received little attention in biomedicine.

It is impossible in the present account to fully cover all aspects and consequences of this discovery. At first, it merely explained many of the conflicting reports on the apparent protecting effects of sulfur amino acids in dietary liver necrosis: they had to be attributed to contamination by selenium. Indeed, only those amino acids were found to be effective that were isolated from animal horn, hoofs, or hair, which contain substantial amounts of selenium. Highly purified or synthetic sulfur amino acids were ineffective. Subsequently, it was found that the role of selenium in protecting the liver was different from that of vitamin E. Vitamin E has a "selenium-sparing" effect, and vice-versa, but neither fully replaces the other. A wide variety of other selenium compounds were tested for biological activity. "Factor 3" was found to be more effective than all others tried thus far, although even inorganic selenium compounds such as sodium selenite were active. Thus, the uncertainties concerning the exact nature of Factor 3 could be circumvented. Later, in a search for biologically active selenium drugs, Schwarz embarked on an extensive research project with Professor Arne Fredga at the Institute of Organic Chemistry, University of Uppsala, Sweden. Between 1958 and 1973, over 850 selenium-containing organic compounds were synthesized and tested, many of these being selenium-containing derivatives of carboxylic acids. A large number of interesting observations were made in these studies. For example, in a series of straight-chain-selenodicarboxylic acids $HOOC-(CH_2)_n-Se-(CH_2)_nCOOH$, activity of compounds with short alkyl chains—i.e., with $n = 2 \rightarrow 4$—was low; the compound with $n = 5$ was five times more potent; from $n = 5 \rightarrow 11$, an alternative effect was observed; and the acids with $n = 9$, 10, and 11 showed optimal activity in preventing dietary liver necrosis in rats (29,33). Perhaps one day, some of these compounds will be used for the treatment of selenium deficiency syndromes in animals.

Selenium deficiency is associated with a number of diseases of laboratory and farm animals. While defects of growth are frequently observed, they are not the most prominent sign and do not appear in all species. Muscular dystrophy, myocardial degeneration, kidney damage, hemorrhages of the lung, pancreatic atrophy, and abnormalities of serum proteins were noted particularly in mice and rats. In chicks and turkeys, many of these symptoms occur in selenium deficiency, but exudative diathesis is more common and has been responsible for many previously unexplained deaths. White muscle disease was recognized to be a selenium deficiency disease. Additional conditions which appear to develop in selenium-deficient laboratory animals are: impaired fertility, diminished sperm motility, unthriftiness, and cataracts.

These developments culminated with the discovery of enzymes that contain selenium. In 1972, glutathione peroxidase was identified to be a selenium enzyme by Hoekstra et al. (10) in Wisconsin and Flohé et al. (4) in Germany; details on the properties and functions of this enzyme are described by Dr. Flohé in this volume. A glycine reductase which converts glycine to acetic acid was isolated from *Clostridium sticklandii* and similarly identified as selenium-dependent (22), as is formate dehydrogenase (1), which is widely distributed in bacteria, including *Clostridium thermoaceticum, C. sticklandii, C. formicoaceticum,* and *Methanococcus vannielli.*

In the late 1960s, I had become interested in demonstrating the cancer-protecting effects of selenium after it was shown that a historical Plasma Cancer Test was, in fact, a crude assay for plasma selenium. What previously was regarded as a positive diagnosis of cancer actually indicated subnormal plasma selenium levels (12).

We concluded from this that selenium could be a natural cancer-protecting agent, as did others at about the same time (31). To prove this hypothesis, we performed animal experiments, and it was in the course of this work that I thought I should consult with Klaus Schwarz. We established contact under somewhat mysterious circumstances: the very moment when I decided to call him, my phone rang and Klaus Schwarz introduced himself—he *also* wanted to meet me personally. From this event in 1969, a deep and intensifying personal- and family friendship developed which was one of the most gratifying experiences in my life. I will not go into details too much here concerning selenium and cancer interrelationships: this matter is treated by a former collaborator of Schwarz, Dr. W. Baumgartner in this volume.

In brief, we have found that subtoxic amounts of selenium in the drinking water or the diet of breast cancer-prone mice causes a dramatic diminution of the tumor incidence. Subsequent epidemiologic studies revealed inverse relationships between the incidence and mortalities from cancer at major sites and the regional availability of selenium. In the United States, certain areas are known to be low in selenium. More women die of breast cancer in these areas than where selenium is more abundant (11). The possibility that selenium may be a cancer-protecting agent to be used at the community level is presently being considered. In addition to cancer, selenium is even believed to have prophylactic effects against heart disease, muscular dystrophy, against premature aging, heart disease, and immune incompetence. These are the more recent developments which more than anything else illustrate the enormous importance of a trace mineral which until 1956 was listed in textbooks only as a poison.

But Klaus Schwarz did not stop with selenium. Two years after the discovery of selenium's essentiality, he together with Walter Mertz was able to report that chromium is also essential (24,25). This element is active at such low concentrations that it was difficult to induce chromium deficiency in laboratory animals. Eventually Schwarz and Mertz succeeded and showed that chromium-deficient animals exhibit a low glucose tolerance which can manifest itself with diabetes-like symptoms. The animals also show shortened lifespans, increased serum cholesterol levels, and a tendency to form "plaques" in the large blood vessels. The work on chromium is presently being continued by Walter Mertz who succeeded in isolating a "glucose tolerance factor" consisting of a complex of chromium(III) with nicotinic acid and the components of glutathione as the ligands (8).

At first, it appeared as if chromium deficiency does not occur in humans.

However, disturbances of glucose metabolism in malnourished infants can sometimes be positively influenced by the addition of trivalent chromium to the formula (5). In other studies with adults, a marked reduction of serum cholesterol levels and improvement of glucose tolerance was observed. Finally, a female patient who had been fed by total intravenous nutrition for more than 5 years developed unexplained weight loss, impaired glucose tolerance, decreased respiratory quotient, pronounced negative nitrogen balance, and peripheral neuropathy, as well as low blood- and hair chromium. The intravenously administered nutrient mix contained only little chromium, furnishing but 8 μg of the element per day. After 45 units of insulin per day failed to improve the condition, 250 μg of chromium per day for 2 weeks normalized all the abnormal signs (6). Efforts are currently underway to determine the human dietary chromium requirement. It is presently estimated to fall within a range of 50–200 μg per day for adults.

The example of chromium, whose activity at very low concentrations seemed equally if not more surprising than that of selenium, prompted Klaus Schwarz in the following years to develop sophisticated methods of maintenance of laboratory animals under "trace-element sterile" conditions. "Ultraclean room" technology was employed together with highly purified diets that were carefully monitored for contaminants. Special "isolators" were built which were made out of plastic; dust in the breathing air was eliminated with special filters.

Animals maintained on highly purified amino acid diets which previously had been doing fine in metal cages developed roughness of coat and seborrhea-like conditions as well as impaired growth and abnormalities of the teeth in the isolators. The addition of the ash of livers or of yeast prevented these symptoms. Klaus Schwarz referred to these nutritional defects as to the "Factor G Effect" (17).

Employing the isolator technique, Schwarz et al. (28) demonstrated positive growth effects of *tin* at levels of 0.5–2 ppm in the diet. Tin is one of those elements which are normally present in foods or introduced as contaminants during food processing. In newborn rats, tin is not detectable immediately after birth, suggesting a placental barrier effect. It becomes detectable a few hours after birth, however, and is believed to be present in colostrum. Although tin deficiency may not be a problem in human adults, Schwarz has suggested that more attention should be given to the tin-status of premature babies. Tin is a most interesting element because it can exist in different oxidation states and catalyze electron transfer reactions. It thus could have very subtle but important effects on growth in early infancy which are still unexplored.

Vanadium was suggested to be essential, without adequate support, as early as in 1903. Schwarz and Milne in 1971 showed that vanadium in form of sodium orthovanadate induced definite growth responses in rats maintained in trace-element sterile isolators (27). Vanadium appears to catalyze the oxidation of certain substrates *in vivo,* e.g., of catecholamines. It also inhibits cholesterol synthesis and lowers the phospholipoid-level in blood.

Fluorine, an element which is presently still in the focus of public attention because of questions concerning the safety of the mandatory fluoridation of public water supply systems, was shown by Schwarz and Milne (26) to be essential in 1972. The addition of fluorine in form of sodium fluoride to the feed of rats at levels of 1.5–2.0 ppm caused a definite enhancement of longitudinal growth of the animals during the first 4 weeks. Apart from stabilizing effects on bones and teeth, fluoride is known to stimulate the growth of cells cultured *in vitro,* to activate the synthesis of citrullin in the liver, and to activate adenyl cyclase. The growth-stimulatory effects of fluoride occur at the same concentrations at which fluoride is added to the drinking water. Schwarz (18) pointed out that the normal dietary fluorine intake of humans is much lower than that of most laboratory animals on commercial feeds. Hence, it is possible that humans may actually not receive enough of this element. He also emphasized that fluoride is not excessively toxic. Diets containing 300–500 ppm of fluorine can be fed to animals for months without adverse effects.

Silicon was also investigated by Schwarz and his co-workers (18). When silicon was added to the feed of rats in form of sodium metasilicate, they were able to demonstrate significant growth effects; animals maintained on low-silicon diets demonstrated impaired growth, disturbances of bone formation and of the pigmentation of the frontal teeth. Edith Carlisle at the UCLA Department of Public Health independently discovered the essentiality of silicon in studies with chicks. The possibility exists that silicon not only promotes growth, bone, and teeth formation, but also has inhibitory effects on coronary heart disease and atherosclerosis (21,30). The human dietary silicon intakes of silicon are in the order of 0.5 g, of which only a fraction is absorbed. The question whether silicon deficiency in pregnancy may contribute to birth-defects was also raised (19).

With the discussion of silicon, we are entering the last phase of Klaus Schwarz's research, dealing with the demonstration of essentiality of the elements cadmium, lead, and arsenic (20). All three are known to be highly toxic at higher concentrations. Applying the isolator technique, statistically significant effects of cadmium on weight gains have been observed at cadmium dosage levels far below the toxicity threshold. The same appear to be true for lead, but this work could not be concluded in time. The addition of sodium arsenite to the feed of rats was also demonstrated to produce positive growth responses of rats maintained in trace-element controlled environments. It took 10 years before a low-As diet could be developed by Schwarz and his co-workers. In the meantime, Nielsen and co-workers in the United States (9), and Anke et al. (2) in East Germany, generated arsenic deficiency diseases in rats, goats, and pigs. Nielsen's rats showed rough fur, increased osmotic fragility of the red blood cells and enlarged spleens containing excess of iron. Anke's As-deficient pigs and goats demonstrated decreased reproduction, low birth rates, and retarded growth. Arsenic-deficient goats died during lactation with myocardial damage. It should be noted that cadmium and arsenic are known to possess selenium-antagonistic properties—i.e., these elements may interact with selenium *in vivo*. Our own studies have shown that arsenic abolishes the cancer-protecting effects of selenium (13); indications are that the same is true for cadmium and lead. Klaus Schwarz pointed out in 1977 that "antagonistic relationships and the need for a close balance are of greater importance in trace element nutrition and toxicity than in any other area" (20). Antagonistic effects of elements in the development of plants have been known to biologists for some time and are beginning to be more fully recognized in animals and man.

In returning to selenium, Klaus Schwarz's first trace element discovery, we have now briefly covered his enormous accomplishments in trace-element research, in which he holds a historically unparalleled record of achievement. Trace element research as we know it began in the 17th century in which it was recognized that iron was essential for the maintenance of health in man. That other elements could be similarly essential did not seem to trouble anyone much in the 18th century, during which no new "essential trace element" was added. This is perhaps not surprising if one considers that many of the elements were just being identified and chemistry had not freed itself from the confinements of phlogiston theory. But even in the 19th century, which saw great advancements in all sciences, only one essential trace element was recognized as such: iodine. Medications containing iodine were used empirically for the treatment of goiter as early as 1819. Boussingault recommended the addition of iodide to table salt as a prophylactic measure shortly thereafter. In 1830, iodine deficiency was clearly shown to be the cause of goiter. However, it took another 100 years until iodized salt was introduced. In Switzerland, a country greatly plagued by goiter and cretinism, the use of iodized salt was "officially encouraged" only as late as 1924.

In the second quarter of the 20th century, at last, five new trace elements were discovered: copper, manganese, zinc, cobalt, and molybdenum. The third quarter (1950–1974) was clearly dominated by Klaus Schwarz's discoveries. Of the eight new elements shown to be essential, selenium, chromium, tin, vanadium, fluorine, silicon, nickel, and arsenic, six were discovered or codiscovered by him (see Table 1). More recently, evidence has begun to accumulate which suggests that cadmium and lead may also be essential (3,20), again largely through the efforts of Klaus Schwarz and his collaborators.

Henry Schroeder has once remarked that *"the trace elements are more important to life than the vitamins, (because) they cannot be synthesized, as can the vitamins, but must be present in the environment within a relatively narrow range of concentration . . . Their only sources are the earth's crust and sea water, and without them life would cease to exist* (14). Yet, the discovery of a new essential trace element does not trigger anywhere near the interest as the discovery of a new vitamin. One reason for this is that acute deficiency syndromes in man that can be attributed to trace mineral deficiencies are rare compared to the many known vitamin deficiency diseases. Another reason is that some of the trace elements are still greatly feared—with some justification—as potent

TABLE 1. *Discovery of trace element requirements*[a]

Element	Year	Discoverers
Iron	17th century	
Iodine	1830–1850	Boussingault, Chatin
Copper	1928	Hart, Steenbrock, Waddell, and Elvehjem
Manganese	1931	Kemmerer and Todd
Zinc	1934	Todd, Elvehjem, and Hart
Cobalt	1935	Underwood and Filmer; Marston; Lines
Molybdenum	1953	deRenzo, Kaleita, Heytler, Oleson, Hutchings, and Williams; Richert and Westerfeld
Selenium	1957	Schwarz and Foltz
Chromium	1959	Schwarz and Mertz
Tin	1970	Schwarz, Milne, and Vinyard
Vanadium	1971	Schwarz and Milne; Hopkins and Mohr
Fluorine	1971	Schwarz and Milne
Silicon	1972	Schwarz and Milne; Carlisle
Nickel	1974	Nielsen and Ollerich; Anke, Grün, Dittrich, Groppel, and Hennig; Kirchgessner and Schnegg
Arsenic	1975	Nielsen, Givand, and Myron; Anke, Hennig, Grün, Partschfeld, Groppel, and Lüdke
Cadmium [b]	1977	Schwarz and Spallholz; Anke, Hennig, Groppel, Partschfeld, and Grün
Lead [b]	1977	Schwarz, Spallholz, and Moran

[a] Updated from Ref. 20, *loc. cit.*
[b] Positive growth effects were observed under carefully controlled experimental conditions; essentiality possible but not yet proved.

poisons rather than as benefactors. Schwarz frequently stressed the great importance of the difference between required, biologically effective concentrations of a trace element and the dose levels producing chronic or acute toxicity symptoms.

For each trace element, a relatively narrow "concentration window" exists, outside of which life cannot be sustained. Lower concentrations cause death via deficiency syndromes, higher concentrations kill by poisoning the organism. However, it is not yet fully appreciated that "toxicity is no counterargument against biological essentiality" (24). Current officially endorsed attempts to remove certain of the "toxic" elements out of the biologic environment should not be carried to the extreme as this could create equal—if not more serious—problems of deficiency.

I can report that Klaus Schwarz remained active to the very end. In spite of adversities of various kinds which even threatened the existence of his laboratory at Long Beach toward the end of 1977, he was determined to go on, to start another round. For 1978, he had planned to continue the study of lead-deficient animals, to use ^{203}Pb as a biological marker to establish the mechanism of uptake. Similarly, the work on cadmium, silicon, and other elements was to be extended. It was not to be. In January 1978 he participated in a "Metals in Biology" Gordon Conference in Santa Barbara where he received an ovation by a prestigeous audience. We who were with him did not realize that this was to be his last.

Returning to Long Beach on January 27, he was back in action. His notebook of this day contains an entry concerning a naturally occurring selenium compound (selenomethionine) isolated from wheat. On January 28, he called me from his home, but I was unavailable, and this last time we could not establish contact. He suffered a heart attack on January 29 and died on January 30. His laboratory was closed and his staff disbanded shortly after his death.

Klaus Schwarz has authored or co-authored over 200 papers. He was a Fellow of the New York Academy of Sciences, the AAAS, the Royal Society of Health (FRSH), an Honorary Fellow of the Association of Clinical Scientists, and the Vice-president of the International Association of Bioinorganic Scientists. In 1961, he received the Borden Award in Nutrition. The International Association of Bioinorganic Scientists has issued a commemorative Klaus Schwarz Medal to be awarded to leading trace element scientists.

Klaus Schwarz ended one of his last publications with the old Greek saying: Παντα ρει [Panta rhei]—everything is in flux (20). This saying is certainly not applicable to his scientific contributions, because these are permanent and of lasting importance.

REFERENCES

1. Andreesen, J. R., and Ljungdahl, L. G. (1973): Formate dehydrogenase of clostridium thermoaceticum: Incorporation of selenium-75, and the effects of selenite, molybdate, and tungstate on the enzyme. *J. Bacteriol.*, 116:867–873.
2. Anke, M., Hennig, A., Grun, M., Partschfeld, M., Groppel, B., and Ludke, H. (1976): Arsen-ein neues essentielles spurenelement. *Arch. Tierernaehr.*, 26:742–743.
3. Anke, M., Hennig, A., Groppel, B., Partschfeld, M., and Grun, M. (1978): The Biochemical Role of Cadmium. In: *Proceedings of the Third International Symposium on Trace Element Metabolism in Man and Animals,* edited by M. Kirchgessner, pp. 540–548. Freising-Weihenstephan, Germany.
4. Flohe, L., Gunzler, W. A., and Schock, H. H. (1973): Glutathione peroxidase: A selenoenzyme. *Fed. Eur. Biochem. Soc. Lett.*, 32:132–134.
5. Hopkins, L. L. Jr., Ransome-Kuti, O., and Majaj, A. S. (1968): Improvement of impaired carbohydrate metabolism by chromium (III) in malnourished infants. *Am. J. Clin. Nutr.*, 21:203–211.
6. Jeejeebhoy, K. N., Chu, R. C., Marliss, E. B., Greenberg, G. R., and Robertson, A. B. (1977): A case of chromium deficiency following total intravenous nutrition. *Am. J. Clin. Nutr.*, 30:531–534.
7. Kuhn, R., and Schwarz, K. (1941): Die Isolierung des Wuchsstoffes H' (p-Aminobenzoeseure). *Ber. Dtsch. Chem. Ges.*, 74:1617–1624.
8. Mertz, W., Anderson, R. A., Wolf, W. R., and Roginski, E. E. (1978): Progress of Chromium Nutrition Research. In: *Proceedings of the Third International Symposium on Trace Element Metabolism in Man and Animals,* edited by M. Kirchgessner, pp. 272–276. Freising-Weihenstephan, Germany.
9. Nielsen, F. H., Givand, S. H., and Myron, D. R. (1975): Evidence of a possible requirement for arsenic by the rat. *Fed. Proc.*, 34:923.
10. Rotruck, J. T., Hoekstra, W. G., Pope, A. L., Ganther, H., Swanson, A. B., and Hafeman, D. G. (1972): Relation of Selenium to GSH Peroxidase. *Fed. Proc.*, 31:691.

11. Schrauzer, G. N., and Ishmael, D. (1974): Effects of selenium and of arsenic on the genesis of spontaneous mammary tumors in inbred C₃H mice. *Ann. Clin. Lab. Sci.,* 4:441–447.
12. Schrauzer, G. N., and Rhead, W. J. (1971): Interpretation of the methylene blue reduction test of human plasma and of selenium. *Experientia,* 27:1069–1071.
13. Schrauzer, G. N., White, D. A., McGinness, J. E., Schneider, C. J., and Bell, L. J. (1978): Arsenic and cancer: effects of joint administration of arsenite and selenite on the genesis of mammary adenocarcinoma in inbred female C₃H/St mice. *Bioinorg. Chem.,* 9:245–253.
14. Schroeder, H. (1965): The biological trace elements. *J. Chronic Dis.,* 18:217–228.
15. Schwarz, K. (1951): Production of dietary necrotic liver degeneration using american torula yeast. *Proc. Soc. Exp. Biol. Med.,* 77:818–852.
16. Schwarz, K. (1952): Casein and Factor 3 in dietary necrotic liver degeneration: Concentration of Factor 3 from casein. *Proc. Soc. Exp. Biol. Med.,* 80:319–323.
17. Schwarz, K. (1970): An agent promoting growth of rats fed amino acid diets (Factor G). *J. Nutr.,* 100:1487–1499.
18. Schwarz, K. (1974): Recent dietary trace element research, exemplified by tin, fluorine and silicon. *Fed. Proc.,* 33:1748–1757.
19. Schwarz, K. (1974): Neuere Erkenntnisse uber den essentiellen Charakter einiger spurenelemente. In: *Spurenelemente in der Entwicklung von Mensch und Tier,* edited by K. Betke, and F. Bindlingmaier, pp. 1–30. Urban and Schwarzenberg, Munich, Germany.
20. Schwarz, K. (1977): Essentiality versus toxicity of metals. In: *Clinical Chemistry and Chemical Toxicology of Metals,* edited by S. S. Brown, pp. 3–22. Elsevier-North Holland, New York.
21. Schwarz, K. (1977): Silicon, fiber and atherosclerosis. *Lancet,* 1:454.
22. Schwarz, K., and Foltz, C. M. (1957): Selenium as an integral part of Factor 3 against dietary necrotic liver degeneration. *J. Am. Chem. Soc.,* 79:3292–3293.
23. Schwarz, K., and Fredga, A. (1974): Biological potency of organic selenium compounds. V. Diselenides of alcohols and amines, and some selenium-containing ketones. *Bioinorg. Chem.,* 3:153–159.
24. Schwarz, K., and Mertz, W. (1959): Chromium (III) and the glucose tolerance factor. *Arch. Biochem. Biophys.,* 85:292–295.
25. Schwarz, K., and Mertz, W. (1961): A physiological role of chromium (III) in glucose utilization (glucose tolerance factor). *Fed. Proc.,* 10:111–114.
26. Schwarz, K., and Milne, D. B. (1972): Fluorine requirement for growth in the rat. *Bioinorg. Chem.,* 1:331–338.
27. Schwarz, K., and Milne, D. B. (1972): Growth promoting effects of silicon in rats. *Nature (Lond.),* 239:333–334.
28. Schwarz, K., Milne, D. B., and Vinyard, E. (1970): Growth effects of tin compounds in rats maintained in a trace element-controlled environment. *Biochem. Biophys. Res. Commun.,* 40:22–29.
29. Schwarz, K., Porter, L. A., and Fredga, A. (1974): Biological potency of organic selenium compounds IV. Straight-chain dialkylmono- and diselenides. *Bioinorg. Chem.,* 3:145–152.
30. Schwarz, K., Ricci, B., Punsar, S., and Karvonen, M. J. (1977): Inverse relation of silicon in drinking water and atherosclerosis in Finland. *Lancet,* 1:538.
31. Shamberger, R. J., and Frost, D. V. (1969): Possible protective effect of selenium against human cancer. *Can. Med. Assoc. J.,* 100:682.
32. Turner, D. C., and Stadtman, T. C. (1973): Purification of protein components of the clostridial glycine reductase system and characterization of protein A as a selenoprotein. *Arch. Biochem. Biophys.,* 154:366–381.
33. Woods, D. D. (1940): The relation of p-aminobenzoic acid to the mechanism of the action of sulphanilamide. *Br. J. Exp. Pathol.,* 2:74–90.

The Glutathione Peroxidase Reaction: A Key to Understand the Selenium Requirement of Mammals*

L. Flohé, W. A. Günzler, and G. Loschen

Grünenthal GmbH, Center of Research, D 5100 Aachen, German Federal Republic

THE ROOTS OF PRESENT KNOWLEDGE

Selenium is known as one of the most toxic trace elements. In particular, the toxic symptoms in grazing animals ingesting seleniferous plants have been so well characterized that it is possible to quote Marco Polo as the first describer of selenium toxicity. In the report on his famous trip from Venice to China, he mentions that certain Asian highlands could be traveled only with inborn pack animals which had learned to avoid eating some poisonous plants, while less experienced animals regularly lost their hoofs (75). Probably they were suffering from "alkali disease" due to ingestion of seleniferous plants.

Several centuries later, it became obvious that selenium may also exert beneficial effects. As a result of his intensive studies on experimental liver necrosis, Klaus Schwarz in 1957 discovered that selenium was an integral constituent of Factor 3 (80) (Table 1). This micronutrient present in crude casein, kidney, liver, and brewer's yeast turned out to be the most effective compound to prevent dietary liver necrosis in rats. On a molar basis, it was superior to α-tocopherol and to sulfur containing amino acids. A variety of inorganic or organic selenium compounds could substitute for Factor 3, although they were less active (81). It thus could be deduced that selenium itself was the active principle of Factor 3 and has since been considered as an essential trace element.

The essentiality of selenium was soon established for many warm-blooded animals (81). The precise physiological role of selenium, however, remained obscure for another decade. The numerous studies on selenium–vitamin E deficiency suggested that these substances might be responsible for the biological defense against oxidative challenge. These considerations culminated in the oversimplified view that selenium and α-tocopherol constitute an antioxidant system (7) protecting either the lipid or the aqueous fraction of living material. It was again Klaus Schwarz who challenged this hypothesis from the beginning. Again and again, he stressed that the extremely low alimentary requirements

---

* In memoriam Klaus Schwarz.

TABLE 1. *The history of GSH peroxidase and related topics of selenium research*[a]

1957–1961	Essentiality of selenium in rats (80)
	Discovery of GSH peroxidase activity (60)
	Mitochondrial swelling *in situ* due to selenium deficiency (73)
	Description of selenium-vitamin E deficiency symptoms in various animal species (81)
1962–1966	Identification of GSH peroxidase as a mitochondrial contraction factor (67)
	Importance of GSH peroxidase for red blood cell integrity (13)
1967–1969	Description of GSH peroxidase deficiency in human red blood cells (65)
	GSH peroxidase can reduce lipid hydroperoxides (10,52)
	GSH peroxidase can reduce hydroperoxides of nucleic acids (12)
	Differentiation between selenium and vitamin E deficiencies (93)
1970–1973	First report on selenium-dependent inhibition of chemical cancerogenesis (84)
	Role of GSH peroxidase in biomembrane protection (18)
	Isolation, molecular weight, and subunit size of GSH peroxidase (21)
	Substrate specificity of GSH peroxidase (22)
	Spectral characteristics of GSH peroxidase (23)
	Subcellular distribution of GSH peroxidase (20)
	Kinetics of GSH peroxidase (24,39)
	First evidence of selenium-dependency of GSH peroxidase (77)
	Crystallization and quantitative determination of selenium stoichiometry of GSH peroxidase (25)
1974–1978	Confirmation of selenium stoichiometry of GSH peroxidases from different sources (2,64,70)
	Substrate-induced redox change of enzyme-bound selenium (95)
	Prevention of lipid peroxidation by selenium and vitamin E *in vivo* (41)
	Molecular dimensions of GSH peroxidase by preliminary X-ray crystallography; intramolecular distances of selenium atoms (49)
	Identification of selenocysteine residues in GSH peroxidase (32,96)

[a] The authors have to apologize to the many researchers who are not mentioned but nevertheless substantially contributed to the present knowledge in this field for the inevitably biased selection that an outline such as this necessitates.

of selenium and α-tocopherol were incompatible with a chemical antioxidant mechanism. Rather could these micronutrients fulfill their biological task only by functioning as a cofactor, a prosthetic group, or a substrate of some unidentified enzymatic system (82). In case of selenium at least, Klaus Schwarz was right.

A brief look into the history of glutathione research (Table 1) reveals some striking similarities to the early selenium research in various respects: Chemical as opposed to biochemical considerations lead to the oversimplification that glutathione is just an antioxidant. However, as shown by Hunter et al. (45), glutathione deprived of its enzymatic environment tends to destroy biomembranes with concomitant lipid peroxidation. That means the antioxidant potential of glutathione per se can also initiate free radical chains instead of protecting the biosystem against oxidative challenge. Many investigations, however, proved that intracellular GSH is indispensable for structural and functional integrity of aerobic organisms. This biological function of glutathione in part coincided with what was claimed to be the biological role of selenium. In retrospect, it is hard to understand why it took so long until the gap between selenium and glutathione research was bridged.

In the early 1970s, three groups of researchers were very close to the solution of the problem of how selenium and glutathione contribute to cell integrity:

1. Some monomaniacs at the University of Tübingen in Germany had spent years in isolating and characterizing glutathione peroxidase (19,29), an enzyme discovered already in 1957 by Mills (60) but badly neglected by the scientific world throughout the 1960s. It had become evident that the GSH peroxidase system was the key for understanding the biological role of most glutathione-dependent redox reactions and the pathology of some glutathione-related deficiency syndromes (26–28,31). But up to 1972, this group, unfortunately, had never read a single paper on trace element research.

2. Klaus Schwarz and his associates meanwhile were well aware of the possible link of selenium and GSH-dependent redox reactions, but unfortunately made a wrong choice in concentrating on glutathione dehydrogenase. In one of his last publications, Klaus Schwarz wrote: "I must blame myself for overlooking glutathione peroxidase. We were at the right spot, so to speak, but we were barking up the wrong tree." (83).

3. Hoekstra's group in Wisconsin, after some trial and error, was the first to arrive at the correct conclusion. Looking for the metabolic background of a selenium deficiency symptom in rats, the H_2O_2-dependent hemolysis, Rotruck et al. (77) found that GSH peroxidase was significantly decreased in selenium-deficient erythrocytes. In addition, if ^{75}Se was administered to selenium-deficient rats, selenium was incorporated into a protein fraction which co-chromatographed with GSH peroxidase activity. These findings have to be considered as the major breakthrough, because they provided strong evidence that GSH peroxidase is a selenium-dependent enzyme.

At that time, however, Hoekstra's group did not yet have purified GSH peroxidase available to run a chemical analysis of the enzyme protein. Alarmed by the Federation Proceedings abstract of Rotruck et al. (77), we subjected some crystals of pure GSH peroxidase to neutron activation analysis and found 4 atoms of selenium per molecule of enzyme (25). The enzyme sample investigated had survived several precipitations and 12 chromatographic purification steps (21), and still the selenium was present in a well-defined stoichiometry. This result revealed that the trace element is tightly if not covalently bound to the enzyme protein. Thus, the hypothesis of Klaus Schwarz that selenium acts as a constituent of an enzymatic system had become scientific reality.

ENZYMOLOGY OF GSH PEROXIDASE

Substrate Specificity

Glutathione peroxidase catalyzes the reduction of hydroperoxides by glutathione according to Eq. 1.

$$ROOH + 2\ GSH \rightarrow GSSG + ROH + H_2O \tag{1}$$

The enzyme is highly specific for GSH. Typical peroxidase substrates such as benzidine or guajacol are not metabolized. Some thiol compounds are accepted as donor substrates, if they show structural similarities to GSH. An extensive specificity study revealed that both carboxylic groups of the GSH molecule are essential for substrate binding, as can be derived from the results shown in Table 2.

With regard to the hydroperoxide substrate, GSH peroxidase is unspecific. With a very few exceptions, all hydroperoxides investigated are reduced by GSH peroxidase (Table 3). The rate constants of the reaction of the enzyme with the hydroperoxides apparently reflect the chemical reactivity of the substrates (see Table 8). The broad specificity range of GSH peroxidase provides a basis for the interaction with multiple metabolic pathways.

Occurrence and Subcellular Distribution

GSH peroxidase activity has been detected in all tissues of warm-blooded animals so far investigated (28). Only one report on GSH peroxidase activity in bacteria has been published so far (37). The homology of this and the mammalian enzyme remains to be established. The selenoenzyme obviously predominates in GSH-dependent hydroperoxide removal in mammals with the possible exception of guinea pigs in which the GSH peroxidase activity of GSH-S-transferase may substitute for the selenoprotein (51). High activities of GSH peroxidase are found in liver, kidney, erythrocytes, stomach, spleen, heart, lung, and lens (59,74).

Figure 1 shows that in several rat tissues, GSH peroxidase activity strongly depends on selenium supply until a plateau is reached. These examples demonstrate that GSH-dependent hydroperoxide metabolism in rats largely depends on the selenoenzyme.

By conventional tissue fractionation of rat liver, GSH peroxidase can be characterized as a soluble enzyme present in the cytosol and the mitochondrial matrix (Tables 4 and 5). These results, however, do not rule out the possibility that GSH peroxidase might be loosely associated with biomembranes *in vivo*. Recent results of Zakowski and Tappel (98) reveal that the mitochondrial enzyme

TABLE 2. Some characteristic examples of thiol oxidation by H_2O_2 catalyzed by GSH peroxidase[a]

A. Variations of the γ-glutamyl residue	
γ-Glu-Cys-Gly (GSH)	100.0%
β-Asp-Cys-Gly	7.6%
Cys-Gly	6.8%
N-Ac-Cys-Gly	2.7%
B. Variations of the glycin residue	
γ-Glu-Cys-Gly	100.0%
γ-Glu-Cys-0-methyl	26.0%
γ-Glu-Cys-NH_2	1.4%

[a] The results were obtained at H_2O_2 concentrations (1 mM) yielding apparent maximum velocity. Data are taken from ref. 22.

TABLE 3. Acceptor substrates of GSH peroxidase

Substrate	Source of enzyme	Reference
H_2O_2	bovine red blood cells	24
Ethyl hydroperoxide	bovine red blood cells	39
Cumene hydroperoxide	rat liver supernatant	52
	bovine lens	44
	bovine red blood cells	39
	pig red blood cells	53
Tert.-Butyl hydroperoxide	rat liver supernatant	52
	bovine lens	44
	bovine red blood cells	39
Linoleic acid hydroperoxide (Hydroperoxyoctadecadienoate)	rat liver supernatant	52
	rat liver supernatant	10
	bovine lens	44
	pig red blood cells	53
	pig aorta	91
Linolenic acid hydroperoxide (Hydroperoxyoctadecatrienoate)	rat liver supernatant	11
Methyl hydroperoxyoctadecadienoates	pig aorta	91
Glyceryl 1-hydroperoxyoctadecadienoates	pig aorta	91
Cholesteryl hydroperoxyoctadecadienoates	pig aorta	91
Ethyl hydroperoxyoctadecatrienoate	rat liver supernatant	52
Prostaglandin G_2	bovine red blood cells	68
15-Hydroperoxyprostaglandine E_1	bovine red blood cells	9
Progesterone 17α-hydroperoxide	pig red blood cells	53
Allopregnanolone 17α-hydroperoxide	pig red blood cells	53
Pregnenolone 17α-hydroperoxide	pig red blood cells	53
Cholesterol 7β-hydroperoxide	pig red blood cells	53
Thymine hydroperoxide	rat liver supernatant	12
Peroxidized DNA	rat liver supernatant	12

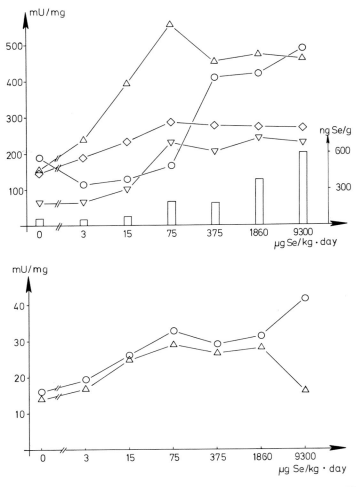

FIG. 1. Dependence of GSH peroxidase activity on selenium supply in rat tissues. The animals received a selenium-deficient torula yeast-based diet and a daily oral selenium supplement as indicated. Seleno bis-(acetyl-glycin) was used as a source of selenium (82). GSH peroxidase activity measurements were performed after a feeding period of 4 weeks. U = Δ lg [GSH] / min was determined at 37°C with a coupled test system and tert.-butyl hydroperoxide as an acceptor substrate. **Top:** ○ red blood cells; △ kidney; ◇ heart; ▽ liver; columns: selenium content of liver tissue per wet weight. **Bottom:** △ plasma; ○ platelets.

is also selenium-dependent and thus, in spite of minor differences, may be considered homologous to cytoplasmatic GSH peroxidase. Interestingly, the subcellular localization of GSH peroxidase does not exactly reflect the distribution pattern of selenium described by Diplock et al. (16,17). This discrepancy suggests that the GSH peroxidase reaction is possibly not the only selenium-dependent metabolic pathway in mammals.

TABLE 4. Subcellular distribution of GSH peroxidase[a] in rat liver

Fraction	Specific activity U_{37}/mg protein	Percent of total activity
Homogenate	0.58	100
Nuclei	0.01 N.S.[b]	
High density mitochondria	0.48	25.9
Low density mitochondria	0.04	
Liposomes	N.D.[b]	
Peroxysomes	0.09	
High density microsomes	0.06 N.S.	
Low density microsomes	0.04 N.S.	
Soluble fraction	1.38	73.3

Data taken from ref. 20.
[a] Tested with H_2O_2 as an oxidizing substrate.
[b] N.S. = not significant; N.D. = not detectable.

Chemical and Physicochemical Properties

GSH peroxidase is unique among the peroxide-reducing enzymes, since it contains neither heme nor flavine but selenium. A stoichiometry of 4 atoms of selenium per molecule of native enzyme has meanwhile been determined by four independent groups of scientists working on GSH peroxidase isolated from four different species (Table 6). Early investigations (40,64) did not reveal any anomalies in amino acid composition, but had to leave open the question of how selenium was bound to the protein. Recently, Wendel et al. (96) and Forstrom et al. (32) supported strong evidence that selenium is present in GSH peroxidase as a selenocysteine residue. This success was essentially furthered by the earlier findings that iodoacetate selectively inactivates substrate-reduced

TABLE 5. Distribution of GSH peroxidase[a] in rat liver mitochondria fractionated according to Parsons et al. (72)

Fraction	Specific activity U_{37}/mg protein	Percent of total activity
Mitochondria	0.49	100
Inner membrane	N.D.[b]	
Outer membrane	N.D.	
Intermembrane space	0.19	4
Matrix space	0.74	92

Data taken from ref. 20.
[a] Tested with H_2O_2 as an oxidizing substrate.
[b] N.D. = not detectable.

TABLE 6. *Selenium content of GSH peroxidase from different sources*

Source of enzyme	Criteria of purity	Molecular weight (method)	Selenium content (gram atoms/mole) (method)	Reference
Bovine erythrocytes	rechromatography (ion exchange and gel filtration), ultracentrifugation; microzone electrophoresis, disc electrophoresis; crystallization	85,000 (gel filtration); 80,000 (disc electrophoresis); 83,000 (ultracentrifugation); 21,000/subunit (SDS electrophoresis)	4.04 (neutron activation analysis)	25
Ovine erythrocytes	rechromatography (gel filtration); disc electrophoresis	22,000/subunit (SDS electrophoresis)	3.8 (fluorimetric analysis)	70
Rat liver	rechromatography (ion exchange and gel filtration); ultracentrifugation	75,000–76,000 (thin layer gel filtration); 76,000 (ultracentrifugation); 19,000/subunit (SDS electrophoresis); 17,000/subunit (amino acid composition)	4.24 (fluorimetric analysis)	64
Human erythrocytes	disc electrophoresis	95,000 (ultracentrifugation); 23,000/subunit (SDS electrophoresis)	3.5 (fluorimetric analysis)	2

TABLE 7. *Methods of identification of a selenocysteine residue in GSH peroxidase*

Sample of enzyme	Reagent	Identified product after derivatization, hydrolysis, and chromatography	Reference
Rat liver GSH peroxidase prelabeled with ^{75}Se, partially purified, and reduced by GSH	$\underset{H_2C\text{———}CH_2}{\overset{H}{\underset{\diagdown\diagup}{N}}}$	COOH | CHNH$_2$ | CH$_2$—*Se—C$_2$H$_5$—NH$_2$	32
	COOH | CH$_2$—I	COOH | CHNH$_2$ | CH$_2$—*Se—CH$_2$COOH	32
Bovine GSH peroxidase; purified and reduced by NaBH$_4$	*COOH | CH$_2$—Cl	COOH | CHNH$_2$ | CH$_2$—Se—CH$_2$—*COOH	96

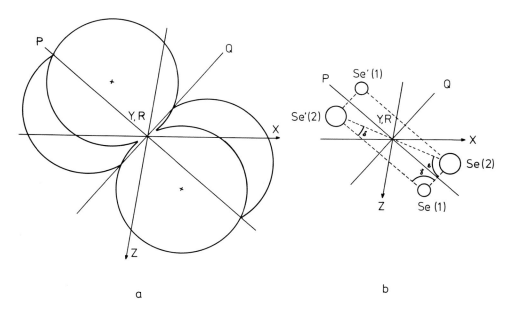

FIG. 2. Arrangement of subunits (**a**) and position of selenium atoms (**b**) in GSH peroxidase according to X-ray crystallographic studies of Ladenstein (49). X, Y, Z, and P, Q, R represent the crystallographic and the molecular coordinate system, respectively. The radius of a subunit is 19 Å, if globular shape is assumed. The angles between the selenium atoms γ, δ, ϵ are 67°, 32°, and 81°, respectively. The following distances between the four selenium atoms were obtained: $\overline{Se(1)\ Se(2)} = \overline{Se'(1)\ Se'(2)} = 21.3$ Å; $\overline{Se'(1)\ Se(2)} = \overline{Se(1)\ Se'(2)} = 39.5$ Å; $\overline{Se(1)\ Se'(1)} = \overline{Se(2)\ Se'(2)} = 36.6$ Å.

GSH peroxidase (26) and by the methodology developed in Thressa Stadtman's laboratory by which selenocysteine was established as a prosthetic group of the clostridial glycine reductase system (14). The methods employed to identify selenocysteine in GSH peroxidase are outlined in Table 7.

Although the amino acid sequence and the complete three-dimensional structure of GSH peroxidase are not yet available, some important structural features may be derived from hitherto published reports. The enzyme is a tetramer with a molecular weight of about 84,000 (Table 6). By means of sodium dodecyl sulfate, it can be dissociated into subunits which appear to be identical or at least very similar. Molecular weight determinations (21,64,70), comparison of amino acid composition with the number of tryptic peptides (29), and preliminary X-ray crystallography (50) did not indicate any unequivalence of the subunits. According to the crystallographic studies of Ladenstein (49), the arrangement of the subunits may be schematized as shown in Fig. 2a. Furthermore, it was possible to localize the selenium atoms within the molecular co-ordinate system (Fig. 2b) by comparing the electron density maps of native and selenium-depleted GSH peroxidase crystals. The resulting selenium distances of 21.3, 36.6, and 39.5 Å, respectively, are considerably too large to allow the formation of intramolecular diselenide bonds. The selenium atoms are each exposed at the surface of a subunit, thus being in an excellent position to react even with hydroperoxy groups attached to bulky organic residues (see Table 3).

The Mechanism of Catalysis

A preliminary but indispensable approach to the reaction mechanism of an enzyme is the evaluation of its initial velocity pattern. In case of GSH peroxidase, typical ping-pong kinetics were observed (24,39) which are indicative of a so-called enzyme-substitution mechanism. This means that different enzyme derivatives (as opposed to complexes) are formed during catalysis. In an enzymatic redox reaction, these different enzyme derivatives most likely represent different redox states of the enzyme, which are formed as a result of consecutive but independent reactions. In other words, the ping-pong kinetics observed with GSH peroxidase were a rather convincing hint that some prosthetic group of the enzyme itself is subjected to a cyclic redox change during catalysis.

Our results of the initial velocity measurements could be fitted into the Dalziel equation (15).

$$\frac{[E_o]}{v} = \frac{\phi 1}{[A]} + \frac{\phi 2}{[B]} \tag{2}$$

This equation stands for a fairly unusual kinetic pattern, because the limiting maximum velocities and the limiting Michaelis constants are infinite. Though uncommon in enzymology, this kinetic behavior appears to be typical for peroxidases. Equation 2 is equivalent to the rate equation of the heme-containing peroxidases (8) which can be transformed into Eq. 3:

$$\frac{d[A]}{dt} = v = [E_o]\left\{\frac{1}{k_{+1}[A]} + \frac{1}{k_{+2}[B]} + \frac{1}{k_{+3}[B]}\right\}^{-1} \qquad (3)$$

k_{+1} represents the rate constant of the reduced enzyme with the peroxide A, k_{+2} and k_{+3} the rate constants of different oxidized enzyme species with the donor substrate B. By comparison of eq. 2 and 3, it is evident that the empirical coefficient ϕ_1 is identical with the reciprocal value of k_{+1}, whereas ϕ_2 specifically describes the two reaction steps of the oxidized enzyme forms. In Table 8, the kinetic coefficients of GSH peroxidase obtained with different hydroperoxides are compiled. Interestingly, the coefficient ϕ_2 describing the GSH-dependent reactions of the oxidized enzyme forms remains constant, when the oxidizing substrate is changed. This provides further evidence that the catalytic cycle consists of largely independent consecutive steps.

Quite a number of additional observations suggest or prove that the enzyme considerably changes its conformation or chemical identity depending on the catalytic state: The UV as well as the CD spectra of GSH peroxidase are substantially altered by addition of GSH (23,96); after preincubation with GSH, the enzyme binds additional pCMB equivalents (23); only after preincubation with GSH, the enzyme is irreversibly inhibited by iodoacetate (26); most surprisingly, no GSH binding could be detected by equilibrium chromatography of GSH peroxidase on [^{14}C] GSH-equilibrated sephadex columns (23). With regard to the pronounced specificity of the enzyme for GSH, this odd result can again be conceivably explained only by assuming different enzyme forms. The reduced enzyme, present at excess concentration of donor substrate, does not possess a measurable affinity to GSH, while the oxidized form or forms of the enzyme specifically bind GSH.

By means of X-ray photoelectron spectroscopy, Wendel et al. (95) unambiguously demonstrated that it is the enzyme-bound selenium that undergoes a substrate-induced redox change. This important finding, the identification of a selenol as the lowest oxidation step, the knowledge of the intramolecular selenium distances, and the other findings mentioned earlier supply the framework for a tentative formulation of the catalytic cycle of GSH peroxidase. In Fig. 3a, the formal description of the mechanism derived by us from kinetic investigations (24,26,39) is recalled. The catalytic cycle essentially consists of three different enzyme forms E, F, and G which may or may not form additional complexes with substrates and products. Initial velocity measurements did not prove the existence of any typical enzyme substrate complex, since the limiting maximum velocity was infinite. As to the first step of the catalytic cycle, the extremely low specificity for the various hydroperoxides does not justify the assumption of a specific enzyme substrate complex either. That is why we formulated the reaction of E with A, as a simple bimolecular redox reaction. The next step, however, shows the pronounced substrate specificity we are used to in enzymatic reactions (22) and therefore must involve the formation of a highly specific complex (FB), which then reacts intramolecularly to yield another enzyme deriv-

TABLE 8. Kinetic constants of GSH peroxidase of bovine erythrocytes[a]

Substrate	Buffer system	pH	$\phi_1[10^{-8}\text{M sec}]$	$\phi_2[10^{-6}\text{M sec}]$	$k_{+1}[10^7\text{M}^{-1}\text{sec}^{-1}]$
Hydrogen peroxide	potassium phosphate (0.05 M)	7.0	0.56	1.27	17.86
Hydrogen peroxide	MOPS (0.25 M)	7.7	0.94	0.83	10.6
Hydrogen peroxide	MOPS (0.25 M)	6.7	1.70	2.19	5.88
Ethyl hydroperoxide	MOPS (0.25 M)	6.7	3.3	2.24	3.09
Cumene hydroperoxide	MOPS (0.25 M)	6.7	7.8	2.24	1.28
T.-Butyl hydroperoxide	MOPS (0.25 M)	6.7	13.5	2.24	0.75

[a] The definition of symbols can be derived from Eq. 2 and 3 and Fig. 3. Data are taken from refs. 24 and 39.

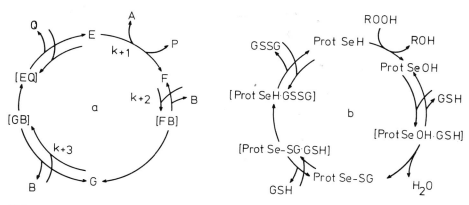

FIG. 3. Hypothetical reaction mechanism of GSH peroxidase. Figure 3a represents a formal description of the reaction sequence as derived from kinetic measurement. A, B, and P, Q are substrates and products, respectively. E, F, and G stand for different enzyme forms, while the expressions in brackets represent enzyme substrate or enzyme product complexes. k_{+1}, k_{+2}, and k_{+3} are the rate constants limiting the overall turnover, thereby determining the rate equation (Equation 3). Figure 3b attempts to correlate our fragmentary knowledge of the enzyme chemistry with the functional studies. The kinetic entities E, F, and G correspond to three different redox states of the enzyme-bound selenium, a selenol, a selenenic acid derivative, and seleno-sulfide formed between the enzyme and GSH, respectively.

ative (G). This intramolecular reaction is obviously very fast, as it does not limit the overall reaction rate under any condition. The final regeneration of E from G may occur in a way analogous to step 2, but according to recent observations, the second reaction with a donor substrate is less specific than the preceding one (99).

In Fig. 3b, some proposals are presented as to the chemical nature of the kinetically identified enzymatic entities E, F, and G. As outlined previously, a selenocysteine residue can be trapped by alkylation in reduced GSH peroxidase. Thus, E, the form which has to react with the hydroperoxide, will contain the highly reactive selenol function. No direct evidence as to the chemical nature of the first intermediate F is available. But taking into account the stoichiometry of the reaction, the extremely high reaction rates, and the isolated position of the selenium atoms within the enzyme, the most adequate assumption for the formation of F appears to be a simple bimolecular reaction between the selenol function of the enzyme and the hydroperoxo group of the substrate yielding a selenenic acid derivative. In the absence of a reducing substrate, such a selenenic acid group would of course be further oxidized. However, this is not likely in the normal catalytic cycle, since GSH peroxidase is easily inactivated by hydroperoxides in the absence of GSH. During regular catalysis, GSH will continuously regenerate the selenol from the selenenic acid function in two consecutive steps. A conceivable intermediate during this selenol regeneration would be a mixed selenosulfide between the enzyme and GSH. An enzyme derivative of this nature could well represent the kinetic entity G. A more detailed knowledge of the three-dimensional structure of GSH peroxidase will eventually provide a better insight into the catalytic mechanism.

THE PHYSIOLOGICAL ROLE OF GSH PEROXIDASE

In a qualitative sense, the importance of the GSH peroxidase reaction is easily estimated by a brief look into the impressive list of acceptor substrates (Table 3). The toxicological risk associated with peroxidized lipids, nucleic acids or xenobiotics, as well as the metabolic role of some hydroperoxy derivatives of arachidonic acid, are most topical issues of today's life sciences, and an enzyme which can handle such exciting compounds certainly deserves our interest.

Unfortunately, the knowledge that a certain reaction is catalyzed by an enzyme *in vitro* does not yet prove its pertinent function *in vivo*. A variety of additional factors such as compartmentation, availability of cosubstrate, rate constants for the specific substrate under consideration, and competing enzymes have to be taken into account, before a reliable prognosis can be made. After all, some kind of an experimental proof for the postulated role is needed. Correspondingly, the facts proving the role of GSH peroxidase *in vivo* are scarce as compared to its eventual potential (31).

Established Functions of GSH Peroxidase

The high reaction rates of GSH peroxidase with H_2O_2 as measured *in vitro* suggested a predominant role of the enzyme in H_2O_2 metabolism at least in tissues or cell compartments in which competing enzymes such as catalase are missing or present only at low activity. This holds true, for instance, in the crystal lens where GSH peroxidase is abundant (44,74), while catalase is absent. It also applies to the cytoplasmatic and mitochondrial compartment of rat liver cells, because catalase is largely, if not exclusively, restricted to the peroxisomes. The resulting idea that the fate of hepatic H_2O_2 largely depends on its site of formation could be corroborated by a series of quasi-*in vivo* measurements—i. e., perfusion studies with rat liver. Addition of H_2O_2 to the perfusion medium results in a decline of the cellular NADPH level and a release of oxidized glutathione (86,88), which are characteristic indicators of GSH peroxidase in action. However, if intracellular H_2O_2 formation is triggered by infusion of substrates of peroxisomal oxidases, the redox state of catalase is changed primarily (88). In contrast, infusion of benzylamine, which is metabolized by mitochondrial monoamine oxidase with concomitant H_2O_2 formation, again induces GSSG release (90). Correspondingly, the effects induced by H_2O_2 and benzylamine are not observed in livers of selenium-deficient rats (90). These results convincingly demonstrate that the selenium-dependent GSH peroxidase is the primary site of extraperoxisomal H_2O_2 metabolism in rat liver. In principle, our conclusion (19) has been recently confirmed by Oshino and Chance (71) and Burk et al. (6).

In red blood cells, the situation is more complicated due to the simultaneous presence of catalase which competes with GSH peroxidase for the common substrate H_2O_2 at comparable reaction rates. However, it can be taken for sure that the GSH peroxidase system of rat and human erythrocytes contributes substantially to the cell membrane integrity. Rat erythrocytes made experimentally deficient in GSH peroxidase by means of restricted selenium supply are particularly prone to peroxide-induced hemolysis (77,78). In case of human erythrocytes, the importance of GSH peroxidase is stressed by the most convincing experiments we can think of—i.e., experiments of nature. Genetic (or possibly nutrition-dependent) deficiency in GSH peroxidase is associated with a nonspherocytic drug-induced hemolytic anemia which very much resembles glucose-6-phosphate dehydrogenase deficiency. It is revealing that all genetic deficiencies resulting in an impaired availability of reduced glutathione, qualitatively at least, are characterized by the same pathological features: The red cells break upon an oxidative challenge (see Table 9). Of course, the severity and the time course of the clinical manifestation varies with the type of genetic defect (31). In principle, however, the common denominator and final cause of the pathophysiological events in all these deficiencies is the functional impairment of the GSH peroxidase reaction. It is still controversial whether the deficient peroxidase reaction affects primarily the H_2O_2 metabolism of the red blood cell or whether additional acceptor substrates of the enzyme play a role in the pathological mechanisms (31). But the significance of GSH peroxidase for the red blood cell integrity can no longer be doubted.

In general, the GSH peroxidase reaction appears to be an important factor in protecting biomembranes against oxidative challenge. A series of *in vitro*

TABLE 9. *Genetic or alimentary deficiencies resulting in nonspherocytic hemolytic anemia*

	Metabolic disorders	Reviews
1.	Impaired synthesis of glutathione from constituent amino acids:	
1.1	Genetic deficiency in γ-glutamylcysteine synthetase (EC 6.3.2.2.)	58
1.2	Genetic deficiency in glutathione synthetase (EC 6.3.2.3.)	58
2.	Impaired recovery of GSH from GSSG:	
2.1	Genetic deficiency in GSSG reductase (EC 1.6.4.2)	3,19,54
2.2	Deficiency in riboflavine (needed as precursor for coenzyme of GSSG reductase)	3,19,54
2.3	Insufficient NADPH supply through genetic deficiency in glucose-6-phosphate dehydrogenase (EC 1.1.1.49) or 6-phosphogluconate dehydrogenase (EC 1.1.1.44)	3,19
3.	Impaired hydroperoxide reduction by GSH:	
3.1	Genetic (?) deficiency in GSH peroxidase (EC 1.11.1.9.)	19,29,66
3.2	Selenium deficiency (experimentally in rats)	29,36,43

experiments revealed that GSH peroxidase can prevent or delay lipid peroxidation in biomembranes exposed to prooxidative agents. Isolated rat liver mitochondria, if exposed to GSH, lose their volume control. This "high amplitude swelling" is due to—or at least accompanied by—a peroxidative destruction of the unsaturated phospholipids (45) and thus can be explained by the initiation of a free radical chain via thiol autoxidation (27,61). Both the degree of swelling and of lipid peroxidation as measured by malone dialdehyde formation are inversely related to the GSH peroxidase activity of the mitochondria (18). The effects can also be inhibited by addition of GSH peroxidase to the incubation medium. This early observation of Neubert et al. (67) resulted in the designation "contraction factor II" for GSH peroxidase, although an adequate explanation of the findings was not yet possible in 1962. If isolated inner mitochondrial membranes are incubated with GSH, addition of GSH peroxidase again significantly delays the peroxidation of the lipids, whereas catalase on a molar basis is considerably less active (27). These results strongly suggest that the protective function is not exclusively due to H_2O_2 removal, but to the metabolism of some hydroperoxylipids catalyzing further membrane destruction. In this context, reference should be made to the histological studies of Piccardo and Schwarz (73) with selenium-deficient rats: The earliest structural anomaly detected in prenecrotic selenium-deficient liver was a swelling of mitochondria *in situ,* a pathological symptom possibly related to our *in vitro* results.

The debate, as to whether lipid peroxidation in biomembranes is a laboratory artefact or possibly a physiological or at least pathological phenomenon *in vivo* could be terminated by an elegant methodology developed in G. Cohen's laboratory (76). *In vivo,* lipid peroxidation can be detected and quantified by measuring ethane evolution of animals. By means of this parameter, Hafeman and Hoekstra (41) were able to prove that *in vivo* lipid peroxidation is increased in selenium-vitamin E-deficient rats, and that each—selenium and vitamin E—contribute to the inhibition of lipid destruction. These effects are even more pronounced if lipid peroxidation is previously stimulated by CCl_4. The relevance of these findings is stressed by the observation that the CCl_4-dependent mortality is similarly reduced by alimentary selenium and/or vitamin E supply.

In conclusion, at least three functions of GSH peroxidase could be experimentally proven:

1. The enzyme regulates H_2O_2 levels in tissues and compartments free of catalase.
2. It protects red cell integrity during oxidative challenge by removal of H_2O_2 and/or other hydroperoxides.
3. It inhibits lipid peroxidation in biomembranes, in particular those of liver cells, by mechanisms which may, but have not to involve H_2O_2 removal.

However, we are not at all inclined to suggest that GSH peroxidase can solve all problems associated with aerobic life. Depending on the species, the tissue and the chemical mechanism of an oxidative stress, GSH peroxidase has to be complemented by, e.g., epoxide hydratase (69), GSH-S-transferase (6,46), catalase (87), superoxide dismutase (33), and α-tocopherol.

Hypothetical Functions

Certainly, the above examples of hydroperoxide-related alterations of mammalian tissues do not cover the whole scope of problems arising from aerobic life. Whenever living organisms gain their metabolic energy from oxygen reduction to water, we have also to face the possibility of an incomplete reduction of O_2(4,29,30,42). The resulting products such as $\cdot O_2^-$, H_2O_2 and $\cdot OH$ may start or sustain free radical chains and eventually form an unforseeable variety of products, in particular, organic hydroperoxides which in turn might attack a variety of targets in the biosystem. As gross biological results, not only acute tissue damage but also changes in the blueprint of the organism could be visualized. Thus the recent paper of Bruyninckx et al. (5) showing that oxygen itself is a mutagen in the classical Ames test is not at all surprising. The formation of reactive oxygen species definitively has to be balanced by the detoxifying enzymatic systems listed above to guarantee the stability of the cellular structure and the genetic material. It seems reasonable to postulate a multiple function of GSH peroxidase in this context. The enzyme could prevent low molecular hydroperoxides from reacting with DNA; it could also reduce mutagenic (92) precursors of nucleic acids; and it could even reduce peroxidized nucleic acids (see Table 3). It remains to be established whether this potential of the enyzme is related to the increasing evidence that an adequate selenium supply antagonizes experimental and spontaneous cancerogenesis (34,79,84).

Although GSH peroxidase is usually considered a detoxifying enzyme, this may not always be correct. Hydroperoxides are fairly reactive, therefore they need care, but they are not poisons. In particular, this applies to the products built from arachidonic acid or other unsaturated fatty acids via the cyclo-oxygenase and the lipoxygenase reactions. The primary products of these reactions are hydroperoxides which have pronounced biological activities by themselves. In addition, the cyclo-oxygenase products—i.e., the prostaglandins of the G-type—are obligatory intermediates in the biosynthesis of other prostaglandins and the thromboxanes. In principle, GSH peroxidase can metabolize the hydroperoxy compounds of the arachidonic acid cascade, although the relevance of this activity has been doubted (9,62). Nevertheless, two possible reactions of GSH peroxidase at least merit further consideration, as long as they are not ruled out by convincing experiments. GSH peroxidase might prevent an accumulation of prostaglandin G, once this compound diffuses away from the cyclo-oxygenase-prostaglandin peroxidase complex (Equation 4).

According to Kuehl et al. (47,48), the regulation of the levels of the G-type prostaglandins and other hydroperoxides might be essential for the prevention of inflammatory responses of tissues, because these hydroperoxides can produce highly aggressive $\cdot OH$ radicals. Besides, Adcock et al. (1) recently demonstrated

[Structure: PGG$_2$ with O-OH group]

Equation [4]   PG peroxidase; GSH peroxidase?

[Structure: PGH$_2$ with OH group]

that fatty acid hydroperoxides can trigger the release of anaphylactic mediators.

GSH peroxidase possibly interacts with another site of the arachidonic acid cascade. Prostacyclin (PGI$_2$) biosynthesis is probably regulated by an endogenous inhibitor which could well be a substrate of GSH peroxidase: 15-hydroperoxy-arachidonic acid (38). Since prostacyclin inhibits aggregation and adhesion of platelets to the endothelial lining of the blood vessels (63), GSH peroxidase might contribute to the homiostasis of platelet aggregation by regulating indirectly prostacyclin biosynthesis.

$$PGH_2 \xrightarrow{T} PGI_2$$
$$\text{15-OOH-arachidonic acid} \xrightarrow{\text{GSH peroxidase}} \text{15-OH-arachidonic acid} \quad (5)$$

Again, circumstantial evidence from epidemiological studies urges investigations in this field, since according to Shamberger (85), the incidence of cardiovascular diseases in the U.S. population might correlate with selenium deficiency.

Considering the numerous selenium deficiency symptoms observed in animals (34,35,81), it is tempting to speculate on the existence of corresponding diseases in man. Kwashiorkor (34) and ceroid lipofuscinosis have been implicated (97) as consequences of insufficient selenium supply. Certainly, these hypotheses need further corroboration. Apart from the hemolytic disorder in GSH peroxidase deficiency mentioned above, no unambiguous correlation of selenium deficiency or decreased GSH peroxidase activity with any clinical disorder could be detected so far, though human GSH peroxidase clearly corresponds to alimentary selenium supply (55–57,94). However, it conflicts with our understanding of evolution that an enzymatic system survives the selection process in spite of being useless to the organism. Therefore, selenium deficiency and the resulting GSH peroxidase deficiency in man should not only be associated with pathological manifestations in the red blood cells. It has to be considered, however, that nearly total and acute selenium deficiency is a highly unlikely event in man due to human eating habits. Only a sporadic or moderate alimentary selenium deficiency could be imagined and, as measured by GSH peroxidase activity in human red blood cells, does occur (57,94). Obviously, a dramatic clinical symptomatology can not be expected under those circumstances. Nevertheless, a subacute deficiency existing for some time might induce a metabolic or histopathological change, which in turn will trigger a disease at some future time when the deficiency itself may not be detectable anymore. Such a mechanism of disease manifestation, though conceivable, is not easily elucidated. This warning should be kept in mind, because it could turn out that the much challenged retrospective epidemiological studies, in spite of all their drawbacks, are the only practicable tool to answer the question: how important is an adequate selenium supply in human health care?

CONCLUSIONS

The discovery of Klaus Schwarz that selenium is an essential trace element has opened up a new field of biochemistry. He was right in his basic assumption that selenium in animals exerts its antioxidant function only as an integral part of an enzyme. This enzyme, GSH peroxidase, is a real selenoprotein. Its functional characteristics have been studied extensively and the structural evaluation is comparatively advanced. Its biological function, as far as established, explains at least some of the activities attributed to the trace element—in particular, the protection of biomembranes against oxidative destruction. It is our understanding, however, that these hitherto well-understood aspects of selenium biochemistry only represent the top of the iceberg. There is a sound theoretical basis for the role of selenium in other most topical and significant fields of biochemistry, such as mutagenesis, platelet function, and others. The present review therefore should not be regarded primarily as a survey of past events, but as a status report indicating where to start off.

REFERENCES

1. Adcock, J. J., Garland, L. G., Moncada, S., and Salmon, J. A. (1978): The mechanism of enhancement by fatty acid hydroperoxides of anaphylactic mediator release. *Prostaglandins*, 16:179–187.
2. Awasthi, Y. C., Beutler, E., and Srivastava, S. K. (1975): Purification and properties of human erythrocyte glutathione peroxidase. *J. Biol. Chem.*, 250:5144–5149.
3. Benöhr, H. Ch., and Waller, H. D. (1974): Hematological manifestations in enzymatic deficien-

cies of glutathione reductase. In: *Glutathione,* edited by L. Flohé, H. Ch. Benöhr, H. Sies, H. D. Waller, and A. Wendel, pp. 184–191. Georg Thieme, Stuttgart, Germany.
4. Boveris, A., Oshino, N., and Chance, B. (1972): The cellular production of hydrogen peroxide. *Biochem. J.,* 128:617–630.
5. Bruyninckx, W. J., Mason, H. S., and Morse, S. A. (1978): Are physiological oxygen concentrations mutagenic? *Nature,* 274:606–607.
6. Burk, R. F., Nishiki, K., Lawrence, R. A., and Chance, B. (1978): Peroxide removal by selenium-dependent and selenium-independent glutathione peroxidases in hemoglobin-free perfused rat liver. *J. Biol. Chem.,* 253:43–46.
7. Caldwell, K. A., and Tappel, A. L. (1965): Acceleration of sulfhydryl oxidations by selenocystine. *Arch. Biochem. Biophys.,* 112:196–200.
8. Chance, B., and Higgins, J. (1952): Peroxidase kinetics in coupled oxidation: an experimental and theoretical study. *Arch. Biochem. Biophys.,* 41:432–441.
9. Christ-Hazelhof, E., Nugteren, D. H., and Van Dorp, D. A. (1976): Conversions of prostaglandin endoperoxides by glutathione-S-transferases and serum albumins. *Biochim. Biophys. Acta,* 450:450–461.
10. Christophersen, B. O. (1968): Formation of monohydroxypolyenic fatty acids from lipid peroxides by a glutathione peroxidase. *Biochim. Biophys. Acta,* 164:35–46.
11. Christophersen, B. O. (1969): Reduction of linolenic acid hydroperoxide by a glutathione peroxidase. *Biochim. Biophys. Acta,* 176:463–470.
12. Christophersen, B. O. (1969): Reduction of X-ray-induced DNA and thymine hydroperoxides by rat liver glutathione peroxidase. *Biochim. Biophys. Acta,* 186:387–389.
13. Cohen, G., and Hochstein, P. (1963): Glutathione peroxidase: The primary agent for the elimination of hydrogen peroxide in erythrocytes. *Biochemistry,* 2:1420–1428.
14. Cone, J. E., Martin del Rio, R., Davis, J. N., and Stadtman, T. C. (1976): Chemical characterization of the selenoprotein component of clostridial glycine reductase: Identification of selenocysteine as the organoselenium moiety. *Proc. Natl. Acad. Sci. USA,* 73:2659–2663.
15. Dalziel, K. (1957): Initial steady state velocities in the evaluation of enzyme-coenzyme-substrate reaction mechanisms. *Acta Chem. Scand.,* 11:1706–1723.
16. Diplock, A. T., Baum, H., and Lucy, J. A. (1971): The effect of vitamin E on the oxidation state of selenium in rat liver. *Biochem. J.,* 123:721–729.
17. Diplock, A. T., Caygill, Ch. P. J., Jeffery, E. H., and Thomas, C. (1973): The nature of the acid-volatile selenium in the liver of the male rat. *Biochem. J.,* 134:283–293.
18. Flohé, L., and Zimmermann, R. (1970): The role of GSH peroxidase in protecting the membrane of rat liver mitochondria. *Biochim. Biophys. Acta,* 223:210–213.
19. Flohé, L. (1971): Die Glutathionperoxidase: Enzymologie und biologische Aspekte. *Klin. Wochenschr.,* 49:669–683.
20. Flohé, L., and Schlegel, W. (1971): Glutathion-Peroxidase, IV. Intrazelluläre Verteilung des Glutathion-Peroxidase-Systems in der Rattenleber. *Hoppe Seylers Z. Physiol. Chem.,* 352:1401–1410.
21. Flohé, L., Eisele, B., and Wendel, A. (1971): Glutathion-Peroxidase, I. Reindarstellung und Molekulargewichtsbestimmungen. *Hoppe Seylers Z. Physiol. Chem.,* 352:151–158.
22. Flohé, L., Günzler, W., Jung, G., Schaich, E., and Schneider, F. (1971): Glutathion-Peroxidase, II. Substratspezifität und Hemmbarkeit durch Substratanaloge. *Hoppe Seylers Z. Physiol. Chem.,* 352:159–169.
23. Flohé, L., Schaich, E., Voelter, W., and Wendel, A. (1971): Glutathion-Peroxidase, III. Spektrale Charakteristika und Versuche zum Reaktionsmechanismus. *Hoppe Seylers Z. Physiol. Chem.,* 352:170–180.
24. Flohé, L., Loschen, G., Günzler, W. A., and Eichele, E. (1972): Glutathione peroxidase, V. The kinetic mechanism. *Hoppe Seylers Z. Physiol. Chem.,* 353:987–999.
25. Flohé, L., Günzler, W. A., and Schock, H. H. (1973): Glutathione peroxidase: A seleno-enzyme. *FEBS Lett.,* 32:132–134.
26. Flohé, L., and Günzler, W. A. (1974): Glutathione peroxidase. In: *Glutathione,* edited by L. Flohé, H. Ch. Benöhr, H. Sies, H. D. Waller, and A. Wendel, pp. 132–145. Georg Thieme, Stuttgart, Germany.
27. Flohé, L., and Zimmermann, R. (1974): GSH-induced high-amplitude swelling of mitochondria. In: *Glutathione,* edited by L. Flohé, H. Ch. Benöhr, H. Sies, H. D. Waller, and A. Wendel, pp. 245–260. Georg Thieme, Stuttgart, Germany.
28. Flohé, L., and Günzler, W. A. (1976): Glutathione-dependent enzymatic oxido-reduction reac-

tions. In: *Glutathione: Metabolism and Function,* edited by I. M. Arias and W. B. Jakoby, pp. 17–34. Raven Press, New York.
29. Flohé, L., Günzler, W. A., and Ladenstein, R. (1976): Glutathione peroxidase. In: *Glutathione: Metabolism and Function,* edited by I. M. Arias and W. B. Jakoby, pp. 115–138. Raven Press, New York.
30. Flohé, L., Loschen, G., Azzi, A., and Richter, Ch. (1977): Superoxide radicals in mitochondria. In: *Superoxide and Superoxide Dismutases,* edited by A. M. Michelson, J. M. McCord, and I. Fridovich, pp. 324–334. Academic Press, London.
31. Flohé, L. (1979): Glutathione peroxidase: fact and fiction. Ciba Foundation Symposium on "Oxygen free radicals and tissue damage." *Excerpta Medica,* Elsevier, Amsterdam pp. 95–122.
32. Forstrom, J. W., Zakowski, J. J., and Tappel, A. L. (1978): Identification of the catalytic site of rat liver glutathione peroxidase as selenocysteine. *Biochemistry,* 17:2639–2644.
33. Fridovich, I. (1978): The biology of oxygen radicals. *Science,* 201:875–880.
34. Frost, D. V. (1974): The two faces of selenium. Can selenophobia be cured? *CRC Crit. Rev. Toxicol.,* 1:467–514.
35. Frost, D. V., and Lish, P. M. (1975): Selenium in biology. *Ann. Rev. Pharmacol.,* 15:259–284.
36. Ganther, H. E., Hafeman, D. G., Lawrence, R. A., Serfass, R. E., and Hoekstra, W. G. (1976): Selenium and glutathione peroxidase in health and disease: A review. In: *Trace Elements in Human Health and Disease,* edited by A. Prasad, pp. 165–234. Academic Press, New York.
37. Grosch, W., Senser, F., and Fischer, K. (1972): Einfluss von auf Heringen wachsenden Mikroorganismen auf die Fettoxidation. III. Glutathion-Peroxidase-Aktivität in einer Candida lipolytica-Art. *Chem. Mikrobiol. Technol. Lebensm.* 1:214–218.
38. Gryglewski, R. J., Bunting, S., Moncada, S., Flower, J. R., and Vane, J. R. (1976): Arterial walls are protected against deposition of platelet thrombi by a substance (prostaglandin X) which they make from prostaglandin endoperoxides. *Prostaglandins,* 12:685–713.
39. Günzler, W. A., Vergin, H., Müller, I., and Flohé, L. (1972): Glutathion-Peroxidase, VI. Die Reaktion der Glutathion-Peroxidase mit verschiedenen Hydroperoxiden. *Hoppe Seylers Z. Physiol. Chem.,* 353:1001–1004.
40. Günzler, W. A. (1974): Glutathion-Peroxidase: Kristallisation, Selengehalt, Aminosäurezusammensetzung und Modellvorstellungen zum Reaktionsmechanismus. *Dissertation,* Tübingen, Germany.
41. Hafeman, D. G., and Hoekstra, W. G. (1977): Protection against carbon tetrachloride-induced lipid peroxidation in the rat by dietary vitamin E, selenium, and methionine as measured by ethane evolution. *J. Nutr.,* 107:656–665.
42. Halliwell, B. (1974): Superoxide dismutase, catalase and glutathione peroxidase: Solutions to the problems of living with oxygen. *New Phytol.,* 73:1075–1086.
43. Hoekstra, W. G. (1975): Biochemical function of selenium and its relation to vitamin E. *Fed. Proc.,* 34:2083–2089.
44. Holmberg, N. J. (1968): Purification and properties of glutathione peroxidase from bovine lens. *Exp. Eye Res.,* 7:570–580.
45. Hunter, F. E., Jr., Scott, A., Weinstein, J., and Schneider, A. (1964): Effects of phosphate, arsenate and other substances on swelling and lipid peroxide formation when mitochondria are treated with oxidized and reduced glutathione. *J. Biol. Chem.,* 239:622–630.
46. Jerina, D. M. (1976): Products, specificity, and assay of glutathione S-transferase with epoxide substrates. In: *Glutathione: Metabolism and Function,* edited by I. M. Arias and W. B. Jakoby, pp. 267–279. Raven Press, New York.
47. Kuehl, F. A., Jr., Humes, J. L., Egan, R. W., Ham, E. A., Beveridge, G. C., and van Arman, C. G. (1977): Role of prostaglandin endoperoxide PGG_2 in inflammatory processes. *Nature,* 265:170–173.
48. Kuehl, F. A., Jr., Humes, J. L., Ham, E. A., and Egan, R. W. (1978): Oxygen free radicals in inflammatory processes. *Intern. Congress of Inflammation,* Bologna, Italy, Abstract, p. 34.
49. Ladenstein, R. (1977): Strukturanalytische Studien an der Glutathion-Peroxidase aus Rindererythrocyten: Isolierung, Kristallisation, Untereinheiten-Symmetrie, 2,6Å-Fourier-synthese. *Dissertation,* Munich, Germany.
50. Ladenstein, R., and Epp, O. (1977): X-ray diffraction studies on the selenoenzyme glutathione peroxidase. *Hoppe Seylers Z. Physiol. Chem.,* 358:1237–1238.
51. Lawrence, R. A., and Burk, R. F. (1978): Species, tissue and subcellular distribution of non-Se-dependent glutathione peroxidase activity. *J. Nutr.,* 108:211–215.

52. Little, C., and O'Brien, P. J. (1968): An intracellular GSH-peroxidase with a lipid peroxide substrate. *Biochem. Biophys. Res. Commun.*, 31:145–150.
53. Little, C. (1972): Steroid hydroperoxides as substrates for glutathione peroxidase. *Biochim. Biophys. Acta*, 284:375–381.
54. Löhr, G. W., Blume, K. G., Rüdiger, H. W., and Arnold, H. (1974): Genetic variability in the enzymatic reduction of oxidized glutathione. In: *Glutathione,* edited by L. Flohé, H. Ch. Benöhr, H. Sies, H. D. Waller, and A. Wendel, pp. 165–173. Georg Thieme, Stuttgart, Germany.
55. Lombeck, I., Kasperek, K., Harbisch, H. D., Feinendegen, L. E., and Bremer, H. J. (1977): The selenium state of healthy children. I. Serum selenium concentration at different ages; activity of glutathione peroxidase of erythrocytes at different ages; selenium content of food of infants. *Eur. J. Pediat.,* 125:81–88.
56. Lombeck, I., Kasperek, K., Harbisch, H. D., Becker, K., Schumann, E., Schröter, W., Feinendegen, L. E., and Bremer, H. J. (1978): The selenium state of children. II. Selenium content of serum, whole blood, hair and the activity of erythrocyte glutathione peroxidase in dietetically treated patients with phenylketonuria and maple-syrup-urine disease. *Eur. J. Pediat.,* 128:213–223.
57. McKenzie, R. L., Rea, H. M., Thomson, C. D., and Robinson, M. F. (1978): Selenium concentration and glutathione peroxidase activity in blood of New Zealand infants and children. *Am. J. Clin. Nutr.,* 31:1413–1418.
58. Meister, A. (1975): Biochemistry of Glutathione. In: *Metabolism of Sulfur Compounds,* edited by D. M. Greenberg, pp. 101–188. Academic Press, New York.
59. Menzel, H. (1973): Untersuchungen zum Glutathion-abhängigen Peroxidstoffwechsel unter *in vivo*-Bedingungen. *Dissertation,* Tübingen, Germany.
60. Mills, G. C. (1957): Hemoglobin catabolism. I. Glutathione peroxidase, an erythrocyte enzyme which protects hemoglobin from oxidative breakdown. *J. Biol. Chem.,* 229:189–197.
61. Misra, H. P. (1974): Generation of superoxide-free radical during the autoxidation of thiols. *J. Biol. Chem.,* 249:2151–2155.
62. Miyamoto, T., Ogino, N., Yamamoto, S., and Hayaishi, O. (1976): Purification of prostaglandin endoperoxide synthetase from bovine vesicular gland microsomes. *J. Biol. Chem.,* 251:2629–2636.
63. Moncada, S., Gryglewski, R. J., Bunting, S., and Vane, J. R. (1976): An enzyme isolated from arteries transforms prostaglandin endoperoxides to an unstable substance that inhibits platelet aggregation. *Nature,* 263:663–665.
64. Nakamura, W., Hosoda, S., and Hayashi, K. (1974): Purification and properties of rat liver glutathione peroxidase. *Biochim. Biophys. Acta,* 358:251–261.
65. Necheles, T. F., Maldonado, N., Barquet-Chediak, A., and Allen, D. M. (1967): Homozygous erythrocyte glutathione peroxidase deficiency. *Blood,* 30:880–881.
66. Necheles, T. F. (1974): The clinical spectrum of glutathione-peroxidase deficiency. In: *Glutathione,* edited by L. Flohé, H. Ch. Benöhr, H. Sies, H. D. Waller, and A. Wendel, pp. 173–180. Georg Thieme, Stuttgart, Germany.
67. Neubert, D., Wojtczak, A. B., and Lehninger, A. L. (1962): Purification and enzymatic identity of mitochondrial contraction-factors I and II. *Proc. Natl. Acad. Sci. USA,* 48:1651–1658.
68. Nugteren, D. H., and Hazelhof, E. (1973): Isolation and properties of intermediates in prostaglandin biosynthesis. *Biochim. Biophys. Acta,* 326:448–461.
69. Oesch, F. (1972): Mammalian epoxide hydrases: Inducible enzymes catalysing the inactivation of carcinogenic and cytotoxic metabolites derived from aromatic and olefinic compounds. *Xenobiotica,* 3:305–340.
70. Oh, S. H., Ganther, H. E., and Hoekstra, W. G. (1974): Selenium as a component of glutathione peroxidase isolated from ovine erythrocytes. *Biochemistry,* 13:1825–1829.
71. Oshino, N., and Chance, B. (1977): Properties of glutathione release observed during reduction of organic hydroperoxide, demethylation of aminopyrine and oxidation of some substances in perfused rat liver, and their implications for the physiological function of catalase. *Biochem. J.,* 162:509–525.
72. Parsons, D. F., Williams, G. R., Thompson, W., Wilson, D., and Chance, B. (1967): Improvements in the procedure for purification of mitochondrial outer and inner membrane. Comparison of the outer membrane with smooth endoplasmatic reticulum. In: *Mitochondrial Structure and Compartmentation,* edited by E. Quagliariello, S. Papa, E. C. Slater, and J. M. Tager, pp. 29–70. Adriatica Editrice, Bari, Italy.

73. Piccardo, M. G., and Schwarz, K. (1958): The electron microscopy of dietary necrotic liver degeneration. In: *Symposium on Liver Function,* edited by R. W. Brauer, pp. 528–533. Amer. Inst. Biol. Sci., Washington, D.C.
74. Pirie, A. (1965): Glutathione peroxidase in lens and a source of hydrogen peroxide in aqueous humor. *Biochem. J.,* 96:244–253.
75. Polo, M. (1295; re-issued 1973): Von Venedig nach China. *Die grösste Reise des 13. Jahrhunderts.* p. 99. Horst Erdmann Verlag, Tübingen and Basel.
76. Riley, C. A., Cohen, G., and Lieberman, M. (1974): Ethane evolution. A new index of lipid peroxidation. *Science,* 183:208–210.
77. Rotruck, J. T., Hoekstra, W. G., Pope, A. L., Ganther, H., Swanson, A., and Hafeman, D. (1972): Relationship of selenium to GSH peroxidase. *Fed. Proc.,* 31:691.
78. Rotruck, J. T., Pope, A. L., Ganther, H. E., Swanson, A. B., Hafeman, D., and Hoekstra, W. G. (1973): Selenium: Biochemical role as a component of glutathione peroxidase. *Science,* 179:588–590.
79. Schrauzer, G. N., and Ismael, D. (1974): Effects of selenium and of arsenic on the genesis of spontaneous mammary tumors in inbred C_3H mice. *Ann. Clin. Lab. Sci.,* 4:441–447.
80. Schwarz, K., and Foltz, C. M. (1957): Selenium as an integral part of factor 3 against dietary necrotic liver degeneration. *J. Am. Chem. Soc.,* 79:3292–3293.
81. Schwarz, K. (1961): Development and status of experimental work on factor 3-selenium. *Fed. Proc., Fed. Amer. Soc. Exp. Biol.,* 20:666–673.
82. Schwarz, K., and Pathak, K. D. (1975): The biological essentiality of selenium, and the development of biologically active organoselenium compounds of minimum toxicity. *Chemica Scripta (Sweden),* 8A:85–95.
83. Schwarz, K. (1976): The discovery of essentiality of selenium, and related topics. In: *Proceedings of the Symposium on Selenium-Tellurium in the Environment,* edited by Industrial Health Foundation, Inc., pp. 349–376. Pittsburgh.
84. Shamberger, R. J. (1970): Relationship of selenium to cancer. 1. Inhibitory effect of selenium on carcinogenesis. *J. Natl. Cancer Inst.,* 44:931.
85. Shamberger, R. J. (1976): Selenium in health and disease. In: *Proceedings of the Symposium on Selenium-Tellurium in the Environment,* edited by Industrial Health Foundation, Inc., pp. 253–267. Pittsburgh.
86. Sies, H., Gerstenecker, C., Menzel, H., and Flohé, L. (1972): Oxidation in the NADP system and release of GSSG from hemoglobin-free perfused rat liver during peroxidatic oxidation of glutathione by hydroperoxides. *FEBS Lett.,* 27:171–175.
87. Sies, H. (1974): Biochemie des Peroxysoms in der Leberzelle. *Angew. Chem.,* 86:789–801.
88. Sies, H., Gerstenecker, C., Summer, K. H., Menzel, H., and Flohé, L. (1974): Glutathione-dependent hydroperoxide metabolism and associated metabolic transitions in hemoglobin-free perfused rat liver. In: *Glutathione,* edited by L. Flohé, H. Ch. Benöhr, H. Sies, H. D. Waller, and A. Wendel, pp. 261–276. Georg Thieme, Stuttgart, Germany.
89. Sies, H., and Moss, K. M. (1978): A role of mitochondrial glutathione peroxidase in modulating mitochondrial oxidations in liver. *Eur. J. Biochem.,* 84:377–383.
90. Sies, H., Bartoli, G. M., Burk, R. F., and Waydhas, C. (1978): Glutathione efflux from perfused rat liver after phenobarbital treatment, during drug oxidations, and in selenium deficiency. *Eur. J. Biochem.,* 89:113–118.
91. Smith, A. G., Harland, W. A., and Brooks, C. J. W. (1973): Glutathione peroxidase in human and animal aortas. *Ster. Lipids Res.,* 4:122–128.
92. Thomas, H. F., Herriott, R. M., Hahn, B. S., and Wang, S. Y. (1976): Thymine hydroperoxide as a mediator in ionising radiation mutagenesis. *Nature,* 259:341–343.
93. Thompson, J. N., and Scott, M. L. (1969): Role of selenium in the nutrition of the chick. *J. Nutr.,* 97:335–342.
94. Thomson, C. D., Rea, H. M., Doesburg, V. M., and Robinson, M. F. (1977): Selenium concentrations and glutathione peroxidase activities in whole blood of New Zealand residents. *Br. J. Nutr.* 37:457–460.
95. Wendel, A., Pilz, W., Ladenstein, R., Sawatzki, G., and Weser, U. (1975): Substrate-induced redox change of selenium in glutathione peroxidase studied by X-ray photoelectron spectroscopy. *Biochim. Biophys. Acta,* 377:211–215.
96. Wendel, A., Kerner, B., and Graupe, K. (1978): The selenium moiety of glutathione peroxidase. *Hoppe Seylers Z. Physiol. Chem.,* 359:1035–1036.

97. Westermarck, T. (1977): Selenium content of tissues in Finnish infants and adults with various diseases, and studies on the effects of selenium supplementation in neuronal ceroid lipofuscinosis patients. *Acta Pharmacol. Toxicol.*, 41:121–128.
98. Zakowski, J. J., and Tappel, A. L. (1978): Purification and properties of rat liver mitochondrial glutathione peroxidase. *Biochim. Biophys. Acta,* 526:65–76.
99. Zakowski, J. J., and Tappel, A. L. (1978): A semiautomated system for measurement of glutathione in the assay of glutathione peroxidase. *Anal. Biochem.,* 89:430–436.

Antioxidants, Cancer, and the Immune Response*

Werner A. Baumgartner

Radioimmunoassay and In Vitro Laboratory, Nuclear Medicine Service, Veterans Administration Wadsworth Hospital Center, Los Angeles, California 90073

The discovery of the biological essentiality of the trace element selenium by Schwarz and Foltz in 1957 was an important turning point in antioxidant biochemistry. For not only did this discovery provide new insights into the causes of certain pathophysiological processes, but it also served to stimulate new ideas on the functions of antioxidants. (17,22,24,36,45,47,64). Klaus Schwarz, for instance, was one of the foremost proponents for the point of view that selenium and vitamin E had more important roles to play than to act merely as protective agents against random lipid peroxidation reactions. He proposed that these alternate functions were produced by these substances as critical components of enzyme systems (63,65). The recent realization that selenium is the active component of glutathione peroxidase (57,58) and the possible involvement of this enzyme in the regulation of certain physiologically significant radical processes such as prostaglandin synthesis (21) and its involvement in immune processes (the subject of this chapter) have given further credence to these ideas. In particular, these findings tend to support the concept that antioxidants could also be involved in the regulation of radical processes occurring as part of normal intermediary metabolism (6). This is an attractive hypothesis, since it contains the most important elements of the other two opposing points of view for the action of selenium and vitamin E.

To date, however, it has not been possible to confirm any of these other proposed functions for vitamin E and selenium. In some instances, such proposals had to be withdrawn, since they were subsequently found to have been based on *in vitro* artifacts (6,53).

Undoubtedly, progress in antioxidant biochemistry has been very much held up by the multitude of artifacts which occur as a consequence of the drastic increase in lipid peroxidation under *in vitro* conditions. For one, it has made the detection of physiologically significant radical processes extremely difficult (6). Progress has also been hampered by the fact that *in vivo* studies have largely been confined to such artificial probes as the induction of pathophysiological conditions by the feeding of antioxidant-deficient diets or by the introduction of toxic chemicals such as carbon tetrachloride.

---

* In memoriam of Klaus Schwarz.

A recent discovery in immunology, however, appears to provide some hope for more effective experimental approaches to the study of vitamin E and selenium effects. This development stems from the recent observation that immune processes depend critically on selenium and vitamin E. Not only has it been found that immunity is depressed under antioxidant-deficient conditions but, more importantly, supplementation of normal diets with antioxidants was found to produce immunostimulatory effects. These latter *in vivo* observations and the fact that immune processes can also be effectively studied under *in vitro* conditions suggests that immune phenomena may provide valuable probes for the action of selenium and vitamin E.

In view of the potential importance of these immunological antioxidant effects, both to antioxidant biochemistry and to immunology, I have restricted the scope of the present chapter entirely to this subject, touching upon cancer-related antioxidant effects only in reference to certain possible applications of antioxidants to tumor immunology. Thus, the present chapter does not cover the by now fairly well established (and extensively reviewed) field of antioxidant protection against chemical carcinogenesis, since this is known to involve nonimmunological mechanisms (81).

Specifically, I will review some of the recent evidence for immunological antioxidant effects. This is followed by a discussion of possible mechanisms for these effects in the hope of stimulating further work in this area. And finally, with the same objective in mind, I will discuss two potential applications of antioxidants to tumor immunology: The enhancement of immune surveillance mechanisms against certain types of neoplasms, and the treatment of tumor-induced depressions of immunity.

Although this chapter focuses to a considerable extent on selenium and vitamin E in honor and memoriam of Klaus Schwarz, there are also reasons to believe that some of the theoretical considerations proposed herein will also apply to some of the other antioxidants such as vitamin C, superoxide dismutase, catalase, and certain sulfhydryl compounds and synthetic antioxidants.

IMMUNOLOGICAL ANTIOXIDANT EFFECTS

The interest in immunological antioxidant effects was given a considerable boost by the discovery of Tengerdy et al. (28,75–78) that supplementation of normal diets with vitamin E enhances the T-cell and macrophage-dependent antibody response against sheep erythrocytes in chicken and mice. Some of the essential features of this discovery are shown in Table 1. Clearly, vitamin E has a beneficial effect on immunity. This was established by two independent assays. The first and more generally used assay was the Jerne-Nordin plaque assay. The other assay employed measures the amount of antisheep-cell antibodies by conventional hemeagglutination procedures.

These investigations were extended to selenium by Spallholz et al. (71,72). It was found that selenium could also produce stimulatory effects on immunity

TABLE 1. *Effect of vitamin E on antibody production against sheep erythrocytes*

Animal	Diet	Immune Response	
		PFC/10^6 Cells	HA $\log_2$ Titer
Mice	Normal	3,386 ± 295	5.3 ± 0.4
	+E	5,565 ± 295	8.3 ± 0.9
Chicken	Normal	3,200 ± 300	—
	+E	6,600 ± 500	—

Modified, with permission, from ref. 77.

(Fig. 1). Once again, the anti-sheep erythrocyte response was used in these studies.

A number of points are illustrated by Fig. 1. For one, we see that a combined dietary deficiency in vitamin E and selenium caused serious depressions of immunity. It was possible to restore immunity to normal levels by addition of either selenium or vitamin E to the diet. At a constant value of 70 ppm vitamin E and increasing selenium concentrations, immunity increased continuously and reached a maximum of approximately four times the normal response at 1.25

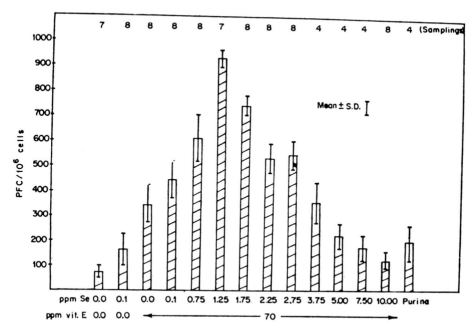

FIG. 1. Effects of vitamin E and selenium on the production of antibody against sheep erythrocytes in Swiss Webster mice. Animals were fed a semi-synthetic Torula yeast diet containing the indicated amounts of antioxidants. (Reprinted, with permission, from ref. 72.)

TABLE 2. *Effect of selenium on phagocytosis and cytotoxicity of neutrophils*

Diet	Ingestion[a]	Killing[b]
−Se	2.0 ± 0.2	6.6 ± 2.1
+Se	2.0 ± 0.1	22 ± 4

Modified, with permission, from ref. 67.
[a] *C. albicans* per PMN.
[b] Stained *C. albicans* as percentage ingested *C. albicans*.

ppm selenium. At higher selenium concentrations, immunity began again to decline due to toxicity effects. These initial observations on the immunostimulatory effects of antioxidants have been confirmed in other laboratories with different antigens and test animals (3,18,54).

Antioxidants have also been found to play an important role in the function

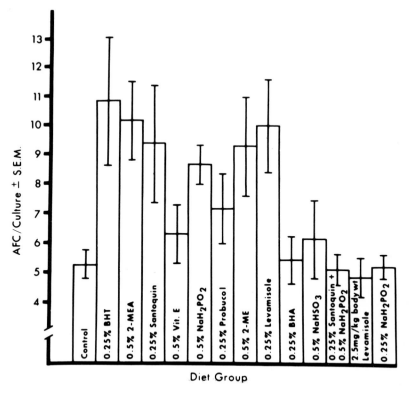

FIG. 2. Humoral response of female C3HeB/FeJ mice spleen cells to sheep erythrocytes: effect of adding levamisole and free-radical inhibitors to the diet. (Reprinted, with permission, from ref. 25.)

of phagocytic cells. For example, Serfass and Ganther (67) reported that a nutritional selenium deficiency reduced the killing capacity of neutrophils in spite of the fact that the deficiency had no effect on their capacity to phagocytize (Table 2). Similar antioxidant effects were observed by other investigators (4).

Harman et al. (25) applied these findings with antioxidants to the problem of age-associated depressions of immunity. He found (Fig. 2) that age-associated depression of immunity in mice could be prevented to a considerable extent by the long-term feeding of diets enriched with a variety of different antioxidants. Once again, the antisheep-erythrocyte response was used in these studies. Similar effects have been reported by Makinodan et al. (37).

Although there is considerable evidence now for immunostimulatory antioxidant effects, very little is known about the underlying mechanisms for these phenomena. This situation is perhaps not surprising, considering how far apart antioxidant biochemistry and immunology have been historically. However, one aspect of this problem appears to be clear even at the present time: the mechanisms for immunological antioxidant effects are likely to differ from the immunological effects produced by other nutrients, as the latter are known to influence immunity largely through effects on protein and nucleic acid metabolism (23,69).

CELLULAR MECHANISMS

In preparation for a discussion of possible biochemical mechanisms for immunological antioxidant effects, it is necessary to consider first some of the underlying cellular mechanisms. We concentrate here on the immune response against sheep erythrocytes, since this has featured so prominently in the demonstration of immunological antioxidant effects. Even then the story is still not complete. However, considerable progress has been made in recent years on cellular mechanisms, mainly because of the rediscovery by immunologists of the importance of the macrophage (5,46). The main features of the cellular mechanisms for T-cell and macrophage dependent antibody responses in immune processes are illustrated in Fig. 3.

Now it must be said that the scheme represented in Fig. 3 is by no means the only cellular mechanism for this particular antibody response. However, for the purposes of a discussion of biochemical mechanisms for immunological antioxidants, we do not have to concern ourselves here with all the possible variations on the presently outlined scheme, since most, if not all, of the other proposed cellular mechanisms have in common the features which are essential for the discussion of biochemical mechanisms. These features are: (a) The close spatial and functional interaction of lymphocytes with macrophages; (b) The elaboration by macrophages of nonspecific mitogenic factors for lymphocytes; and (c) (not shown in the diagram) the production by macrophages and other phagocytes of lymphocyte-inhibitory factors. Little is known about the chemical nature of these macrophage factors.

Essentially, Fig. 3 summarizes the findings that virtually no antibody response

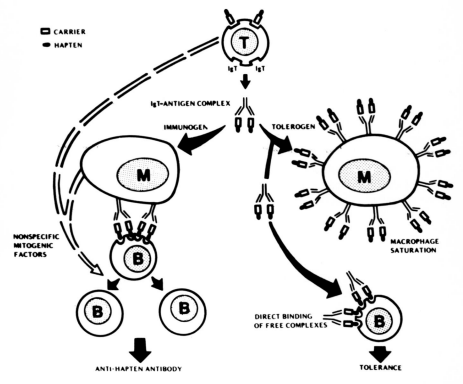

FIG. 3. Mechanism for T-cell and macrophage dependent antibody response and development of tolerance. (Reprinted, with permission, from ref. 5.)

against sheep erythrocytes can occur in the absence of macrophages and T-lymphocytes. To account for this observation, it has been proposed that activated T-cells (T) release an as yet unidentified Ig-T-antigen complex (or related factor) which is cytophilic for macrophages (M). The macrophages then present these complexes to B-cells (B) as an immunogenic matrix that cross-links haptenic receptors (signal 2), thereby triggering the cells into antibody production. Nonspecific factors from T-cells and/or macrophages act at the proliferative stage of the immune response to increase the number of antibody-forming cells. If a high dose of antigen is used, the macrophage surface becomes saturated with Ig-T-antigen molecules, thereby allowing direct binding of antigen-complexes to the B-cell receptors resulting in immunological tolerance.

Now, by looking at this mechanism for cell interactions, one gets very few clues regarding the possible mode of action of antioxidants, and this rightfully so, for the scheme illustrated in Fig. 3 deals only with cellular and not with biochemical mechanisms. Obviously, however, the scheme provides us with insights as to how macrophages are involved in the production of adjuvant effects.

Thus the immunostimulatory effects of such classical adjuvants as Freund's Complete Adjuvant are largely due to their macrophage-attracting properties, since this produces immune granulomas at the site of antigen injection. I would like to designate adjuvants which operate according to this basic mechanism as immunobiological adjuvants.

What I wish to propose is that the immunostimulatory effects of selenium, vitamin E, and certain other substances are also due to an adjuvant effect. The reason for this is that antioxidants—although not causing the cellular interactions of the classical adjuvants—nevertheless appear to be able to optimize these cellular interactions by providing the proper biochemical environment. Accordingly, one should distinguish between the immunobiological adjuvant effects of the type produced by the classical adjuvants and the immunobiochemical adjuvant effects as produced by antioxidants.

BIOCHEMICAL MECHANISMS

Having established the cellular basis for the antisheep cell response, we are now ready to speculate on how these processes can be affected beneficially by selenium and vitamin E.

This story had its beginning with what is generally considered to be only a tissue culture effect—i.e., the ability of mercaptoethanol to stimulate virtually every conceiveable type of immune process when these are carried on in tissue culture. The following immune processes are known to be beneficially affected by mercaptoethanol: antisheep erythrocyte response (16); B-lymphocyte colony formation (42); cytotoxic product formation (19); response to T- and B-cell mitogens (20); lymphocyte replication (10); and mixed leukocyte reaction (2). A considerable number of other sulfhydryl compounds have given similar results (10,31).

There is now a vast literature in existence on the possible mechanisms for these immunological sulfhydryl effects. This is partly due to the fact that sulfhydryl groups, in addition to acting as antioxidants or deactivators of metal peroxidation catalysts, can influence numerous enzyme and membrane processes via sulfhydryl/disulfide exchange reactions.

One critical discovery with respect to the mechanism of sulfhydryl effects has been that sulfhydryl compounds can replace the function of macrophages in the immune response (35,51). The puzzle of how a chemical could replace the antigen presentation function of macrophage was solved when it was found that the original experiments suggesting this effect had still contained a sufficient number of macrophages as impurities to carry out the antigen presentation function. This led to the realization that sulfhydryl compounds actually only replaced the nonspecific lymphocyte-mitogenic factors of macrophages, produced in adequate amounts only by high concentrations of macrophages.

This was followed by another critical discovery, made by Campbell et al. (13), namely that vitamin E can also substitute for the mitogenic factors of

TABLE 3. Effect of macrophages, 2-mercaptoethanol and vitamin E on antibody production

Culture composition	Immune response (PFC/culture)
Lymphocytes + macrophages	1,800
Lymphocytes + macrophages + E	5,054
Lymphocytes	91
Lymphocytes + E	3,700
Lymphocytes + 2-Me	1,854

Modified, with permission, from ref. 13.

macrophages in tissue culture. Table 3 contains representative data of these studies, which suggest that sulfhydryl compounds and vitamin E stimulate the *in vitro* antierythrocyte response by similar mechanisms.

These observations with vitamin E have simplified the mechanistic picture of *in vitro* sulfhydryl effects considerably, since they showed that the mercaptoethanol effect, at least as far as the antierythrocyte response was concerned, did not involve one of the myriad of possible disulfide/sulfhydryl exchange reactions, but that it was instead due to an antioxidant effect. It also demonstrated that at least some of the nonspecific mitogenic factors of macrophages possess antioxidant properties. To my knowledge, the above studies have so far not been repeated with some of the other antioxidants such as with the selenium-dependent glutathione peroxidase or with superoxide dismutase.

I do not consider that an antioxidant mechanism indicates that mercaptoethanol and vitamin E protect only against certain *in vitro* artifacts i.e., against lipid and/or sulfhydryl group oxidation. No doubt, protection against such artifacts is one of their functions in tissue culture, and may indeed be the only function for the many immunological processes for which beneficial sulfhydryl effects have so far only been observed under *in vitro* conditions. One reason for believing this is that in tissue cultures, lipid and sulfhydryl-group oxidations are greatly enhanced by the prevailing elevated oxygen tensions and by the presence of trace metal peroxidation catalysts released from cellular debris. It must be remembered that it is mainly due to the release of these catalysts that lipid peroxidation under *in vitro* conditions is several orders of magnitude greater than that occurring in the intact animal. However, the mere prevention of such tissue culture artifacts cannot be the only role for antioxidants in the case of the antisheep cell response, since here both *in vivo* as well as *in vitro* effects have been obtained.

IN VIVO IMMUNOLOGICAL ANTIOXIDANT EFFECTS

We are thus confronted with the question: What is the nature and source of the *in vivo* oxidative processes against which the antierythrocyte response is protected by antioxidants? Here we are provided with clues from inflammation biochemistry, for it has been known for quite some time now that the phagocytes present in immune granulomas, notably the neutrophils and macrophages, generate during phagocytosis considerable quantities of oxidants, such as lipid hydroperoxides (67,74), superoxide (48), singlet oxygen (1), hydrogen peroxide (29,34), OH radicals (39), and via oxidative radical processes also prostaglandins (30,79). All these substances are known as potent lymphocyte deactivators (52). As a matter of fact, lymphocytes distinguish themselves by their exquisite sensitivity to such radical processes. In contrast to this, macrophages are known for their extreme resistance to these agents—possibly as protection against their own cytotoxic radicals.

This suggests that antioxidants such as selenium, vitamin E, and sulfhydryl compounds are needed in immune processes under *in vivo* conditions for the protection of lymphocytes against the inhibitory products of phagocytes. Antioxidants also inhibit prostaglandin synthesis and, therefore, can modulate also this mode of suppression of lymphocyte activity.

It appears plausible, therefore, that some of the lymphocyte inhibitory factors of macrophages are nothing else but some of the above named cytotoxic agents leaking from macrophages. By the same token, some of the lymphocyte mitogenic factors of macrophages may be antioxidants which macrophages probably require for their own protection, e.g., glutathione peroxidase, superoxide dismutase, catalase, vitamin E and vitamin C. These protective agents may be transferred to lymphocytes by various release mechanisms. Alternatively, dietary supplementation with antioxidants or their precursors (sodium selenite) may result in their optimum accumulation in macrophages, thereby preventing leakage of lymphocyte-inhibiting factors.

In general then, I would like to propose that antioxidant-adjuvants act to optimize the environment for the immunobiological interaction of the biochemically somewhat incompatible lymphocytes and phagocytes. I believe that this view can also, in part, explain why excessive use of classical adjuvants, such as Freund's Adjuvant, produces immunodepressive rather than immunostimulatory effects (32,33). In other words, immunity apparently becomes depressed whenever too much inflammatory activity is generated. Similarly, I believe that the theory can provide an explanation for some of the immunodepressive effects of autoimmune inflammatory diseases such as rheumatoid arthritis, and also for some of the immunodepressive effects of cancer. Interestingly, all of these conditions are also accompanied by a depression in serum sulfhydryl levels. There are good reasons to believe that some of the other antioxidants are also depressed under these conditions.

ANTIOXIDANT EFFECTS IN TUMOR IMMUNITY

A recent study by Schrauzer and Ishmail (60) has shown that the incidence of mammary tumors afflicting over 80% of aged C_3H mice can be reduced to

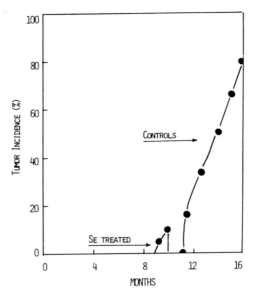

FIG. 4. Appearance of spontaneous tumors in selenite-treated mice and controls during 16 months of observation. (Reprinted, with permission, from ref. 60.)

10% by the long-term feeding of subtoxic levels of selenium in the form of sodium selenite. These results are illustrated in Fig. 4. Although the mechanism for this protective effect is still unclear, there are good reasons to believe that immunological antioxidant effects are involved in the prevention of this type of cancer. For one, it is known that these cancers in C_3H mice are caused by the horizontal transmission of mammary tumor virus via mothers milk (43). Consequently, one cannot invoke any of the mechanisms which are known to operate in the protective action of selenium and other antioxidants against chemical carcinogens (81). Even the operation of such chemical mechanisms against some as yet unidentified chemical co-carcinogen in the environment appears unlikely, since these tumors in C_3H mice arise with clockwork-like regularity with advancing age in many different laboratory environments.

More directly supportive of a beneficial effect of selenium on tumor immunity is the established importance of immune surveillance mechanisms in the prevention of virally caused cancers (41). This is due to the fact that virally caused tumors—in contrast to those of chemical etiology—express strong and constant viral transplantation antigens against which an effective immune response can be readily launched, provided of course that immunity has not been compromised by age, viruses (including tumor virus), or by a chemical or some other agent. Since the mammary tumors in C_3H mice arise only in old animals, it is tempting to speculate that the etiology of these tumors is linked to the usual age-associated depression of immunity (38), with possible further depression arising from the

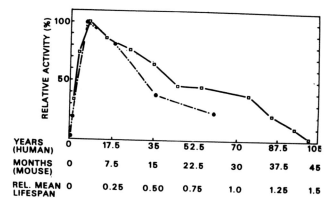

FIG. 5. Effects of age on serum agglutinin titers in humans and mice. (□), Natural serum anti-A isoagglutinin titers in the human. (●), Peak serum agglutinin response titer to RBC stimulation by intact long-lived mice. (Reprinted, with permission, from ref. 38.)

action of the tumor virus. That such age-associated changes in immunity can be quite severe is illustrated in Fig. 5. Consequently, it appears plausible that the anticancer action of selenium may be due to a protective effect against age-associated and/or virally caused depressions of immunity similar to that already observed with other antioxidants (7,25).

The relevance of these findings to the human situation is still unclear, since the question of a viral etiology in human cancers has not been settled to date (73). It may, however, be that too little attention has been given in the virus controversy to age-dependent factors, that is, to the possibility that viral cancers are relatively rare in young and middle-aged individuals, but that such cancers become of increasing importance in older members of the population suffering from depressed immunity.

ANTIOXIDANT EFFECTS IN PARANEOPLASTIC SYNDROMES

Finally, I would like to discuss some evidence which suggests that the adjuvant properties of antioxidants, including those of the water-soluble antioxidant vitamin C, may also produce beneficial effects under conditions of established neoplastic disease.

We first became interested in this subject when we noted some time ago that states of depressed immunity—caused either by excessive uses of adjuvants, chronic inflammation, cancer, or aging—were associated with rather drastic depressions in serum sulfhydryl levels (9,59). We found that a major factor in this loss of protective sulfhydryl groups is a drop in albumin level.

Although the exact mechanism for these sulfhydryl changes is still unclear, there are reasons to believe that they are at least in part caused by enhanced oxidative activity. In the case of cancer, this could arise from the oxidative

TABLE 4. *Liver vitamin E deficiencies in tumor hosts*

Tumor	Liver-vitamin E content[a]	
	Tumor host	Control
Walker Carcinoma	14.0	23.8
Hepatoma XXII	32.0	107.4
Ehrlich ascites carcinoma	23.1	66.0
Ascites sarcoma	12.5	66.0

Modified, with permission, from ref 11.
[a] μg/g tissue.

products generated by the phagocytizing cells engaged in the clean-up of necrotic tumor tissue, i.e., from enhanced inflammatory activity. Incidentally, this link with inflammation may explain why mixed function oxidase activity is depressed and why ceruloplasmin levels are elevated in both cancer and inflammatory disorders (9,56). Another possibility for enhanced peroxidation is the release from necrotic tumor tissue of trace-metal lipid peroxidation catalysts. As indicated previously, the release of such catalysts is the main reason why lipid peroxidation under *in vitro* conditions is several orders of magnitude higher than those occurring normally in the intact organism. Although the release of toxic substances from tumor tissue has long been suspected by pathologists (40), their nature or mode of action has remained unclear.

Further evidence in support of the postulated increase in lipid peroxidation comes from a recent Russian report which shows that the vitamin E content of livers in tumor-bearing animals can decrease by up to 80% (11). The results in Table 4 show that this was the case for all the tumors studied by these Russian investigators.

We decided to test our hypothesis of tumor-enhanced peroxidation processes further by investigating whether tumors could also cause selenium deficiencies, since the resulting combined deficiencies in both vitamin E and selenium could possibly lead to serious consequences for the tumor host. For instance, it was Klaus Schwarz who first showed that liver necrosis can only occur under conditions of a combined nutritional deficiency in vitamin E and selenium. Now, obviously we do not expect that tumor-induced antioxidant deficiencies can produce such drastic lesions in the tumor-hosts. However, what we did consider as a distinct possibility in view of the dependence of immunity on antioxidants is that such antioxidant deficiencies could contribute to malignant anergy, that is, to a general tumor-induced depression of immunity. It even appears plausible that the antioxidant deficiencies could contribute to certain other paraneoplastic syndromes such as disturbed liver functions (56,80) which in turn may contribute to a general condition of wasting known as cachexia. It should be pointed out that malignant anergy is a serious complication in neoplastic disease, for not only is it a serious obstacle to immunotherapy of cancer, but it is also a major cause of death in cancer patients (40)—that is, cancer patients frequently die of viral or bacterial infections rather than from any direct effect of the tumor itself.

We investigated the possibility of tumor-induced selenium deficiencies by means of selenium turnover experiments rather than by direct chemical analysis of selenium, since we were interested in determining whether lower antioxidant levels in tumor-hosts were caused by a reduction in nutrient intake (due to loss of appetite of the tumor host), or by the faster utilizations of the antioxidant as a result of an increase in oxidative processes (8). These experiments were carried out by pre-labeling the various selenium pools of Swiss Webster mice with Se^{75}-labeled sodium selenite. It was Klaus Schwarz who first showed that selenite is rapidly transformed into the biologically active form of selenium, i.e., into what we now know to be glutathione peroxidase. Half of the mice were inoculated with Ehrlich Ascites carcinoma 96 hr after selenite injection, that is, with one of the tumors for which vitamin E deficiencies have already been demonstrated.

The results in Table 5 show that Ehrlich ascites tumors have increased the rate of turnover of selenium from liver, the body's main selenium store. Only a marginally significant increase in turnover was observed in kidney; whereas that observed in thymus was found to be due to a decrease in organ weight. No such weight changes were observed for kidney or liver.

In another series of experiments, we compared the selenium-75 loss caused by tumor growth to one caused by a vitamin E and selenium-deficient diet capable of producing fatal tissue necrosis (Fig. 6). To our surprise, we found that tumor growth caused the same increase in selenium-turnover as a vitamin

TABLE 5. *Effect of Ehrlich ascites tumor on selenium-75 distribution in Swiss Webster mice.*[a] *Percent of total body selenium-75 in different tissue.*[b]

Tissue	Animal group			
	6 days tumor growth	6 day control (no tumor)	10 days tumor growth	10 day control (no tumor)
Liver	28.7 ± 4.4[c]	34.1 ± 3.3	22.8 ± 2.7[c]	33.9 ± 3.4
Kidney	6.1 ± 0.8[c]	7.7 ± 0.7	6.2 ± 0.7[c]	7.2 ± 0.4
Spleen	1.05 ± 0.14[c]	0.78 ± 0.14	0.73 ± 0.06	0.80 ± 0.11
Thymus	0.13 ± 0.04[d]	0.16 ± 0.03	0.10 ± 0.02[c]	0.18 ± 0.03
Lymph nodes	0.043 ± 0.007	0.047 ± 0.015	0.037 ± 0.008	0.043 ± 0.006
Tumor cells	3.5 ± 1.6	—	6.0 ± 1.5	—
Tumor cells + ascites fluid	5.1 ± 2.8	—	14.2 ± 5.1	—

Modified, with permission, from ref. 8.
[a] Each value represents the mean of 10 animals ± standard deviation.
[b] Total body selenium-75 not including the selenium-75 in tumor or ascites fluid.
[c] Average values differing by $p < 0.01$, as compared to control group.
[d] Average values differing by $p < 0.05$, as compared to control group.

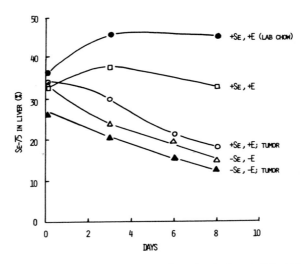

FIG. 6. The effects of tumor growth and vitamin E and selenium-deficient diet on selenium-75 turnover in mouse liver. Results are expressed as percentage of total body selenium-75: (●), normal mice on lab chow containing normal antioxidant levels; (□), normal mice on Torula yeast containing normal antioxidant levels: (△), normal mice on antioxidant deficient Torula yeast; (○), tumor bearing mice on Torula yeast containing normal antioxidant levels; (▲), tumor-bearing mice on antioxidant deficient Torula yeast. (Modified, with permission, from ref. 8.)

E- and selenium-deficient diet capable of producing fatal liver necrosis. Moreover, no futher increase in selenium turnover rate could be achieved by the combined application of these two stresses, i.e., by growing tumors in animals which had been fed a vitamin E- and selenium-deficient diet. This result suggests that tumor growth in normal animals was able to exert a maximal detrimental effect on liver as far as selenium turnover was concerned.

As an interesting sideline, this experiment allowed us also to investigate how tumor growth was affected by a nutritional vitamin E and selenium deficiency. We performed this experiment, since the naturally necrotizing tendencies of tumor tissue suggested that a vitamin E- and selenium deficient-diet could induce selective necrosis in neoplastic cells.

The data in Fig. 7 show that this was indeed the case. It is evident that the tumor in vitamin E- and selenium-deficient animals grew initially at the same rate as those in normal animals. However, after six days of normal growth, the tumors in the antioxidant-deficient animals contracted abruptly or disappeared completely—i.e., we could not detect any significant volume of packed tumor cells in the small amount of remaining ascites fluid. The extent to which tumors can be caused to regress is actually much greater than indicated in the figure, since the last point on the graph is the average of two complete remissions and three contractions in tumor volume. We consider it rather likely that the three shrunken tumors would have also disappeared if sufficient time

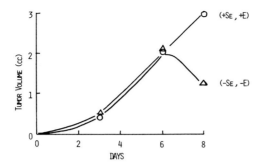

FIG. 7. Selective induction of necrosis in tumor cells. Growth rate of Ehrlich ascites carcinoma in Swiss Webster mice fed a Torula yeast diet which was either deficient in (△), or contained normal levels of (○) selenium and vitamin E.

had been given for the full effects of the antioxidant deficiency to manifest themselves. This, of course, presupposes that the animals would not have died in the interim from fatal tissue necrosis.

Interestingly, tumor regression occurred in spite of the fact that the selenium-sequestering capacity of the tumor increases drastically under the conditions of the nutritional antioxidant deficiency. These changes are shown in Table 6. No such changes occurred in normal tissue such as liver which exhibited the highest selenium-sequestering capacity of all organs. It is interesting to speculate whether this increase in the selenium-sequestering capacity of cancer cells in response to an oxidative stress is related to the nonsenescing qualities of neoplastic cell lines (27,49,50) or even to certain aspects of their radiation sensitivities (e.g., oxygen effect).

Further support for our suggestion of tumor-induced antioxidant deficiencies comes from Schamberger's and Schrauzer's (61,62,68) laboratories, namely from their observations that serum-selenium levels are depressed in human cancer patients. These results, however, have so far been discussed mainly from a causal point of view. Although we believe that such deficiencies may in large measure be a consequence of the disease (8), we do not feel that these two

TABLE 6. *Effect of diet and antioxidant content on the selenium sequestering capacity of liver and tumor tissue*

Diet	Tissue	SE-sequestering capacity (Specific activity)
Lab chow	Tumor	1.6
Torula yeast,+E,+Se	Tumor	5.6
Torula yeast,−E,−Se	Tumor	14.0
Lab chow	Liver	19.4
Torula yeast,−E,−Se	Liver	19.0

Modified, with permission, from ref. 8.

interpretations are necessarily in conflict with one another, since it is quite easy to envisage mechanisms by which causes and effects become linked through a vicious cycle. Such cycles obviously exist for depressed immunity and decreased mixed-function oxidase activity in tumor hosts. For instance, low activities in these two functions predispose towards cancer. However, once neoplastic disease has become established, immunity and mixed-function oxidase activity become further depressed by the disease, thereby completing the cycle. Interestingly, antioxidants play an important role in both of these functions (14,15).

In conclusion: On the basis of the demonstrated tumor-induced antioxidant deficiencies, we suggest that cancer patients could derive considerable benefit from supplemental amounts of antioxidants. We suggest this approach not only because such treatment can potentially reverse any existing tumor-induced antioxidant deficiencies—deficiencies which could play an important role in the development of malignant anergy—but also simply because of the inherent adjuvant properties of antioxidants, which are expected to enhance immunity even under normal conditions, i.e., under conditions where no such deficiencies exist.

Evidence for such beneficial antioxidant effects in cancer patients may already be in hand. This is suggested by the work of Cameron and Pauling (12) which shows that megadoses of vitamin C have a beneficial effect on the survival rate of terminal cancer patients.

Additional reasons for administering antioxidants to cancer patients are that antioxidants may reduce some of the undesirable side effects of chemotherapy. For example, vitamin E has been shown to reduce the cardiotoxicity of Adriamycin without interfering with its chemotherapeutic effectiveness (44). Also, since most chemotherapeutic agents are carcinogens to varying degrees (55,70) as well as known depressants of immunity (26), there is the possibility that antioxidants could also offer some protection against the undesirable side effects of these agents.

ACKNOWLEDGMENT

The author is greatly indebted to Annette Baumgartner and Joel Miller for their expert advice and assistance with this research. The work was supported in part by Grant 5RO1-CA-14221 from the National Institutes of Health.

REFERENCES

1. Allen, R. C., Stjernholm, R. L., and Steel, R. H. (1972): Evidence for the generation of an electronic excitation states in human polymorphonuclear leukocytes and its participation in bactericidal activity. *Biochem. Biophys. Res. Commun.*, 47:679–684.
2. Axelsson, J. A., Kallen, B., Nilsson, O., and Trope, C. (1976): Effect of 2-mercaptoethanol on the mixed leukocyte reaction in man. *Acta Path. Microbiol. Scand. Sect. C.*, 84:390–396.
3. Barber, T. L., Nockels, C. F., and Jochim, M. M. (1977): Vitamin E enhancement of Venezuelan equine encephalomyelitis antibody response in guinea pigs. *Am. J. Vet. Res.*, 38:731–734.
4. Bass, D. A., De Chatelet, L. R., Burk, R. F., Shirley, P., and Szejda, P. (1977): Polymorphonuclear leukocyte bactericidal activity and oxidative metabolism during glutathione peroxidase deficiency. *Infect. Immun.*, 18:78–84.

5. Basten, A., and Mitchell, J. (1976): Role of macrophage in T-cell-B-cell collaboration in antibody production. In: *Immunobiology of the Macrophage,* edited by D. S. Nelson, pp. 43–90. Academic Press, New York.
6. Baumgartner, W. A., Hill, V. A., and Wright, E. T. (1978): Anomalous Vitamin E effects in mitochondrial oxidative metabolism. *Mech. Aging Develop.,* 8:311–328.
7. Baumgartner, W. A., Makinodan, T., and Blahd, W. H. (1978): Comparison of isotopic and conventional assays for delayed type hypersensitivity in old and young mice. *J. Nucl. Med.,* 19:722.
8. Baumgartner, W. A., Hill, V. A., and Wright, E. T. (1978): Antioxidant effects in the development of Ehrlich ascites carcinoma. *Am. J. Clin. Nutr.,* 31:457–465.
9. Baumgartner, W. A., Beck, F. W. T., Lorber, A., Pearson, C. M., and Whitehouse, M. W., (1974): Adjuvant disease in rats: Biochemical criteria for distinguishing several phases of inflammation and arthritis. *Proc. Soc. Exp. Biol. Med.,* 145:625–630.
10. Broome, J. D., and Jeng, M. W. (1973): Promotion of replication in lymphoid cells in comparison with splenic lymphocytes. *J. Exp. Med.,* 138:547–592.
11. Burobina, S. A., and Nejfakh, E. A., (1970): Investigation of a natural antioxidant (vitamin E) during malignant growth. *Moscow Soc. Naturalists,* 32:56–61.
12. Cameron, E., and Pauling, L. (1976): Supplemental ascorbate in the supportive treatment of cancer: Prolongation of survival times in terminal cancer. *Proc. Natl. Acad. Sci. USA,* 73:3685–3689.
13. Campbell, P. A., Cooper, H. R., Heinzerling, R. H., and Tengerdy, R. P. (1974): Vitamin E enhances *in vitro* immune response by normal and nonadhering spleen cells. *Proc. Soc. Exp. Biol. Med.,* 146:465–469.
14. Carpender, M. P., and Howard, C. N. (1974): Vitamin E steroids, and liver microsomal hydroxylations. *Am. J. Clin. Nutr.,* 27:966–979.
15. Carpender, M. P. (1972): Vitamin E and microsomal drug hydroxylation. *Ann. N.Y. Acad. Sci.,* 203:81–92.
16. Click, R. E., Benck, L., and Alter, B. J. (1972): Enhancement of antibody synthesis *in vitro* by mercaptoethanol. *Cell. Immunol.,* 3:155–160.
17. Diplock, A. T. (1974): Possible stabilizing effect of vitamin E on microsomal, membrane-bound, selenide-containing proteins and drug-metabolizing enzyme systems. *Am. J. Clin. Nutr.,* 27:995–1004.
18. Ellis, R. P., and Vorhies, M. W. (1976): Effect of supplemental dietary vitamin E on the serologic response of swine to an *Escherichia coli* bacteria. *J. Am. Vet. Med. Assoc.,* 168:231–232.
19. Engers, H. D., MacDonald, H. R., Cerottini, J. C., and Brunner, K. T. (1975): Effect of delayed addition of 2-mercaptoethanol on the generation of mouse cytotoxic T-lymphocytes in mixed leukocyte cultures. *Eur. J. Immunol.,* 5:223–225.
20. Fanger, M. W., Hart, D. A., Wells, J. V., and Nisonoff, D. (1970): Enhancement by reducing agents of the transformation of human and rabbit peripheral lymphocytes. *J. Immunol.,* 105:1043–1045.
21. Flohe, L., Gunzler, W. A., and Ladenstein, R. (1976): Glutathione Peroxidase. In: *Glutathione: Metabolism and Function,* edited by I. M. Arias and W. B. Jakoby, pp. 115–138. Raven Press, New York.
22. Gilbert, J. J. (1974): Effect of tocopherol on the growth and development of rotifers. *Am. J. Clin. Nutr.,* 27:1005–1015.
23. Good, R. A., Jose, D., Cooper, W. C., Fernandes, G., Kramer, T., and Yunis, E. (1977): *Malnutrition and the Immune Response.* edited by R. M. Suskind, pp. 169–183. Raven Press, New York.
24. Green, J. (1969): Vitamin E and the biological antioxidant theory. In: *The Fat Soluble Vitamins,* edited by H. F. DeLuca and J. W. Suttie pp. 293–305. University of Wisconsin Press, Madison, Wisconsin.
25. Harman, D., Heidrick, M. L., and Eddy, D. E. (1977): Free radical theory of aging: Effect of free-radical-reaction inhibitors on the immune response. *J. Am. Geriat. Soc.,* 9:400–407.
26. Haskell, C. M. (1977): Immunological aspects of cancer chemotherapy. *Ann. Rev. Pharmacol. Toxicol.,* 17:179–195.
27. Hayflick, L. (1965): The limited *in vitro* lifetime of human diploid cell strains. *Exptl. Cell. Res.,* 37:614–636.
28. Heinzerling, R. H., Tengerdy, R. P., Wick, L. L., Lueker, D. C. (1974): Vitamin E protects mice against *Diplococcus pneumoniae* type I infection. *Infect. Immun.,* 10:1292–1295.

29. Homan-Muller, J. W. T., Weening, R. S., and Roos, D. (1975): Production of hydrogen peroxide by phagocytizing human granulocytes. *J. Lab. Clin. Med.,* 85:198–207.
30. Hope, W. C., Dalton, C., Machlin, L. J. M., Filipski, R. J., and Vane, F. M. (1975): Influence of dietary vitamin E on prostaglandin biosynthesis in rat blood. *Prostaglandins,* 10:557–571.
31. Jackson, J. F., and Lindahl-Kiessling, J. F. (1964): Action of sulfhydryl compounds on human leukocyte mitosis *in vitro. Exp. Cell. Res.,* 34:515–524.
32. Jamkovic, B. D. (1963): Adjuvant-induced immunological non-reactivity in guinea-pigs. *Colloques Intern. C.N.R.S. (Tolerance Immunol.).,* 116:187–198.
33. Jollés, P., and Paraf, A. (1973): *Chemical and Biological Basis of Adjuvants,* pp. 1–3. Springer-Verlag, New York.
34. Karnovsky, M. L. (1968): The metabolism of leukocytes. *Sem. Hematol.,* 5:156–165.
35. Lemke, H., and Opitz, H. G. (1976): Function of 2-mercaptoethanol as a macrophage substitute in the primary immune response *in vitro. J. Immunol.,* 117:388–395.
36. Lucy, J. A. (1972): Functional and structural aspects of biological membranes: a suggested structural role for vitamin E in the control of membrane permeability and stability. *Ann. N.Y. Acad. Sci.,* 203:4–12.
37. Makinodan, T., Deitchman, J. W., Stoltzner, G. H., Kay, M. M., and Hirokawa, K. (1975): Restoration of the declining normal immune functions of aging mice. *Proc. 10th Internat. Cong. Gerontol.,* 2:23.
38. Makinodan, T., and Adler, W. (1975): The effects of aging on the differentiation and proliferation potentials of cells of the immune system. *Fed. Proc.,* 34:153–158.
39. McCord, J. M. (1974): Free radicals and imflammation: Protection of synovial fluid by superoxide dismutase. *Science,* 185:529–531.
40. Meissner, W. A., and Diamandopoulos, G. T. H. (1977): Neoplasia. In: *Pathology, Vol. 1 (7th ed.),* edited by W. A. D. Anderson and J. M. Kissane, pp. 640–691. C.V. Mosby, St. Louis.
41. Melief, C. J. M., and Schwartz, R. S. (1975): Immunocompetence and malignancy. In: *Cancer, Vol. 1: Chemical and Physical Carcinogenesis.,* edited by F. F. Becker, p. 128. Plenum Press, New York.
42. Metcalf, D. (1976): Role of mercaptoethanol and endotoxin in stimulating B lymphocyte colony formation *in vitro. J. Immunol.,* 116:635–638.
43. Moore, D. H. (1975): Mammary tumor virus. In: *Cancer, Vol. 2, Viral Carcinogenesis.,* edited by F. F. Becker, p. 133. Plenum Press, New York.
44. Myers, C. E., McGuire, W. P., Liss, R. H., Frim, I., Grotzinger, K., and Young, R. C. (1977): Adriamycin: The role of lipid peroxidation in cardiac toxicity and tumor response. *Science,* 197:165–167.
45. Nair, P. P., (1972): Vitamin E and metabolic regulation. *Ann. N.Y. Acad. Sci.,* 203:53–61.
46. Nelson, D. S. (1976): Macrophages: Perspectives and prospects. In: *Immunobiology of the Macrophage.,* edited by D. S. Nelson, pp. 617–621. Academic Press, New York.
47. Olson, R. E. (1974): Creatine kinase and myofibrillar proteins in hereditary muscular dystrophy and vitamin E deficiency. *Am. J. Clin. Nutr.,* 27:1117–1129.
48. Oyanagui, Y. (1976): Inhibition of superoxide anion production in macrophages by anti-inflammatory drugs. *Biochem. Pharmacol.,* 25:1473–1480.
49. Packer, L., and Smith, R. (1977): Extension of the life-span of cultured normal human diploid cells by vitamin E: A reevaluation. *Proc. Natl. Acad. Sci. USA,* 74:1640–1641.
50. Packer, L., and Smith, R. (1974): Extension of the life-span of cultured normal human diploid cells by vitamin E. *Proc. Natl. Acad. Sci. USA,* 71:4763–4767.
51. Pierce, C. W., Kapp, J. A., Wood, D. D., and Benacerraf, B. (1974): Functions of macrophages. *J. Immunol.,* 112:1181–1189.
52. Plescia, O. J., Smith, A. H., and Grinwich, K. (1975): Subversion of immune system by tumor cells and role of prostaglandins. *Proc. Nat. Acad. Sci. USA,* 72:1848–1851.
53. Pollard, C. J., and Bieri, J. G. (1959): Studies of the biological functions of vitamin E. *Biochim. Biophys. Acta,* 34:420–430.
54. Renoux, M. L., de Montis, G., Roche, A., and Hemon, D. (1976): Effect of a polyvitamin preparation on immunological antibody response. *Nouv. Presse. Med.,* 5:2053–2056.
55. Rosner, F. (1978): Is chemotherapy carcinogenic? *Ca—A Cancer Jr. for Clinicians.,* 28:57–59.
56. Rosso, R., Donelli, M. G., Franchi, G., and Garattini, S. (1971): Impairment of drug metabolism in tumor-bearing animals. *Eur. J. Cancer.,* 7:565–577.

57. Rotruck, J. T., Pope, A. L., Ganther, H. E., Swanson, A. B., Hafeman, D., and Hoekstra, W. G. (1973): Selenium: Biochemical role as a component of glutathione peroxidase. *Science,* 179:588–590.
58. Rotruck, J. T., Hoekstra, W. G., Pope, A. L., Ganther, H., Swanson, A., and Hafeman, D. (1972): Relationship of selenium to GSH peroxidase. *Fed. Proc.,* 31–691.
59. Schoenbach, E. B., Wiessman, N., and Armistead, E. B. (1951): The determination of sulfhydryl groups in Serum II. Protein alterations associated with disease. *J. Clin. Invest.,* 30:762–777.
60. Schrauzer, G. N., and Ishmael, D. (1974): Effects of selenium and arsenic on the genesis of spontaneous mammary tumors in inbred C_3H mice. *Ann. Clin. Lab. Sci.,* 4:441–447.
61. Schrauzer, G. N., Rhead, W. J., and Evans, G. A. (1973): Selenium and cancer: Chemical interpretation of a plasma cancer test. *Bioinorgan. Chem.,* 2:329–340.
62. Schrauzer, G. N., and Rhead, W. J. (1971): Interpretation of the methylene blue reduction test of human plasma and the possible cancer protecting effect of selenium. *Experientia,* 27:1069–1071.
63. Schwarz, K., and Pathak, K. D. (1975): The biological essentiality of selenium and the development of biologically active organo-selenium compounds of minimum toxicity. *Chemica Scripta. (Sweden),* 8A:85–95.
64. Schwarz, K., and Baumgartner, W. A. (1969): Kinetic studies on mitochondrial enzymes during respiratory decline relating to the mode of action of tocopherol. In: *The Fat Soluble Vitamins,* edited by F. H. De Luca and J. W. Suttie, pp. 317–346. University of Wisconsin Press, Madison, Wisconsin.
65. Schwarz, K. (1965): Role of vitamin E, selenium and related factors in experimental nutritional liver disease. *Fed. Proc. Fed. Am. Soc. Exp. Biol.,* 24:58–67.
66. Schwarz, K., and Foltz, C. M. (1957): Selenium as an integral part of factor 3 against dietary necrotic liver degeneration. *J. Am. Chem. Soc.,* 79:3292–3293.
67. Serfass, R. E., and Ganther, H. E. (1975): Defective microbicidal activity in glutathione peroxidase-deficient neutrophils of selenium deficient rats. *Nature,* 255:640–641.
68. Shamberger, R. J., Rukovena, E., Longfield, A. K., Tytko, S. A., Deodhar, S., and Willis, C. E. (1973): Antioxidants and cancer: Selenium in the blood of normals and cancer patients. *J. Natl. Cancer Inst.,* 50:863–870.
69. Sheffy, B. E., and Schultz, R. D. (1978): Nutrition and the immune response. *Cornell Vet. Suppl.,* 48–61.
70. Sieber, S. R. (1977): The action of antitumor agents, a double edged sword? *Med. Pediatr. Oncol.,* 3:123–131.
71. Spallholz, J. E., Martin, J. L., Gerlach, M. L., and Heinzerling, R. H. (1973): Enhanced immunoglobulin M and immunoglobulin G antibody titers in mice-fed selenium. *Infect. Immun.,* 8:841–842.
72. Spallholz, J. E., Martin, J. L., Gerlach, M. L., and Heinzerling, R. H. (1973): Immunologic responses of mice-fed diets supplemented with selenite selenium. *Proc. Soc. Exp. Biol. Med.,* 143:685–689.
73. Spiegelman, S. (1974): Molecular evidence for viral agents in human cancer and its chemotherapeutic consequences. *Cancer Chemotherapy Report,* 58:595–613.
74. Stossel, T. P., Mason, R. J., and Smith, A. L. (1974): Lipid peroxidation by human blood phagocytes. *J. Clin. Invest.,* 54:638–645.
75. Tengerdy, R. P., and Nockels, C. F. (1975): Vitamin E and vitamin A protects chickens against *E. coli* infections. *Poultry Science,* 54:1292–1296.
76. Tengerdy, R. P., and Nockels, C. F. (1973): The effect of vitamin E on egg production, hatchability and humoral immune response of chickens. *Poultry Science,* 52:778–783.
77. Tengerdy, R. P., Heinzerling, R. H., Brown, G. L., and Mathias, M. R. (1973): Enhancement of humoral immune response by vitamin E. *Int. Arch. Allergy, Applied Immunol.,* 44:221–232.
78. Tengerdy, R. P., Heinzerling, R. H., and Nockels, C. F. (1972): Effect of vitamin E on the immune response of hypoxic and normal chickens. *Infect. Immun.,* 5:987–989.
79. Velo, G. P., Dunn, C. J., Giroud, J. P., Timsit, J., and Willoughby, D. A. (1973): Distribution of prostaglandins in inflammatory exudate. *J. Pathol.,* 111:149–157.
80. Waterhouse, C. (1974): How tumors affect host metabolism. *Ann. N.Y. Acad. Sci.,* 230:86–93.
81. Wattenberg, L. W., Loub, W. D., Lam, L. K., and Speier, J. L. (1976): Dietary constituents altering the responses to chemical carcinogens. *Fed. Proc.,* 35:1327–1331.

Subject Index

A

Acrodermatitis enteropathica, 167, 172
Actinomycetes
  chromium and, 10
  soil and isolation of, 4
Aging
  aluminum and, 131-132
  calcium and, 202-203
  metal ions and, 128-129
    concentrations of, 131
Alopecin, 167
Aluminum, 131-132
Alveolar macrophage, 76-77
Anemia, nonspherocytic hemolytic, 277
Anesthetics
  calcium antagonists versus, 229
  verapamil and, 237
Angina, see Cardiovascular disease
Animal cells, chromium and in vitro, 118
Antibiotics, siderophores and, 34-38
Antimetabolite drugs, 32-36
Antioxidant immunological effects, 287-302
  biochemical mechanisms in, 293-294
  cellular mechanisms in, 291-293
  in vivo, 294-295
  paraneoplastic syndromes and, 297-302
  in tumor immunity, 295-297
Arsenic, 113, 257-258
Atomic microscopy, 155-165; see also Scanning transmission electron microscope

B

B_{12}-dependent methionine synthesis, 50
B_{12}-dependent methyl-transfer, 44-51
Bacteria
  chromium and spore-forming, 8-10
  enteric, 29
  mutagenicity of chromium and, 115-116
  siderophores and, 28
  soil and isolation of, 4
  spore-forming, 8-10
Benzimidazole, methylation and, 49-50
Biomethylation of metals, see Methylation of metals
Brain implants of carcinogens, 85

C

Cadmium, 257-258
Calcium
  age and, 202-203
  homeostatic system, 201-204
  transport in intestine and vitamin D, 209-210
  in vitamin D metabolism, 199-204
Calcium antagonists, 217-222
  anesthetics versus, 229
  and cardiovascular disease, 227-232
    angina, 231, 245-247
    arrhythmia, 221-222, 227-229, 236-238
    myocardial ischemia, 221, 224-227
  classification of, 217-218
  depressant action of, 230-231
  drugs included in concept of, 227-228
  investigative uses of, 217
  MDI compounds as, 218-222
    applied pharmacology of, 220-222
    basic pharmacology of, 219-222
    chemistry and structure-activity relationships of, 218-219
    toxicology of, 222
  mechanisms of action of, 227-232

Calcium antagonists (*contd.*)
  myocardium and, 227-228
  pharmacological effects of, 231
  vasodilators versus, 230-231
  verapamil and, 235
Cancer, *see also* Carcinogenesis; Carcinogens
  antioxidants and, 295-302
  essential trace elements and, 100-104
  selenium and, 255
  siderophores and, 38
Cancer-related metals, 113-115
Carbanion methyl-transfer, 45
Carbon metabolism, 11-14
Carcinogenesis; 83-89; *see also* Carcinogens
  cancer-related metals and, 113-115
  metal mutagenic inhibitors of, 109-120
  chromium as, 115-120
  model of, 110
  mutagenesis and, 63-66, 109-120, 135-136
  models for, 150
  physical form of implanted material in, 85
  route of administration of carcinogen in, 84-88
    brain, 85
    dermal, 85-86
    ingestion, 87
    inhalation, 86
    intramuscular, 86
    intrarenal, 85
    intrathoracic, 87
    intratracheal, 87
    orthopedic, 88
    parenteral, 84
    subcutaneous, 84-85
  trace element interaction in, 93-104
    antagonism as, 95-97
    parameters of, 94
    synergism as, 97-100
Carcinogens, 110; *see also* Carcinogenesis
  animal and human, 112-113
  mutagens and, 55-56, 109-120
  as occupational hazard, 86, 89
  route of administration of, *see* Carcinogenesis, route of administration in
Cardiovascular disease, 280-281
  angina
    calcium antagonists and, 231
    verapamil and, 245-247
  calcium antagonists and, 227-232
  cardiac arrhythmia, 221-222
    calcium antagonists and, 227-229
    verapamil and, 236-238
  GSH peroxidase and, 280-281
  myocardial ischemia, 221
    verapamil and, 244-247
  verapamil and, 241-247
Cell cultures, toxicity of metal in, 71-80
  early work on, 72
Cell cycle and zinc, 177-187; *see also* Zinc in the cell cycle
Cellular aging and metal ions, 128-129, 131
Cellular lethality, mutagen-mediated, 60-61
CH_3^+, 46-47
Chemical form and toxicity, 1-3
Chemical mutagens, 61-63
Chinese hamster ovary (CHO), 56
CHO (Chinese hamster ovary), 56
CHO genetic toxicity assay, 56
  characteristics of, 60
  development of, 58-64
  mutagenicity and carcinogenicity in, 63
  validation of, 63-64
CHO/HGPRT assay, *see* CHO genetic toxicity assay
Cholylhydroxyamic acid, 38
Chromium, 255-256
  DNA repair assays and, 117-119
  in vitro animal cells and, 118
  metabolism and, 118-119
  as mutagenic initiator of carcinogenesis, 109, 112, 115-120
  mechanism of, 119-120

Chromium (*contd.*)
  soil microbial cultures and, 8-25
    biochemical analyses of, 20-24
    location of chromium in, 20-21
    mechanism of effect of, 11-14
    modification of chromium in, 21-24
    soil respiration and, 10-15
  toxicity of, 14-20
Chromosome aberration
  CHO genetic toxicity assay for, 63-64
  chromium and, 117-118
cis-DDP, 64-65
Clonal growth assay for toxicity, 71
  advantages for toxicity testing of, 72-73
  alveolar macrophage and W138 viability tests versus VERO, 77
  application of, 72
  cytotoxicity of various metals and, 76
  in vitro, 71-80
  metal ions and RPE of VERO cells in, 75
  objective, 72
  RPE_{50} and LD_{50} in, 78
  VERO cytotoxicity assay and, 79
Cloning efficiency, 63
Co-C bond of methyl-B_{12}, 44-46
  electrophilic attack on, 44-46
  free radical attack on, 46-48
  nucleophilic attack on, 50
Congenital manifestations, 167
Copper ions, 126-127
Cytotoxicity
  CHO genetic toxicity assay and, 63-64
  clonal growth assay and *in vitro*, 76
  mutagenicity and, 61, 64-66
  selenium and, 290
  toxicity and *in vitro*, 79
  of various metals, 76

D
DDP, *cis*- and *trans*-, 64-65

Deferration drugs, 36-38
Demetallation drugs, 36-38
Depressants
  calcium antagonists and, 230-231
  verapamil and, 239
Dermal absorption of carcinogens, 85-86
Dermatitis, 167
Desferal, 36-39
Dietary liver necrosis, 251-254
Dietary supplements including minerals, 87
DNA, 64-65; *see also* Nucleic acids and metal ions
  aluminum and crosslinking in, 132
  chromium and, 117-119
  fidelity of synthesis of, 138-149
    activating metal ions and, 140-142
    mechanism of metal ions decrease of, 146-149
    natural DNA templates and, 144-146
    nonactivating metal cations and, 142-143
    polynucleotide templates and, 139-140
    screening for mutagens and carcinogens and, 143-144
  polymerase miscoding and, 117, 135-151
    mechanism of polymerization and, 136-138
  repair assays, 117-119
  templates and, 139-140, 144-146
  zinc and cell content of, 178-180
  zinc deficiency and, 169-170, 185-187
Drinking water, 87-88

E
E. coli, mutagenicity of chromium and, 116
EDTA (ethylene-diamine-tetraacetic acid), zinc deficiency and, 171
Electrophilic attack on Co-C bond of methyl-B_{12}, 44-46
EMS and mutagenesis, 60-62
Enteric bacteria, siderophores and, 29
Enterobactin, 29, 38

Enzymes and metal ions, 130
Euglena gracilis cell cycle, 177-187; see also Zinc in the cell cycle
Eukaryotes, chromium and chromosome damage in, 117-118

F

Ferredoxin, 39
Ferric-specific ligands, 28
Fluorine, 257
Food and siderophores, 39-40
Food chain metal transfer research areas, 2-3
Free radical attack on Co-C bond of methyl-B_{12}, 46-48
Fungi
  culturing metal-resistant, 5-7
  siderophores and, 28
  soil and isolation of, 4

G

Genetic basis of mutagenesis, 60
Genetic information transfer, 128-132, 135-151; see also DNA synthesis, fidelity of; Mutagenesis
  genetic miscoding, 135-151
    mechanism of, 146-149
  genetic toxicity of metals, 63-64, 110-115
    measuring, 110-112
    metal carcinogens exhibiting, 109
Glucose
  measurement of soil, 4-5
  metabolism, 11-14, 255-256
Glutathione peroxidase reaction, 263-281; see also GSH peroxidase
Glycoproteins, STEM studies of, 162-165
GSH peroxidase
  cardiovascular disease and, 280-281
  chemical and physicochemical properties of, 269-272
  enzymology of, 266-276
  inflammation and, 279-280
  mechanisms of catalysis of, 272-276
  initial velocity measurement in, 272-273
  kinetic constants in, 274
  mutagenesis and, 279
  physiological role of, 276-281
    established, 277-279
    hypothesized, 279-281
  subcellular distribution in, 266-269
  substrate specificity in, 266, 267

H

Heart ailments, see Cardiovascular disease
Heavy atom labeling, 155-165; see also Scanning transmission electron miscroscope
Heavy metals in biosphere, 43-51
HGPRT (hypoxanthine-guanine phosphoribosyl transferase), 56
Hydronamic acids, 39
Hyperferremia, 36-38
Hypoxanthine-guanine phosphoribosyl transferase (HGPRT), 56

I

Immune response and antioxidants, see Antioxidant immunological effects
Immunity and iron, 31-32
Implantation of carcinogens, 84-85
Infection and immunity, iron and, 31-32
Inflammation and hydroperoxides, 279-280
Ingestion of carcinogens, 87
Inhalation of carcinogens, 86
Intramuscularly implanted carcinogens, 85
Intrarenal administration of carcinogens, 85
Intrathoracic injection of carcinogens, 87
Intratracheal instillation of carcinogens, 87
Ions in transuranic series, 38
Iproveratril, see Verapamil
Iron assimilation, 27-40
  antimetabolite drugs and, 32-36
  high affinity iron transport in, 28-31

Iron assimilation
  high affinity iron transport in (contd.)
    siderophores in, 28-31
  infection and immunity and, 31-32
  iron deficiency anemia and, 31
  low affinity iron transport in, 27-28
  mechanisms of, 27
  nutrition and, 39
  siderophores and, 28-40
Iron deficiency anemia, 31
Iron poisoning, 36-38

L

LD (lethal dose), 77-78
LD_{50}, 78
Lead, 257-258
  biomethylation of, 45-46
  salts, 46
Lecithin, 168
Leghemoglobulin, 39
Lethal dose (LD), 77-78
Ligands, ferric-specific, 28
Liver necrosis, dietary, 251-254
Lungs, 168

M

Magnesium ions, 125
Manganese (Mn), 113-115
  cytotoxicity and mutagenicity of, 64-66
  $MnCl_2$, 65-67
MDI compounds (2-substituted methylenedioxyindenes), 218-222; see also Calcium antagonists, MDI compounds
Membrane receptors
  "cross test" for competition for, 35
  siderophores and, 28, 30, 33-36
Mercuric ion, 45
Mercury, 44-45
Metabolism
  carbon, 11-14
  chromium and, 118-119
  glucose, 11-14
  soil, 1-25; see also Soil metabolism of heavy metals
Metal carcinogenesis, see Carcinogenesis
Metal interactions in carcinogenesis, see Carcinogenesis, trace element interactions in
Metal ions
  copper, 126-127
  enzymes and, 130
  magnesium, 125
  mercuric, 45
  methyl-B_{12} and inorganic, 44
  nucleic acids and, see Nucleic acids and metal ions
  platinum, 126-127, 132
  RNA synthesis and, 129-132
  RPE and, 75
  in transuranic series, 38
Metal-resistance
  bacteria and, 3-7
  chemical form and, 3
  fungal cultures and, 5-7
Metal salts, 71-80
Methionine synthesis, 50
Methylated metals, 43-51
  lead as, 45-46
  mercury as, 44-45
  methyl-B_{12} as
    electrophilic attack on Co-C bond of, 44-46
    free radical attack on Co-C bond of, 46-48
    nucleophilic attack on Co-C bond of, 50
    oxidation of, 51
    platinum and, 48-50
  platinum as, 48-50
  tin as, 47
  toxicity of, 43-44
Methylation of metals
  B_{12}-dependent methyl transfer in, 44-51
  benzimidazole and, 49-50
  carbanion methyl-transfer in, 45
  dynamic aspects of, 44
  lead and, 45-46
  mechanisms of methyl transfer in, 44-51

Methylation of metals
  mechanisms of methyl transfer in (*contd.*)
    standard reduction potential and, 47-48
    mercury and, 44-45
    platinum and, 48-50
    thiols and, 47
    tin and, 47
Milk, cow versus human, 172-173
Mn, *see* Manganese
Mutagenesis, 55-67; *see also* Genetic information transfer
  carcinogenesis and, 63-66, 109-120, 135-136
    models for, 150
  cellular lethality and, 60-61
  chemical mutagens and, 62-63
  CHO genetic toxicity assay for, 58-64
  of chromium, 115-120
  cytotoxicity and, 61, 64-66
    platinum compounds and, 64-65
  EMS and, 60-62
  genetic basis of, 60
  hydroperoxide and, 279
  ionic composition and, 66-67
  quantifying specific gene, 58-60
Mutagens
  carcinogens and, 55-56
  cellular lethality and, 60-61
Mutation induction, 60-61
Mycobactin, 32
Myocardial disease, *see* Cardiovascular disease

N

Neonatal development and zinc deficiency, 167-173
  neonatal survival and, 170-171
Neoplastic disease, 297-302
Nifedipine, 227-232
Nitrogen fixation in biosphere, 39
Nitrogenase, 39
Nonspherocytic hemolytic anemia, 277

Nucleic acids, STEM studies of, 157-159
Nucleic acids and metal ions, 123-132
  aluminum and DNA and, 32
  cellular aging and, 128-129
  concentration of metal ions and, 129
  DNA synthesis and
    crosslinks in, 124, 132
    DNA-polylysine complex and, 124-128
    DNA-polypeptide binding and, 124-128
    metal ions and *in vitro*, 136
  enzymes and, 130
  nucleic acid structure and, 123-124
  RNA synthesis and, 129-130
Nucleophilic attack on Co-C bond of methyl-B_{12}, 50
Nucleosome structure, STEM analysis of, 161-162
Nutrition
  cow versus human milk, 172-173
  dietary supplements and, 87
  drinking water and, 87-88
  iron, 39
    iron deficiency anemia and, 31
    siderophores in food products and, 39-40
  vitamin D and, 191-193

O

Occupational hazards, 86, 89
Orthopedic implantation of carcinogens, 88
Osteoporosis, 202-205
Oxidation of methyl-B_{12}, 51

P

Parenteral administration of carcinogens, 84
Phosphorus and vitamin D, 204-206
Platinum, 48-50, 113-115
  cytotoxicity and mutagenicity of, 64-66
  ions, 126-127, 132

Polysaccharides, STEM studies of, 162-165
Prenatal development and zinc deficiency, 167-173
　in early stages of pregnancy, 168-169
　mechanisms involved in, 169-170
　plasma zinc of mother in, 169
Proteins, STEM studies of, 159-161

R

Receptors, *see* Membrane receptors
"Redox-Switch," 44, 48-49
Reduction potential, 47-48
Renal osteodystrophy, 206-208
Respiration rate of microorganisms in soil, heavy metals and, 10-20
Rhodotorulic acid, 37
RNA metabolism
　base composition of, 184-185
　metal dependence studies for, 181-183
　Mn concentration and, 186
　zinc and, 180-187
RNA synthesis and metal ions, 129-132
RPE (relative plating efficiency), 75
RPE_{50}, 76
　LD_{50} and, 78
　lethal dose (LD) and, 77-78

S

Scanning transmission electron microscope (STEM), 155-157
　artifacts resulting from observation in, 159
　in studies of
　　glycoproteins, 162-165
　　nucleic acids, 157-159
　　nucleosome structure, 161-162
　　polysaccharides, 162-165
　　proteins, 159-162
Schwarz, Klaus, research, 251-260
　on arsenic, 257-258
　on cadmium, 257-258
　on chromium, 255-256
　on dietary liver necrosis, 251-254
　on fluorine, 257
　on lead, 257-258
　on selenium, 254-255
　on silicon, 257
　on tin, 256
　on vanadium, 256
Selenium, 113-115, 254-255, 258
　history of research in, 263-265
　immunological effects of, 287-291, 293-301
Siderophore(s), 28-40
　agriculture and, 39
　antibiotics and, 34-38
　antimetabolite drugs and, 32-35
　cancer and, 38
　as chemical entities, 30
　distribution among bacteria and yeast and fungi, 28
　enteric bacteria and, 29
　food and, 39-40
　infection and immunity and, 31-32
　iron in nutrition and, 39
　membrane receptors and, 30
　molecular mechanics of transport of, 30
　mycobactin as first, 32
　surface receptors for, 33-36
　types of, 29
Silicon, 257
Silver, 113-115
Sister-chromatid exchanges, CHO genetic toxicity assay for, 63-64
Soil metabolism of heavy metals, 1-25
　carbon metabolism in, 11-14
　chemical form of heavy metals and, 1-3
　chromium in, 14-20
　　soil respiration and, 10-15
　glucose levels and, 11-14
　measurement of, 4-5
　isolation of representative soil microbiota in, 4
　metal-resistant bacteria and, 3-7
　metal-resistant fungal cultures and, 5-7
　soil respiration and, 10-15
　　respiration rate of microorganisms and, 10

Soil metabolism of heavy metals (*contd.*)
   toxicity of heavy metals and, 1-3
   ranking for, 14-20
STEM, *see* Scanning transmission electron microscope
Stillbirths, 170-171
Subcutaneous implantation of carcinogens, 84-85
Surfactant, 168

T
Titanium, 113
Toxicity, 55
   chemical form and, 1-3
   CHO genetic toxicity assay for, *see* CHO genetic toxicity assay
   clonal growth assay for, 71-80; *see also* Clonal growth assay
   advantages of, 72-73
   genetic, *see* Genrtic information transfer, genetic toxicity of metals
   in vitro cytotoxicity and, 79
   in vitro toxicity assays compared, 76-77
   of metal salts, 71-80
      in vitro clonal growth assay for evaluating, 71-80
   methylation of metals and, 43-44
   ranking of metal, 14-20
   trace metal essentiality and, 259
   VERO cytotoxicity assay and, 79
Trace elements
   cancer patients and homesotatic levels of, 100-104
   discovery of
      research in, by Klaus Schwarz, 251-260
      table of, 258-259
   interaction in carcinogenesis, 93-104
   toxicity and essentiality of, 259
Transuranic series of ions, 38
Tumor induction, 84-88; *see also* Carcinogenesis

V
Vanadium, 256
Vasodilators
   calcium antagonists and, 230-231
   verapamil and, 241
Verapamil, 235-248; *see also* Calcium antagonists
   anesthetics and, 237
   angina and, 245-247
   arrhythmia and, 236-238
   calcium antagonism and, 235
   depressant effect of, 239
   effects of oral, 246-247
   electrophysiological effects of, 236-240
   hemodynamic effects of, 241-244
   myocardial ischemia and, 244-247
   structural formula for, 235
VERO cell clonal growth assay, *see* Clonal growth assay
Vitamin D, 189-210
   calcium and, 199-204, 209
   chemical synthesis of antagonists of, 199
   diet and, 191-193
   disease and defects in metabolism of, 206-208
   function of, 189-191
   intestine and function of, 208-210
   mechanism of hydroxylation of, 196-198
   metabolism of, 199-204, 206-208
   metabolites, 191-196
   phosphorus and, 204-206
Vitamin E, 287-289, 294, 300

W
Water, 87-88
WI38 cell assay, 76-77

Y
Yeast
   mutagenicity of chromium and, 116

Yeast (*contd.*)
   siderophores and, 28

Z

Zinc-binding ligand, 176
Zinc deficiency
   absorption of zinc and, 172-173
   cow versus human milk and, 172-173
   DNA synthesis and, 169-170, 185-187
   in eraly stages of pregnancy, 168-169
   EDTA and, 171
   genetic factors in, 171
   hari zinc level and, 173
   milder, 170
   neonatal survival and, 170-171
   other metals and, 185-186
   prenatal and neonatal development and, 167-173
   RNA content per cell and, 186
   RNA metabolism and, 185-187
   severe, 167-170
   stillbirths and, 170-171
   zinc-binding ligand and, 176
Zinc in the cell cycle, 177-187
   DNA content and, 178-180
   growth and, 180
   other metals and, 178
   RNA metabolism and, 180-187

THE LIBRARY
UNIVERSITY OF CALIFORNIA
San Francisco
(415) 476-2335

THIS BOOK IS DUE ON THE LAST DATE STAMPED BELOW

Books not returned on time are subject to fines according to the Library Lending Code. A renewal may be made on certain materials. For details consult Lending Code.

RETURNED
APR 2 3 1987

14 DAY
MAR 2 5 1993

RETURNED
MAR 3 1 1993

28 DAY
JUN 1 9 1995

RETURNED
MAY 3 0 1995